Biomedical Optical Phase Microscopy and Nanoscopy

Biomedical Optical Phase Microscopy and Nanoscopy

Editors:

Natan T. Shaked
Zeev Zalevsky
Lisa L. Satterwhite

AMSTERDAM • BOSTON • HEIDELBERG • LONDON
NEW YORK • OXFORD • PARIS • SAN DIEGO
SAN FRANCISCO • SINGAPORE • SYDNEY • TOKYO
Academic Press is an imprint of Elsevier

Academic Press is an imprint of Elsevier
The Boulevard, Langford Lane, Kidlington, Oxford OX5 1GB, UK

First edition 2013

Copyright © 2013 Elsevier Inc. All rights reserved

No part of this publication may be reproduced, stored in a retrieval system or transmitted in any form or by any means electronic, mechanical, photocopying, recording or otherwise without the prior written permission of the publisher

Permissions may be sought directly from Elsevier's Science & Technology Rights Department in Oxford, UK: phone (+44) (0) 1865 843830; fax (+44) (0) 1865 853333; email: permissions@elsevier.com. Alternatively you can submit your request online by visiting the Elsevier web site at http://elsevier.com/locate/permissions, and selecting *Obtaining permission to use Elsevier material*

Notice

No responsibility is assumed by the publisher for any injury and/or damage to persons or property as a matter of products liability, negligence or otherwise, or from any use or operation of any methods, products, instructions or ideas contained in the material herein. Because of rapid advances in the medical sciences, in particular, independent verification of diagnoses and drug dosages should be made

British Library Cataloguing-in-Publication Data
A catalogue record for this book is available from the British Library

Library of Congress Cataloging-in-Publication Data
A catalog record for this book is availabe from the Library of Congress

ISBN: 978-0-12-415871-9

For information on all Academic Press publications visit our web site at books.elsevier.com

Typeset by MPS Limited, Chennai, India
www.adi-mps.com

Printed and bound in the US

12 13 14 15 16   10 9 8 7 6 5 4 3 2 1

Working together to grow
libraries in developing countries

www.elsevier.com | www.bookaid.org | www.sabre.org

ELSEVIER   BOOK AID International   Sabre Foundation

Contents

Preface ix

PART 1:   Phase Contrast Microscopy and Differential Interference Contrast Microscopy

Chapter 1:   Phase Contrast Microscopy 3
  1.1   Introduction 3
  1.2   How Phase Contrast Works 4
      1.2.1   Basic Overview 4
      1.2.2   What Do You Actually See in a Phase Contrast Image? 10
      1.2.3   Positive and Negative Phases 11
      1.2.4   Halo Artifacts 12
  1.3   How to Acquire Phase Contrast Images 12
      1.3.1   What Components Are in a Phase Contrast System? 12
      1.3.2   General Alignment 13
      1.3.3   Phase Ring Alignment 13
      1.3.4   Adjustability 14
  1.4   Experimental Uses 15
      1.4.1   Cell Culture Models 15
      1.4.2   Image Analysis 15
      1.4.3   Fluorescence Overlay 17
  1.5   Limitations of Phase Optics 17
    References 18

Chapter 2:   Differential Interference Microscopy 19
  2.1   Introduction 19
  2.2   Measuring Shear Angle of DIC Prism 22
  2.3   Bias Optimization 27
  2.4   A Quantitative OI-DIC Microscope with Fast Modulation of Bias and Shear Direction 31
  2.5   Combination of OI-DIC and Orientation-Independent Polarization Imaging 37
  2.6   Combination of OI-DIC and Fluorescence Imaging 39
    References 40

Chapter 3: Long-Term Recordings of Live Human Cells Using Phase Contrast Microscopy 43
3.1 Introduction 43
3.2 The Microscope 44
  3.2.1 Illumination 44
  3.2.2 Temperature Control 45
  3.2.3 Other Considerations 47
  3.2.4 Observation Chambers 48
  3.2.5 Preparing Chambers 50
3.3 Correlative Live Cell–Fixed Cell Observations 51
References 52

Chapter 4: Phase Imaging in Plant Cells and Tissues 53
4.1 Out of Focus Imaging 54
4.2 Dark Field 54
4.3 Phase Contrast Techniques 54
4.4 Becke Line in Optical Microscopy 57
4.5 DIC Microscopy 58
4.6 Hoffmann Modulation Contrast 60
4.7 Adaptive Optics Microscopy 61
4.8 Interferometric Microscopy 62
4.9 Second Harmonic Imaging Microscopy 65
4.10 Magnetic Resonance Imaging of Plants Showing the Flow Component 65
References 67

PART 2: Digital Holographic Phase Microscopy

Chapter 5: Digital Holographic Microscopy for Measuring Biophysical Parameters of Living Cells 71
5.1 Introduction 71
5.2 Technical Introduction 74
  5.2.1 Classical Holography 74
  5.2.2 From Classical to Digital Holography 74
  5.2.3 Digital Holography Methods 75
  5.2.4 Cell Imaging and Quantitative Phase Signal Interpretation 76
5.3 Biological Applications 78
  5.3.1 Extended Depth of Focus and 3D Tracking 79
  5.3.2 Dry Mass and Cell Cycle 79
  5.3.3 Biophysical Parameters in Hematology 80
  5.3.4 Cell Membrane Fluctuations 82
  5.3.5 Neuroscience 84

    5.3.6   Neuronal Cell Death 85
    5.3.7   Future Perspectives 86
  References 88

Chapter 6: Holographic Microscopy Techniques for Multifocus Phase Imaging of Living Cells **97**
  6.1   Introduction 97
  6.2   DHM Setups for Live Cell Imaging 98
    6.2.1   Fiber Optic Modular DHM 99
    6.2.2   Self-Interference DHM 100
  6.3   Recording and Numerical Evaluation of Digital Holograms 101
    6.3.1   Intensity Distribution in the Hologram Plane 101
    6.3.2   Spatial Phase Shifting-Based Reconstruction of Digital Holograms 102
    6.3.3   Quantitative Phase Imaging 103
    6.3.4   Subsequent Refocusing and Autofocusing 106
  6.4   Refractive Index Determination of Cells in Suspension 109
  6.5   DHM Analysis of Morphology and Thickness Changes 111
    6.5.1   Cell Reaction on Chemical Substances 112
    6.5.2   Imaging of Vacuole Formation 114
    6.5.3   Imaging of Apoptosis and Necrosis 114
  6.6   Quantitative Cell Division Monitoring 116
  6.7   Cell Tracking in 3D Environments 121
  6.8   Conclusions 124
  References 124

Chapter 7: Phase Unwrapping Problems in Digital Holographic Microscopy of Biological Cells **129**
  7.1   Introduction 129
  7.2   Experimental Techniques 130
    7.2.1   Dual-Wavelength Digital Holography Apparatus 130
    7.2.2   Angular Spectrum Method 132
    7.2.3   Curvature Correction 133
    7.2.4   Additional Phase Background Removal 134
  7.3   Varying Reconstruction Distance Method 137
  7.4   Dual-Wavelength Phase Imaging 142
    7.4.1   Synthetic Wavelength 142
    7.4.2   Linear Regression Method 145
  7.5   Conclusions 151
  References 152

Chapter 8: On-Chip Holographic Microscopy and its Application for Automated Semen Analysis 153
- 8.1 Introduction 153
- 8.2 Partially Coherent In-Line Holography on a Chip 156
- 8.3 Reconstruction of Microscopic Images and the Phase Recovery Process 161
- 8.4 On-Chip Imaging of Sperm Using Partially Coherent Lensfree In-Line Holography 162
- 8.5 High-Throughput On-Chip Semen Analysis Results 165
- 8.6 Conclusions 168
- References 168

Chapter 9: Synthetic Aperture Lensless Digital Holographic Microscopy (SALDHM) for Superresolved Biological Imaging 173
- 9.1 Introduction 173
- 9.2 SALDHM Inside the Gabor's Regime 177
- 9.3 SALDHM Outside the Gabor's Regime 182
- 9.4 Conclusions 189
- References 189

Chapter 10: Combining Digital Holographic Microscopy with Microfluidics: A New Tool in Biotechnology 193
- 10.1 Introduction 193
- 10.2 Drive and Analyze 194
  - 10.2.1 Trapping Theory 195
  - 10.2.2 Experimental Setup 196
  - 10.2.3 Experimental Results 198
- 10.3 Particle Tracking in 3D 202
  - 10.3.1 Modeling for 3D Tracking 202
  - 10.3.2 Experimental Results 205
- 10.4 Conclusions 207
- References 207

Chapter 11: Holographic Motility Contrast Imaging of Live Tissues 211
- 11.1 Introduction and Review 211
- 11.2 Optical Coherence Imaging 213
- 11.3 Motility Contrast Imaging 216
- 11.4 Dynamic Light Scattering (DLS) Spectroscopy 219
- 11.5 Tissue Dynamics Spectroscopy 223
- References 227

PART 3: Advanced Interferometric and Polarization Techniques

Chapter 12: Tomographic Phase Microscopy (TPM) 231
- 12.1 Introduction 231
- 12.2 Theory—Optical Diffraction Tomography 234
    - 12.2.1 Inverse Radon Transform 234
    - 12.2.2 Optical Diffraction Tomography 234
- 12.3 Experimental Implementation 236
    - 12.3.1 Experimental Setup 236
    - 12.3.2 Data Processing by Inverse Radon Transform 237
    - 12.3.3 Data Processing by the ODT 240
- 12.4 Video-Rate TPM 247
    - 12.4.1 Experimental Scheme 247
    - 12.4.2 Results 250
- 12.5 Biological Applications 253
    - 12.5.1 Study Malaria-Infected Red Blood Cells 253
    - 12.5.2 Assessing Light Scattering of Intracellular Organelles in Single Intact Living Cells 255
- 12.6 Conclusions 258
- References 259

Chapter 13: Phase-Sensitive Optical Coherence Microscopy (OCM) 261
- 13.1 Introduction 261
- 13.2 OCM Principles 262
- 13.3 Phase OCM 266
- 13.4 Phase OCM Applications 267
    - 13.4.1 Imaging of the Outer Retina 268
    - 13.4.2 Determining Elastic Properties of Skin 269
    - 13.4.3 Motion Correction of Rodent Cerebral Cortex 270
    - 13.4.4 Visualizing the Microvasculature within the Retina 271
    - 13.4.5 Refractive Index Measurements 272
    - 13.4.6 Nerve Displacement during Action Potential 272
    - 13.4.7 Measurements of Red Blood Cells 273
    - 13.4.8 Detection of Gold Nanoparticles 275
- 13.5 Conclusions 277
- References 278

Chapter 14: Polarization and Spectral Interferometric Techniques for Quantitative Phase Microscopy 281
- 14.1 Introduction 281
- 14.2 Dual-Interference Channel Quantitative Phase Microscopy 282

- 14.3 Spectral-Domain Differential Interference Contrast Microscopy 286
  - 14.3.1 Principles of Spectral Domain-DIC Microscopy 287
  - 14.3.2 Characterization of USAF Resolution Target 289
  - 14.3.3 Live Cell Imaging with Rat Cardiomyocytes 290
- 14.4 Spectral Multiplexing by RGB Color Channels 292
  - 14.4.1 RGB Camera Multiplexing for Phase Unwrapping 293
  - 14.4.2 Spectral Multiplexing of Quantitative Phase and Fluorescence Biomarkers 296
- 14.5 Nonlinear Phase Dispersion Spectroscopy 299
  - 14.5.1 Fluorescent and Nonfluorescent Polystyrene Beads 301
  - 14.5.2 Red Blood Cells 304
- 14.6 Conclusions 306
- References 307

Chapter 15: Polarization Microscopy 311

- 15.1 Introduction 311
- 15.2 Traditional Polarized Light Microscopy 314
  - 15.2.1 Basic Setup 315
  - 15.2.2 Birefringence, Retardance, and Slow Axis 317
  - 15.2.3 Quantitative Analysis of Specimen Retardance 318
- 15.3 The Liquid-Crystal Polarization Microscope (LC-PolScope) 320
- 15.4 Practical Considerations 322
  - 15.4.1 Choice of Optics 322
  - 15.4.2 Specimen Preparation 324
  - 15.4.3 Combining Polarized Light with DIC and Fluorescence Imaging 325
- 15.5 Polarized Light Imaging in Three Dimensions 327
- 15.6 Conclusion 329
- 15.7 Glossary of Polarization Optical Terms 331
  - 15.7.1 Analyzer 331
  - 15.7.2 Azimuth 331
  - 15.7.3 Birefringence 331
  - 15.7.4 Compensator 332
  - 15.7.5 Dichroism 332
  - 15.7.6 Extinction 332
  - 15.7.7 Fast Axis 333
  - 15.7.8 Optic Axis 333
  - 15.7.9 Polarized Light 333
  - 15.7.10 Linearly Polarized Light 333
  - 15.7.11 Circularly Polarized Light 333
  - 15.7.12 Elliptically Polarized Light 333
  - 15.7.13 Polarizer 334

	15.7.14	Retardance	334
	15.7.15	Retarder	335
	15.7.16	Slow Axis	336
	15.7.17	Waveplate	336

References 336

PART 4: Phase Nanoscopy

Chapter 16: Is There a Fundamental Limit to Spatial Resolution in Phase Microscopy? 341

16.1 Introduction 341
16.2 Where Was Abbe's Theory Incomplete? 342
16.3 Parallel Full-Field Linear Imaging 343
16.4 Reconstruction from Diffraction Patterns 343
16.5 Structured Illumination Microscopy 345
16.6 Imaging Three-Dimensional Phase Objects 349
16.7 Conclusions 350
References 350

Chapter 17: Nano-Holographic Interferometry for In-Vivo Observations 353

17.1 Introduction: Theoretical Basis of Nano Fourier Transform Holography 353
17.2 Gratings Illuminated by Evanescent Waves 355
17.3 Basic Setup Utilized for the Observation of Nano-Objects 356
17.4 System of Fringes Contained in the Recorded Image: Multi-k Vector Fields 359
17.5 Determination of the Pitch of the Gratings 364
17.6 Formation of the Holograms at the Nanoscale 365
17.7 Procedures to Extract Metrological Information from the Recorded Images 368
    17.7.1 Image Shift 368
    17.7.2 Change of Optical Path Through the Observed Object 369
17.8 Analysis of the Scales 370
17.9 Observation of Nano-Sized Objects 372
    17.9.1 Prismatic NaCl Nanocrystals 372
    17.9.2 Determination of Thickness of Nanocrystals Through Evaluation of the Optical Path Change 374
    17.9.3 Observation of the Polystyrene Nanospheres 377
17.10 Conclusions 382
References 384

Chapter 18: Fluorescence Phase Microscopy (FPM) and Nanoscopy ... 387
- 18.1 Introduction ... 387
- 18.2 The Fluorescence Self-Interference Process ... 389
- 18.3 Experimental Setups of FPM ... 390
  - 18.3.1 Time-Domain Fluorescence Phase Microscopy ... 391
  - 18.3.2 Spectral-Domain Fluorescence Phase Microscopy ... 392
  - 18.3.3 Time Domain/Spectral Domain-FPM—Capabilities and Limitations ... 394
- 18.4 Applications of FPM ... 395
  - 18.4.1 Experimental Characterization of the Fluorescence Phase Microscopes ... 396
  - 18.4.2 Optical Sectioning Imaging with Mesoscopic Resolution by FPM ... 397
- 18.5 Conclusions and Outlook ... 401
- References ... 402

Index ... 405

Preface

Living biological specimens, such as cells, tissues, or microorganisms, are microscopic dynamic objects, continuously responding to their environments and performing multiple processes by adjusting their three-dimensional sizes, shapes, and other biophysical features. Microscopy and nanoscopy of living specimens can provide a powerful tool for basic biological and biophysical studies, and for medical diagnosis and monitoring of disease progression.

Many living biological specimens such as cells in-vitro are transparent objects, and imaging them with conventional bright-field light microscopy fails to provide adequate contrast in the microscope image. For this reason, exogenous contrast agents such as fluorescent labels are widely used in biomedical microscopy. However, these exogenous agents are often cytotoxic and may influence the specimen behavior, especially in the long run. Additionally, fluorescent agents tend to photobleach which might limit imaging time and make signal quantification difficult.

Phase microscopy proposes a unique solution to the contrast problem. Phase is proportional to optical path delays of the light passing through the sample, and thus it captures information on the specimen structure and dynamics without using exogenous labeling.

This book presents a cutting-edge review of a variety of phase microscopy and nanoscopy techniques with emphasis on biomedical and clinical applications. The authors of each chapter are internationally renowned scientists, either phase microscopy and nanoscopy technology experts, or researchers who are interested in the biological and medical applications of these technologies. Based on this heterogeneous nature of the contributing authors, this book will not only be useful for researchers in the areas of biomedical engineering, electro-optics engineering, and nanotechnology engineering, but can also help biologists and clinicians who are interested in understanding the underlying principles and in learning about new biological and medical applications of conventional and novel phase microscopy methods.

As editors, we have invested efforts to order the book as a whole and the structure of each chapter in such as a way that the book is self-contained, with introductory subjects, followed by specific biomedical applications. Therefore, we expect that this book will be useful for researchers at all levels including advanced undergraduate students, graduate students, postdoctoral fellows, and established researchers.

The book is divided into four parts based on the historical order of the technology inventions, starting with the older and widely-used technologies and finishing with the newer ones. The four part of the books are: Phase Contrast Microscopy and Differential Interference Contrast (DIC) Microscopy, Digital Holographic Phase Microscopy, Advanced Interferometric and Polarization Techniques, and Phase Nanoscopy. The first chapters in each of the four parts of the book contain an introduction to the relevant basic optical principles, followed by more advanced chapters that describe state-of-the-art advances and new applications.

Part 1 of this book deals with conventional and widely used phase microscopy methods: phase contrast and DIC microscopy. Chapter 1 introduces the basic principles of phase contrast microscopy, the first phase microscopy method, proposed by Zernike in the third decade of the nineteenth century, an invention that gained him the Nobel Prize in Physics in 1953. The typical artifacts of this technique and some common applications are also reviewed in this chapter. Chapter 2 presents another widely used phase microscopy technique: DIC microscopy, invented by Francis H. Smith in 1947. Chapter 3 describes a practical method that is based on phase contrast microscopy for conducting long-term time-lapse observations of living cells. Chapter 4 reviews phase imaging methods for plant cells and tissues.

Part 2 of this book introduces quantitative phase microscopy performed by digital holographic microscopy (DHM). Chapter 5 explains the principles of DHM used for measuring the quantitative phase maps and their interpretation for the calculation of biophysical parameters of biological cells. Chapter 6 presents new DHM setups and various applications. Chapter 7 focuses on the problem of 2π ambiguities in the quantitative phase profile and the possible solutions. Chapter 8 presents compact and portable phase microscopic designs and demonstrates the analysis of living sperm. Chapter 9 focuses on super-resolution methods using synthetic aperture for a simplified lensless DHM setup, which is used as examining tools for red blood cells and sperm cells. Chapter 10 presents the application of DHM for analyzing particles or live cells in microfluidic devices. Finally, Chapter 11 presents low-coherence DHM-based methods for tracing the internal motion in live cells and tissues, and demonstrates its applications for mapping functional motion in tumors and assessing drug effects.

Part 3 of this book focuses on advanced polarization and interferometric methods of phase microscopy, which are based on DHM introduced in Part 2 of this book, polarization microscopy, or a combination of both. Chapter 12 presents the tomographic phase microscopy technique in which the sample phase is captured using multi-viewpoint interferometric method, and then the data are processed into the three-dimensional map of the refractive index of the sample. Several demonstrations of cell imaging are presented. Chapter 13 discusses the phase-sensitive optical coherence tomography (OCT), an interferometric method for measuring the quantitative phase for a single-point or a line of points on the sample, and several possible applications. Chapter 14 presents various advanced interferometric and DIC methods that use polarized light to capture the phase of the sample. Finally, Chapter 15 presents the polarization microscopy, a label-free technique that exploits the inherent birefringence in living cells to visualize dynamics of cell organelles.

Last but not the least, Part 4 of this book introduces phase nanoscopy used to visualize nanoscale objects. Chapter 16 discusses the idea of breaking the spatial resolution limit in phase microscopy, including definitions of the theoretical limit of resolution. Chapter 17 presents a method for combining total internal reflection (TIR) and DHM for visualizing nanoscale objects. Finally, Chapter 18 presents a unique nanoscopy approach that is based on interferometric imaging of fluorescence molecules using self-interference.

We strongly believe that phase microscopy and nanoscopy in the fields of biomedical, electro-optical and nanotechnology engineering will continue to develop rapidly and become one of the most important and widely used tools in biology and medicine, while offering unique opportunities for new and exciting applications. We hope that this book will contribute to the development of the field by providing a balanced and self-contained presentation of the field from different perspectives.

Natan T. Shaked, Ph.D., Tel–Aviv University, Israel

Zeev Zalevsky, Ph.D., Bar–Ilan University, Israel

Lisa L. Satterwhite, Ph.D., Duke University, USA

PART 1

Phase Contrast Microscopy and Differential Interference Contrast Microscopy

CHAPTER 1

Phase Contrast Microscopy

Sam A. Johnson

Light Microscopy Core Facility, Duke University and Duke University Medical Center, Durham, NC

Editor: Lisa L. Satterwhite

1.1 Introduction

Since the mid-1600s microscopes of increasing sophistication have opened a world invisible to the naked eye and are now ubiquitous and essential biomedical research tools. Magnification allowed observation of some objects and structures that were previously unknown and the capture of details impossible without the use of optics. This alone was transformative and opened entire fields of study.

But magnification is nothing without contrast—defined as the difference in intensity between two points divided by their average intensity—and many samples of considerable biological interest provide very limited contrast in homogeneous brightfield transmitted light microscopy. Phase contrast provides a means of extracting additional contrast from a variety of samples which provide little absorption-based contrast. Contrast can also be added by preparation and manipulation of the sample but this often involves fixing the sample, which can confer a considerable disadvantage in the study of biology and the dynamic processes it involves. Therefore, a method of observation of intact, live, and unlabeled samples is key.

Figure 1.1 shows the difference between a brightfield image and a phase contrast image of a typical thin transparent biological sample, a flat mammalian cell.

As seen in Fig. 1.1, phase contrast [1–5] provides additional clarity and detail in live biological samples: more contrast is provided and finer structures can be visualized. Today, these abilities make it an indispensable tool for many who are interested in imaging diverse samples such as cultured cells, bacteria, thin tissue samples, and yeast. The significance and advance that this method of contrast formation provides merited a Nobel Prize in 1953 to Zernike for its development in the 1930s. Today, phase contrast is a very common feature of thousands of microscopes worldwide and can be considered an absolutely core technique

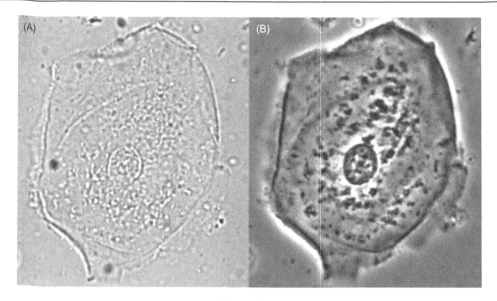

Figure 1.1
Comparison of a mammalian cell imaged with (A) brightfield and (B) phase contrast light microscopy.

of microscopy. This chapter will explain how phase contrast works and how it is done practically in standard microscopes; some typical uses of the technique are also presented.

1.2 How Phase Contrast Works

1.2.1 Basic Overview

Brightfield transmitted light microscopy has a fundamental weakness in the study of thin transparent biological specimens: they generally absorb little light to provide limited contrast by this means alone. The samples do however interact with light in ways other than absorption, and phase contrast is a method to exploit this and turn these interactions into observable contrast.

What does happen to light as it passes through transparent biological material such as the cell in Figure 1.1? Though transparent, the cell will cause diffraction and scatter of some of the light that passes through it. This process causes deviation and a phase shift of $\lambda/4$ between the small amounts of light that is deviated relative to the undeviated wave. A cell is far from homogeneous. The variety of components of the cell and the molecular, macromolecular, and organelle level structures into which they are arranged provide a complex landscape of changes in refractive index, and this variation will also have an effect on the light passing through the different parts of the cell.

Refractive index of a medium is defined as the ratio of speed of light in a vacuum to the speed of light in a medium.

$$n = c_{vac}/c_{medium} \qquad (1.1)$$

A vacuum therefore has a refractive index of 1.0000, air 1.0003, water 1.33, cytosol 1.35, and the glass most commonly used to make optics 1.52. The refractive index of a medium is an aspect of the interaction between the light and the electrical and magnetic susceptibilities of the medium. Overall the cell has a different refractive index than the surrounding medium. Finer scale differences within the cell provide observable subcellular resolution: different constituents such as lipids, proteins, and nucleic acids and different concentrations of these in an organelle or region produce local changes in refractive index. In a transmitted light image, some of the light will have passed through a cell and will have interacted with this complex refractive index topology and have been affected. Both the occurrence of diffraction of the light and the refractive index differences produce phase shifts which can be exploited to produce contrast.

This and the consequent means of exploiting the effects in a phase contrast microscope are most easily explained pictorially with a series of annotated diagrams.

Figure 1.2 shows schematically a phase shift by a sample. Reducing the sample to a gray box that represents an object, such as a cell, of a different refractive index to its surroundings and having a propensity to diffract and deviate some of the light we see light

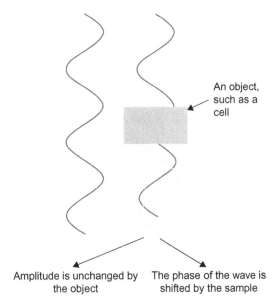

Figure 1.2
Phase shift introduced by a sample.

that passes through it becomes phase shifted relative to a reference wave that passes around the object unaffected. The phase shift has two components: (1) diffraction causes a λ/4 phase shift and (2) the difference in refractive index and optical path length changes the speed of the wave and so the transit time across the object relative to the surround which causes a typically small phase shift between the deviated wave passing through the sample and reference surround wave. The object here has a higher refractive index, a lower speed of transmission of the wavefront, so the phase is retarded slightly. It is noted that no difference in amplitude is produced—there is no absorption—and so no contrast is provided by this means alone. Our eyes and digital cameras are insensitive to phase or differences in phase of light, and we are unable to distinguish the object from the surrounding background without further action. (Objects such as our limiting case example which produce a phase shift but no absorption are called "phase objects"; at the other extreme, objects which produce strong absorption but little phase interaction are termed "amplitude objects" but of course nearly everything really lies somewhere between these two extremes.)

The change in phase presents some possibility for interference between the light that is shifted and that which is not. An observable difference in intensity could be produced in this way. To provide a very simple example consider the wavefronts occurring in Figure 1.2. Of the light illuminating the sample and surrounding, most of the light is undeviated (the zeroth-order light) and has no phase shift. Some of the light passes through the sample and is diffracted and phase shifted, by an amount of around λ/4. The precise magnitude of the phase shift is the sum of the phase shift from diffraction (λ/4) and the effect of a difference in the optical path length—the product of refractive index and thickness.

$$\text{Optical Path Length} = n \times t \tag{1.2}$$

$$\text{Optical Path Difference, } \Delta = (n_{\text{sample}} - n_{\text{surround}}) \times t \tag{1.3}$$

The relationship between the optical path difference and the phase shift of a wave in radians is

$$\text{Phase shift} = 2\pi\Delta/\lambda \tag{1.4}$$

Using examples of refractive index of the cytosol (1.35) and culture medium (1.33) and a range of thickness of the cell up to 10 µm, it can be calculated that the phase shift from optical path length differences are generally smaller than λ/4. Figure 1.3 shows the phase shift between the undeviated surround wave and the deviated wave shifted by around λ/4.

Because of the limited extent of the phase shift (shown in Figure 1.3A) and the difference in amplitude of the two waves (which to underscore is due to only a small fraction of the light being diffracted rather than this light being attenuated) the difference between the observed interfered output (the particle wave) and unaltered surround wave is very minimal.

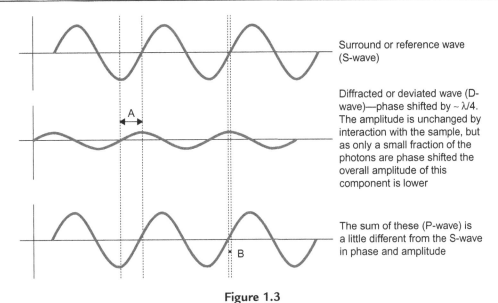

Figure 1.3
Phase shift between the diffracted wave and the surround wave.

The small phase difference between the S- and the P-waves is shown in Figure 1.3B and the amplitudes of the S- and the P-waves are essentially the same so we are still not provided with contrast to distinguish the object from background. This is essentially the state in brightfield illumination. How could a greater magnitude of interference be produced?

If we are able to produce an additional relative phase shift of $\lambda/4$ by advancing the S-wave, the total will be $\sim\lambda/2$ (Figure 1.4C), thus presenting the maximal destructive interference potential. If we are also able to make the two waves of more equal amplitude, the efficiency of extinction through interference will be higher. This situation providing very high extinction is shown in Figure 1.4.

Should you prefer these concepts in vector diagram form where phase is presented by angle (retardation is clockwise, an advance is anticlockwise) and the amplitude by length of the line—the two situations of modest phase shift (Figure 1.3) and very high extinction (Figure 1.4) in phasor diagrams are shown in Figure 1.5.

How do we construct a physical optical arrangement to perform this contrast formation? The key point is the spatial separation of the phase shifted diffracted light and the reference waves. This is achieved by a structured pattern of illumination produced by a ring annulus in the condenser and a matching phase plate in the objective. Figure 1.6 shows a simplified schematic cross-section diagram of the arrangement in an inverted phase microscope. (The diagram simplifies the light path slightly to make it clear how this is achieved; the illumination in reality is not precisely a hollow cone but broader parallel wavefronts that are focused by the objective back to a ring pattern in the back focal plane.)

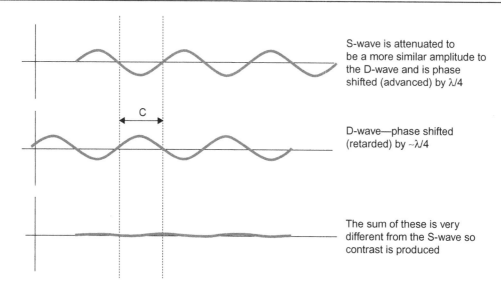

Figure 1.4
Phase shift between the diffracted wave and the phase shifted surround wave.

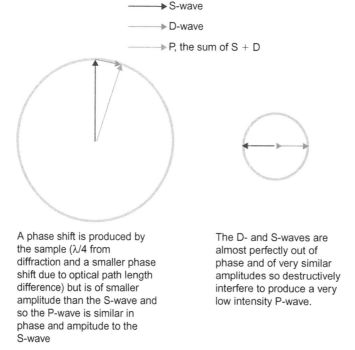

A phase shift is produced by the sample ($\lambda/4$ from diffraction and a smaller phase shift due to optical path length difference) but is of smaller amplitude than the S-wave and so the P-wave is similar in phase and amplitude to the S-wave

The D- and S-waves are almost perfectly out of phase and of very similar amplitudes so destructively interfere to produce a very low intensity P-wave.

Figure 1.5
Phasor diagrams of the phase shifts in Figures 1.3 and 1.4.

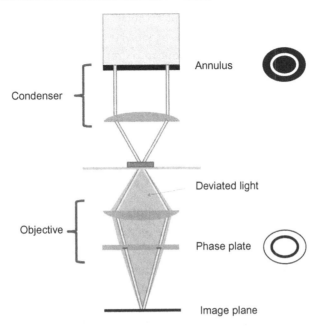

Figure 1.6
Schematic diagram of the arrangement of a phase contrast microscope.

A light source giving homogeneous illumination is used. This is typically a halogen bulb or LED providing Köhler illumination of the sample. The light that passes straight through without being diffracted (the zeroth order, undeviated, or reference waves) passes through the phase plate in a specific region (the conjugate region). The light that is diffracted changes angle and when collected by the objective passes through a different region of the phase plate (the complementary region). The phase plate is of a different thickness in these two regions and the difference in the optical path lengths (the product of refractive index (n) and thickness) produces a phase shift between surround and diffracted waves of $\lambda/4$. The undeviated light in Figure 1.6 passes through a thinner region of the phase plate and is advanced relative to the diffracted waves. This addition of a phase shift between the two waves adds to the difference in phase shift produced by the sample to enable intensity changes at the image plane through interference of the two waves.

The second function of the phase plate is to balance the intensity of the diffracted and reference waves, again achieved by means of them being spatially separated and passing through a different part of the phase plate. The unaltered reference wave passes through the ring on the phase plate, which is typically made partially opaque with a neutral density coating to attenuate this light to make it more comparable in amplitude to the diffracted light. Typically, around 75% of the surround light is absorbed at the phase plate to provide balance.

The two waves then interfere and a high contrast image is produced at the image plane that may be captured by CCD or CMOS camera, or viewed with extra magnification through eyepieces. Maximum extinction will be produced by regions or objects which produce a deviated wave that has a phase shift which when summed with the phase shift introduced by the phase plate produce a total shift of $\lambda/2$ and is of intensity comparable to the attenuated surround wave. A heterogeneous biological sample will also produce a range of phase shifts and extents of diffraction to have a range of contrast.

1.2.2 What Do You Actually See in a Phase Contrast Image?

The contrast added by exploiting phase effects within a sample clearly reveals many details not otherwise visible. With understanding of the method of phase contrast, it is easy to remember that darker regions of the image are not accounted for by absorption. The edge of a culture cell is clearly defined because of the difference between the propensity for diffraction and the phase shift in the cell and space surrounding it. But what are the biological structures or materials that give contrast within a sample? What of the organelles inside a cell—what is visible, what is not? With knowledge of the structure of a cell, it is clear that many familiar structures are visible—the nucleus, the nucleolus, and the vacuoles are clearly defined and identifiable. Other organelles are typically less clear but can be seen (e.g., mitochondria).

The simplest general explanation of the difference in contrast across a sample is the difference in optical path length (as defined earlier, this is the product of refractive index and thickness) at different points. Organelles and fine structures within the cell produce small local differences in refractive indices but because the thickness of the cell is also involved in the optical path length, the contrast distribution cannot be simply interpreted in terms of the refractive index. Often ruffles are visible in cells (see Figure 1.1), and these could be interpreted as differences in thickness of the cell with a relatively homogeneous refractive index of the cytosol but the precise contribution of the two components of optical path length are never known for sure.

The regions that appear darkest in a positive phase contrast image are those that present the extent of diffraction and phase shift from optical path length difference to the surround that produce a total phase shift offering highest extinction. Regions with a total phase shift differing from this maximum will appear lighter. It is not always trivial to attribute differences in intensity to particular directions of phase shift from optical path length differences since large, say, retardations may produce the same effect as a small advance. Indeed, thick objects with large optical path differences could produce a phase shift of more than one wavelength.

Even in the absence of linear direct quantitative attribution between levels of contrast observed, the magnitude of the phase shift that produces the contrast and the precise biological materials or subcellular structures that account for the contrast, the definition produced by phase contrast is extremely useful.

1.2.3 Positive and Negative Phase

The principles discussed earlier were explained in terms of positive phase contrast which is the most popular implementation in biological applications. Positive phase contrast involves retardation by the sample, and advance of the surround wave by the phase plate to produce λ/2 total relative phase shift for maximum extinction. A negative phase plate can be used and the principle of contrast formation is essentially the same except the phase plate produces a relative shift between diffracted and surround light in the opposite direction. In negative phase contrast, the surround wave is retarded relative to the diffracted wave and so regions of phase retardation within the sample produce positive interference and appear brighter than the background. It is somewhat a matter of sample-specific taste whether positive or negative phase gives better results. Generally, positive phase contrast is found to be more clear and these objectives are more commonly available. Figure 1.7 shows the

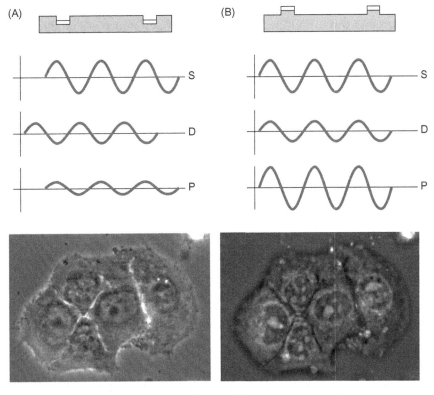

Figure 1.7
(A) Positive and (B) negative phase contrast.

difference in the phase plates, directions of phase shift of the S-waves, and examples of images of the same mammalian cell produced by positive and negative phase contrast.

1.2.4 Halo Artifacts

Phase contrast images have a ring around the larger objects—for positive phase contrast this is a ring of bright light that surrounds the object, for negative phase contrast it is a dark ring that is between object and background. What causes the phase halo and why is it only seen around the outer edge of large objects? The halo is caused by light which is diffracted and phase shifted but not diffracted by a magnitude or direction that causes it to pass through the central part of the phase plate. The phase ring is normally made to be slightly larger than the conjugate size of the annulus so it efficiently accepts the surround light after the slight aberrations in the optics act to expand the size of the illumination ring. This makes the system slightly more susceptible to halo than would be expected. Large objects with lower spatial frequencies diffract the light at a lesser angle than finer scale structures represented by high spatial frequencies, so the light that comes from large objects accounts for the majority of this phase shifted light that passes "incorrectly" through the ring of the phase plate. Because this light is phase shifted by the phase plate in the wrong direction relative to the surround wave, no means of destructive interference are produced for this light and so a high intensity region is produced. The halo has both good and bad properties: good because it provides a strong intensity change around the edge of the cell, highlighting this region; bad because the spatial accuracy of the definition of the edge is reduced by the overwhelming contrast and because the relationship between intensity and optical path length differences is lost in this part of the image.

Shade-off is a related artifact in large homogeneous objects. The center of such objects appears brighter in positive phase contrast because this light is little deviated and so, just as in the halo, not acted upon by the phase contrast system to produce attenuation.

1.3 How to Acquire Phase Contrast Images

1.3.1 What Components Are in a Phase Contrast System?

Good news: only two extra parts are required over brightfield and the general layout remains compatible with other standard transmitted and epi-illumination imaging techniques and modalities. As described earlier, phase contrast is based on a ring-shaped illumination and a matching phase plate, most commonly in the objective itself. The illumination can be achieved by introducing a mask in the aperture plane in the condenser. These are simple metal disks with different sized rings; larger rings are used for higher power lenses. Most commonly manufacturers refer to phase condenser elements with Arabic numbers such as phase 1, 2, and 3 (Roman numerals are generally used for DIC).

The objective needs to be one specifically designed for phase contrast, but this typically now adds little cost to the optic. If an objective without a phase plate is used even with the phase condenser element, an image like brightfield will be obtained and represent the equivalent of Figure 1.3. Objectives capable of phase contrast generally have green writing and indicate which condenser phase ring position offers compatibility. The physical diameter of the ring in the phase plate decreases with increasing magnification and Numerical Aperture (NA) of the lens and generally covers about 10% of the area of the objective.

1.3.2 General Alignment

The components are easy to align. No rotary alignment between components or the samples is necessary in setting up the scope and only a few adjustments are necessary to get a phase image of optimal quality making the technique quick and simple to use.

Nearly all modern microscopes have a transmitted light path employing an adjustable condenser giving Köhler illumination and this is the starting point for phase contrast. It is important to establish Köhler illumination under brightfield for the objective to be used to ensure the condenser is centered and positioned so as the image and illumination planes are physically optimally arranged and homogeneous illumination be provided.

The correct phase ring within the condenser needs to be selected for the chosen phase objective. Commonly, when a phase ring position on a condenser is selected, the condenser aperture is moved out of the path or fully opened. If this is not the case and the condenser aperture is not opened to the diameter of the phase ring, all illumination will be blocked.

A color filter (most commonly green) may be included in the path to ensure the illumination reaching the sample is monochromatic or of a small spectral range. This helps to produce maximum contrast by ensuring optimal phase shift by the phase plate; if white light is used a range of phase shifts around $\lambda/4$ will be produced as described in Eq. (1.4). Defined color illumination also reduces the degradation of the contrast and resolution in the image from chromatic aberrations and allows inexpensive achromat lenses to be used.

1.3.3 Phase Ring Alignment

The most critical alignment in phase contrast is the coincidence of the illumination ring and phase plate in conjugate planes at the condenser aperture and back focal plane of the objective, respectively. Figure 1.8 shows the impact of misalignment of the phase ring and the consequent degradation of image contrast and quality. The image formed with misaligned phase optics is less clear because the misaligned illumination is not adequately attenuated by the phase plate, so it is bright and the imperfect contrast formation overwhelmed.

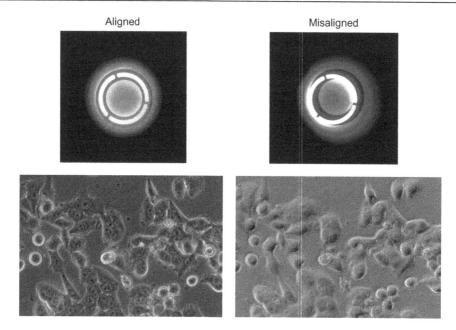

Figure 1.8
Phase ring alignment and an example of an image produced with a misaligned phase ring.

The phase ring can be aligned by viewing the back focal plane of the objective where the phase ring is present, and the annulus giving ring illumination is conjugate. This can be most easily viewed by removing one of the eyepieces. Magnification can be provided by using a Bertrand telescope but it is not impossible to align without this tool, especially for low power lenses. Removing the sample may make this alignment easier so the diffracted rays are not present to lessen the definition of the illumination ring. The sample was present when the images of the back focal plane of the objective were captured for Figure 1.8. The diffracted rays can be seen across the complementary region, but the illumination ring is still visible in the presence of these waves for this sample.

The condenser will have at least two screws giving fine adjustment of the position of the ring to allow superimposition. It is important to remember that adjustment screws in the condenser are not the same as those providing centration adjustments for Köhler alignment. Often once a ring is aligned for an objective, it tends to be fairly stable and may not need adjusting with every use but the alignment is very quick to test to make sure it is correct. The phase plate in the objective is virtually always fixed and nonadjustable.

1.3.4 Adjustability

Phase contrast requires fewer setup steps than many other contrast methods for transmitted light imaging. In most cases for a particular objective, once it is properly set up, there essentially are not any adjustments to make. Different samples present a substantial range of extent of phase shifts and diffraction. To cope with this range, different objectives can be used to find an optimal balance of the amount of surround light that needs to be blocked by the phase plate (based on the extent of diffraction by the sample). But these are generally different complete objectives so the cost of acquiring this variability is fairly significant.

1.4 Experimental Uses

Phase contrast is one of the most popular, versatile, and widely used transmitted light contrast techniques used today. It is difficult to represent adequately the breadth of uses in the biomedical research community, but this will be attempted by presenting some general classes of examples.

1.4.1 Cell Culture Models

Cell culture models are a very common research tool, and phase contrast is ideally suited to the routine observation of these cells. Simple and relatively inexpensive inverted microscopes with low magnification or higher magnification long working distance dry lenses with phase contrast optics are situated in nearly all tissue culture facilities. They provide a quick and effective means of assessing the health and density of the culture cells in the plastic culture dishes and flasks used. The key advantage of phase microscopy for this purpose (as well as simplicity and low cost) is the ability to image through relatively thick optically imperfect plastic. The plastic used in culture vessels tends to present relatively little absorption and local differential phase shift of the light but significant interaction (principally birefringence) with polarized light meaning techniques such as DIC are not achieved with good results without specialized versions of these optical contrast methods. See Chapter 3 for excellent description of how phase can be used in live cell time-lapse studies.

1.4.2 Image Analysis

Threshold-based image analysis routines allow quantification of a broad range of features of samples. Examples include automated counting of cells, measurement of areas, and tracking the movement of live cells in a time lapse. Thresholding involves physically identifying an object by a difference in intensity between the object of interest and background. Phase contrast has two advantages for certain uses of this. The first advantage is that it provides large contrast—the difference in intensity between the cell edge and the background—making identification easier and more robust than with the use of brightfield images. Anybody who has ever attempted any image analysis with biological samples using threshold-based identification will know that with more contrast available, the easier and more accurately one can pick apart the objects of interest and the task can range enormously in difficulty with, among other factors, the level of contrast available. Typically, transmitted light microscopy does not provide as high a level of contrast as does fluorescence because the background is bright, whereas it is dark in fluorescence and so the ratio of intensities is much larger. But the enhanced edge contrast provided by phase is adequate for identification of the edge of cells thus providing a very useful tool. The second advantage of phase contrast is the enhanced edge definition is provided symmetrically around the cell. With DIC (see Chapter 2), enhanced edge contrast is also provided but the shadow effect means it is bright on one side of the cell and dark on the other. Our visual system has no problem understanding this arrangement but an image analysis process looking only at differences in intensity needs to be smarter than one analyzing a symmetrical pattern. Figure 1.9 shows the ability of easy identification of a cell by phase

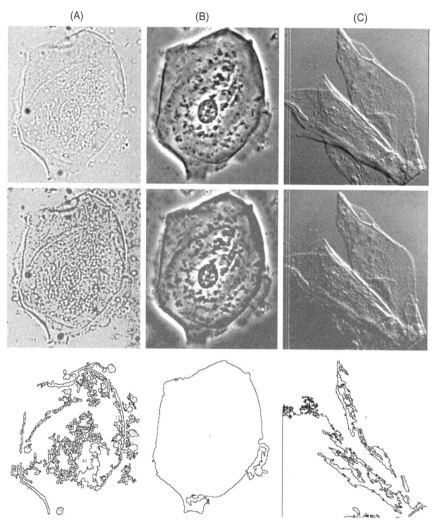

Figure 1.9
Threshold-based cell identification in (A) brightfield, (B) phase contrast, and (C) DIC.

contrast making it ideal for uses such as counting and tracking cells. The automated outlining (third row) of thresholded regions (second row) of a certain size range is produced. The lower contrast in brightfield does not allow easy contiguous identification of the cell perimeter. With DIC, the shadow gradient presents a significant issue.

1.4.3 Fluorescence Overlay

As with other transmitted light microscopy techniques, phase contrast images can easily be overlaid with fluorescence images providing a general morphology of the sample on which the specific fluorescence labels can be interpreted. The transmitted and fluorescence images are taken sequentially using the same camera and digitally overlaid. As with many transmitted light and fluorescence overlays, the best results are often produced by scaling the transmitted image so that it is relatively dark so as not to overwhelm the fluorescence that is to be overlaid. Phase also provides a high contrast method to focus on and select live samples avoiding the use of phototoxic intense fluorescence illumination when no information is obtained with it.

Phase optics has a minor effect on the fluorescence image. Obviously, the ring annulus in the condenser is irrelevant for epi-fluorescence configuration and hence presents no issue. The phase plate is normally fixed in the objective and so fluorescence emission light passes through it. The light that passes through the phase ring itself will be attenuated, but the overall effect is quite minor but can be significant in very demanding applications where the loss of a few percent of the signal is noticeable. The light will also be partially phase shifted by the plate; since this is not coupled with spatially arranged illumination this is of limited consequence.

Confocal microscopes are rarely equipped with phase optics. Point scanning confocals could be affected a little more seriously than widefield as the beam pivots near the phase plate and a coherent light source is used so the aberrations could be more apparent. Also, the means of phase contrast formation described here is not compatible with a point-based assembly so a phase contrast transmitted correlate to a confocal image cannot be built with a nondescanned transmitted photo multiplier tube (PMT). DIC is typically used for transmitted light contrast overlay on a confocal, and this works well in point scanning with laser illumination that provides polarized light. DIC also offers the advantage of allowing the removal of the analyzer and Wollaston prism offering maximum photon collection in fluorescence imaging mode.

1.5 Limitations of Phase Optics

The phase contrast method presents enormous utility but does have some limitations. The interpretation of the image is not trivial—it is really optical path length difference not refractive index variation that is visualized, very large phase shifts can produce contrast inversion, and halo and shade-off artifacts produce phase shift independent contrast. The halo effect also limits the accuracy of edge definition.

Thick objects with large optical path length differences to the surround wave may produce reduced contrast because the total phase shift of a particular region may be far from the $\lambda/4$ produced by diffraction and around which the phase plate is designed. Very thick samples may also present aberrations from out-of-focus regions and so contrast is reduced. Again the nonquantitative nature of phase contrast is apparent.

The effective numerical aperture of phase optics tends to be lower than other contrast methods (but this does give good depth of field which is convenient for uses such as inspection of tissue culture cells). This limits the sectioning ability of phase contrast which is frequently seen to be less good than in DIC. The fixed nature of the condenser ring annulus also means the condenser aperture is smaller than in brightfield or DIC and the condenser cannot be adjusted to balance contrast and resolution.

The phase plate in the objective does have a small effect on light such as fluorescence when the lens is used for other purposes than transmitted light phase contrast formation. The overall attenuation of the light is quite small, but the ring tends to be relatively close to the edge of the objective in a region of the Fourier transform of the image that contains the higher spatial frequencies which are normally present at low amplitudes.

The fixed nature of a particular phase objective does not provide great variability for samples that behave very differently than a typical range of biological samples the objective is designed for.

Other transmitted light contrast methods present different advantages and disadvantages. Understanding and using these methods offers a wealth of possibilities for visualizing interesting biological objects. The ideas presented in the other chapters of this book present great promise for expanding these tools.

References

[1] D.B. Murphy, Fundamentals of Light Microscopy and Electronic Imaging, Wiley-Liss, New York, 2001.
[2] F. Zernike, How I discovered phase contrast, Science 121 (1955) 345–349.
[3] S. Inoué, K.R. Spring, Video Microscopy: The Fundamentals, 2nd Ed., Springer-Verlag, New York, 1997.
[4] Phase contrast microscopy, in: E.M. Slayter (Ed.), Optical Methods in Biology, Wiley-Interscience, New York, 1970.

Further Reading

MicroscopyU—<http://www.microscopyu.com/articles/phasecontrast/index.html>—a detailed coverage of phase contrast and other microscopy. The Interactive Java tutorials are great educational tools.

CHAPTER 2
Differential Interference Microscopy

Michael Shribak
Marine Biological Laboratory, Woods Hole, MA

Editor: Lisa L. Satterwhite

2.1 Introduction

Differential interference contrast (DIC) light microscopy is widely used to observe structure and motion in unstained living cells and isolated organelles. DIC microscopy is a beam-shearing interference system in which the reference beam is sheared by a small amount, generally less than the diameter of an Airy disk. The technique produces a monochromatic shadow-cast image that displays the gradient of optical paths. Those regions of the specimen where the optical paths increase along a reference direction appear brighter (or darker), while regions where the path differences decrease appear in reverse contrast. As the gradient of optical path grows steeper, image contrast is significantly increased. Another important feature of the DIC technique is that it produces effective optical sectioning. This is particularly obvious when high numerical aperture (NA) objectives are used together with high NA condenser illumination.

The DIC technique was invented by Smith in 1947 [1,2]. He placed one Wollaston prism between a pair of polarizers at the front focal plane of the condenser and another Wollaston prism in the back focal plane of objective lens (Figure 2.1). The first Wollaston prism splits the input beam angularly into two orthogonally polarized beams. The condenser makes the beam axes parallel with a small shear. Then the objective lens joins them in the back focal plane where the second Wollaston prism introduces an angular deviation into the beam and makes them parallel. The splitting angles ε_1 and ε_2 are connected with the focal distances of the condenser and objective f_c and f_{ob} and the shear amount d by the following relation:

$$f_c \varepsilon_1 = f_{ob} \varepsilon_2 = d \qquad (2.1)$$

This optical configuration creates a polarizing-shearing interferometer, by which one visualizes phase nonuniformity of the specimen. Smith originally called this device the

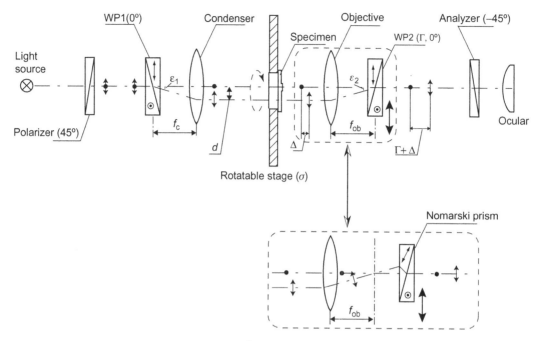

Figure 2.1
DIC microscope setup. Polarizer at 45° azimuth; WP1: first Wollaston prism at 0° azimuth; ε_1: splitting angle; f_c: condenser lens focal distance; d: shear amount; Δ: optical path difference introduced by specimen under investigation; σ: azimuth of rotatable stage; f_{ob}: objective lens focal distance; WP2: second Wollaston prism at 0° azimuth (the second prism introduces bias Γ); ε_2: splitting angle; analyzer at −45° azimuth; Wollaston prism can be replaced by Nomarski prism.

"Smith double-refracting interference shearing microscope system." However, he also gave credit [2] to Lebedeff [3], who constructed the first double-refracting interference shearing microscope.

In addition, Smith [1,2] proposed another DIC microscope with the vertical shear direction, which he called the "Smith double-focus system." The system employed lenses made from birefringent crystals associated with the condenser and objective of the microscope, to impart a double-focus effect on two beams generated by the birefringence. The two lenses must be in conjugate positions, preferably at the front focal plane of the condenser and the back focal plane of the objective, respectively. However, Smith's double-focus DIC system did not find wide use.

The both Smith's approaches suffered from the same problem. In conventional medium to high numerical aperture (NA) objective lenses, the back focal plane is located inside the lens system and therefore not available for insertion of a Wollaston prism or a birefringent lens. In particular, in the shearing DIC microscope system, if the Wollaston prism is placed far from the back focal plane, the prism only makes the rays parallel. But those beams are spatially displaced and hence are not recombined. Therefore, the Smith-shearing DIC scheme requires special design for the microscope objective lenses so that the Wollaston prism can be incorporated within.

Nomarski [4,5] took another approach and proposed in 1952 the use of a special polarization prism. This Nomarski prism (see insert in Figure 2.1) introduced spatial displacement and angular deviation of orthogonally polarized beams simultaneously. The prism can therefore be placed outside of the objective lens. By using crystal wedges with appropriately oriented axes, the Nomarski prism recombines the two beams that were separated by the condenser Wollaston as though a regular Wollaston prism were located at the back aperture plane in the objective lens. This feature enables the use of the Nomarski DIC scheme with regular high NA microscope objectives.

A DIC image can be modeled as the superposition of one image over an identical copy that is displaced by a small amount d and phase shifted by bias Γ. For simplicity, consider a phase nonbirefringent specimen, which is described by Cartesian coordinates XOY in the object plane. The specimen is illuminated by monochromatic light with wavelength λ. The intensity distribution $I(x,y)$ in the images depends on specimen orientation and varies proportionally with the cosine of the angle made by the gradients azimuth θ and the relative direction of wavefront shear σ [6]:

$$I(x,y) = \tilde{I} \sin^2\left(\frac{\pi}{\lambda}(\Gamma + d\gamma(x,y)\cos(\theta(x,y) - \sigma))\right) + I_c(x,y) \quad (2.2)$$

where $\tilde{I}$ is the initial beam intensity, $\gamma(x,y)$ and $\theta(x,y)$ are the gradient magnitude and azimuth, and $I_c(x,y)$ corresponds to a constant offset of the intensity signal.

It follows from formula (2.2) that if the shear direction is parallel to the optical path gradient ($\theta - \sigma = 0°$ or $\theta - \sigma = 180°$) the image contrast is maximal. Where the shear direction is perpendicular to the gradients ($\theta - \sigma = 90°$ or $\theta - \sigma = 270°$) the contrast equals zero. Thus, the regular DIC technique shows the two-dimensional distribution of optical path gradients encountered along the shear direction. It is therefore prudent to examine unknown objects at several azimuth orientations [5,7].

The DIC microscopy demonstrates remarkable optical sectioning capability, like confocal microscopy. The depth in specimen space that appears to be in focus within the image, without readjustment of the microscope focus, is the depth of field. In a regular bright-field microscope, the total depth of field d_{tot} is given by the sum of the diffraction-limited wave and geometrical optical depths of field as [8,9]:

$$d_{tot} = \frac{\lambda n}{NA^2} + \frac{n}{M \times NA} e \quad (2.3)$$

where n is the refractive index of the specimen, M is the combined lateral magnification of the objective and zoom lenses, and e is the smallest distance that can be resolved by the image detector (measured on the detector's face plate). The wave optical depth of field is determined by one quarter of the distance between the first two diffraction minima, above and below focus. The geometrical optical depth is a result of the "circle of confusion." For example, a bright-field microscope using a 40×/0.95NA objective lens and a CCD camera with 6.45 μm square pixels would have the field depth in water 1.0 μm at wavelength 546 nm. As it was shown by Inoué [10], the thickness of optical section (depth of field) of a conventional DIC microscope equipped with the same objective lens could be as little as 0.25 μm.

The contrast in DIC is produced by the optical path difference in a small in-focus volume where two interfering beams are spatially separated. Here, the beams travel through the different areas of the specimen under investigation. The out-of-focus object introduces practically the same phase disturbance in both the beams because the beams go through almost the same area of the specimen. Therefore, the out-of-focus disturbance is suppressed by optical subtraction. The optical section depth becomes thinner if the shear amount is smaller, and the objective and condenser NAs are larger. The narrow optical sectioning DIC phenomenon is similar to removing an out-of-focus haze in the structured illumination microscopy (SIM) [9]. The SIM employs a single-spatial-frequency grid pattern, which is projected onto the object under investigation. Raw images are taken at three spatial positions of the grid. The out-of-focus picture of the object does not depend on the pattern position. As a result, the out-of-focus haze in SIM is subtracted computationally.

Application of computation subtraction in the DIC would be expected to improve its sectioning capability even further. The computation subtraction of images with different biases is employed in various techniques, such as polarization modulation DIC (PM-DIC) [11,12], differential detection DIC (D-DIC) with polarizing beamsplitter [13], phase-shifting DIC (PS-DIC) [14–16], retardation modulation DIC (RM-DIC) [17–19], and orientation-independent DIC (OI-DIC) [6,20,21]. In particular, the PM-DIC removes a background contribution that is insensitive to defocus [12]. It is shown theoretically and confirmed experimentally that an RM-DIC microscope has stronger optical sectioning than a conventional DIC microscope, and the optical section depth is thinner if a Nomarski prism with smaller shear amount is used [19]. Our experiments with the OI-DIC microscope using 100×/1.3NA oil immersion objective lens demonstrated the optical section depth about 0.1 μm. The corresponding field depth of a bright-field microscope would be 0.5 μm.

2.2 Measuring Shear Angle of DIC Prism

Shear amount (distance) is the critical parameter of a DIC microscope that determines its contrast, sensitivity, resolution, and optical section depth. Another issue with DIC microscopy is that, to derive quantitative information, one must know the amount of image shear. Generally, however, microscope manufacturers do not make that information available. As a result, one must measure this parameter. Munster et al. [22] determined the lateral shift by measuring the distance between the center of the bright spot and the center of the dark spot in an image of a submicroscopic transparent latex sphere with bias set at $\pi/2$. Mehta and Sheppard [23] measured shear by studying intensity distribution in the back focal plane of the microscope objective lens. Müller et al. [24] used a combination of fluorescence correlation spectroscopy and dynamic light scattering to determine shear. Duncan et al. [25] described a measurement setup with a standard optical wedge.

The technique described later for determining shear distance is simpler, faster, and more accurate than just mentioned. We measure a shear angle and then compute the shear amount by employing formula (2.1), as it is explained in the end of this section.

In order to find the shear angle of a DIC prism, we do not need to know how the passing beam is transformed inside the prism. A DIC prism splits an incident monochromatic plane wave into two orthogonally polarized plane waves with wavefronts of slightly different direction. Figure 2.2 illustrates splitting the incident linearly polarized beam into two separate output beams with shear angle ε. Here, the shear plane is parallel to the X-axis. As one can see in Figure 2.2, the shear angle ε (in radians) is equal to the derivative of the optical path difference (bias) Γ with respect to the coordinate x:

$$\varepsilon = \frac{d\Gamma}{dx} \qquad (2.4)$$

The optical path difference Γ is connected with retardance δ, written in degrees, and wavelength λ in the simple way:

$$\Gamma = \frac{\delta}{360°} \lambda \qquad (2.5)$$

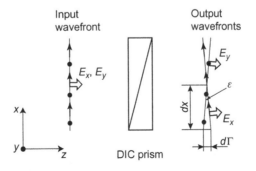

Figure 2.2
Splitting the incident beam with two orthogonal polarizations E_x and E_y by DIC prism into two separate output beams with shear angle ε.

Thus, the shear angle could be found by measuring a derivative of the retardance with respect to the coordinate along the shear direction:

$$\varepsilon = \frac{\lambda}{360°} \frac{d\delta}{dx} \qquad (2.6)$$

Retardance can be determined by a number of techniques. In particular, we employed the Senarmont compensator [26]. A schematic of the used setup is shown in Figure 2.3. A laser radiates the collimated narrow beam with horizontal linear polarization. Here, angle 0° corresponds to the horizontal direction (the X-axis). Then, the first quarter-wave plate QWP1, with fast axis oriented at 45°, transforms the linear beam polarization into the circular polarization. Polarizer, the second quarter-wave plate QWP2 and analyzer are mounted into rotatable holders with angular scale each. The DIC prism under investigation is placed on a linear stage with a Vernier micrometer. The stage allows movement of the prism in the horizontal direction. The minimum intensity of the passed laser beam was visually determined by observing the laser spot brightness on a screen.

The measuring procedure of retardance derivative consists of the following steps. Before placing the DIC prism under investigation in the setup, the prism is positioned between the crossed polarizers. The shear plane is perpendicular to the observed color fringes or/and the black fringe. The shear plane orientation is noted.

Next, the second quarter-wave plate, QWP2, is removed from the setup, cross polarizer and analyzer, and the DIC prism is positioned in the optical path. By rotating simultaneously

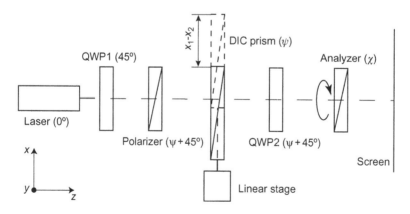

Figure 2.3
Setup for measuring shear angle of DIC prism by employing the Senarmont compensator. Laser with horizontally linearly polarized beam; QWP1: first quarter-wave plate at 45° azimuth; polarizer at $\psi + 45°$ azimuth; DIC prism under investigation with shear plane azimuth at angle ψ; linear stage for shifting DIC prism in the horizontal direction; QWP2: second quarter-wave plate at $\psi + 45°$ azimuth; rotatable analyzer at χ azimuth.

the polarizer and the analyzer from 0° to 180°, we achieve two extinctions of the laser spot on the screen. These extinctions differ by 90°. One extinction occurs when the polarizer transmission axis is parallel to the shear plane of the DIC prism. In this case, the polarizer angular scale points the shear plane orientation ψ. The other extinction corresponds to direction, which is perpendicular to the shear plane.

Next, the DIC prism is removed, and crossed polarizer and analyzer are rotated by 45°. If the orientation of the polarizer is $\psi + 45°$, orientation of the analyzer is $\psi - 45°$. The analyzer orientation is noted: $\chi_0 = \psi - 45°$.

The second quarter-wave plate, QWP2, is then placed into the optical path. The fast axis of the quarter-wave plate must be parallel to the polarizer transmission axis. The laser spot on the screen extinguishes if the fast axis is precisely aligned.

Finally, the DIC prism under investigation is placed in the measuring setup. The laser spot becomes bright. Rotating the analyzer allows to extinguish the spot on the screen. The DIC prism retardance δ in a point of the laser beam incidence is determined by the simple formula:

$$\delta = 2(\chi - \chi_0) \tag{2.7}$$

where χ is the orientation angle of the analyzer.

Using a linear stage, the DIC prism is moved in the horizontal direction and retardances δ_1 and δ_2 are measured at two points x_1 and x_2. The retardance derivative along with the shear direction X could be found by the following formula:

$$\frac{d\delta}{dx} = \frac{\delta_1 - \delta_2}{(x_1 - x_2)\cos\psi} = \frac{2(\chi_1 - \chi_2)}{(x_1 - x_2)\cos\psi} \tag{2.8}$$

Then, we compute the shear angle (in radians):

$$\varepsilon = \frac{\lambda}{180°} \frac{(\chi_1 - \chi_2)}{(x_1 - x_2)\cos\psi} \tag{2.9}$$

Instead of finding retardance values in two points only, we can measure dependence of the analyzer angle χ on the laser beam position x and use its derivative in formula (2.9):

$$\varepsilon = \frac{\lambda}{180°} \frac{1}{\cos\psi} \frac{d\chi(x)}{dx} \tag{2.10}$$

In the described setup, we utilized generic green and red laser pointers. Wavelength of the green laser pointer was 532 nm and the red laser pointer was 641 nm. The achromatic quarter-wave plates were made by Bolder Vision Optik (Boulder, CO, USA, http://www.boldervision.com). Regular grade polarizers for the visible spectral range and the employed

Table 2.1: Measured Shear Angles, Computed Shear Distances, and Ratios of Shear Distance to Airy Disk Radius for Various Combinations of Olympus DIC Prisms and Objective Lenses at Wavelength 532 nm

DIC Prism Type	Shear Angle, ε	Shear Distance, d in μm (Ratio d/r_{Airy})						
		UplanFl 10×/0.30	UplanFl 20×/0.50	UplanSApo 30×/1.05 Sil	UplanFl 40×/0.75	UplanSApo 60×/1.20 W	UplanSApo 60×/1.30 Sil	UplanFl 100×/1.30 Oil
U-DICTHR	40 μrad	0.72 (0.67)	0.36 (0.55)	0.24 (0.78)	0.18 (0.42)	0.12 (0.44)	0.12 (0.48)	0.072 (0.29)
U-DICT	74 μrad	1.32 (1.23)	0.67 (1.03)	0.44 (1.44)	0.33 (0.77)	0.22 (0.82)	0.22 (0.89)	0.13 (0.53)
U-DICTHC	143 μrad	2.57 (2.38)	1.29 (1.98)	0.86 (2.78)	0.64 (1.49)	0.43 (1.59)	0.43 (1.72)	0.25 (1.03)

mechanical components were purchased from Thorlabs (Newton, NJ, USA, http://www.thorlabs.com).

Table 2.1 summarizes representative results of the shear angle measurements of various DIC prisms currently manufactured by Olympus (Tokyo, Japan, http://www.olympus.com). The high-resolution DIC prism U-DICTHR has the smallest shear angle. The prism enables observations with high resolution but with less glare even for thick specimens used in developmental and genetic research, such as finely structured diatoms, embryos, zebrafish and *Caenorhabditis elegans*. The general-use prism U-DICT with the intermediate shear angle is suitable for observing a wide range of general specimens, such as tissue. The high-contrast DIC prism U-DICTHC has the largest shear angle. Using this prism, high contrast can be obtained even in high-magnification observations of thin specimens, such as culture cells.

Table 2.1 shows also the corresponding computed shear distance d in the object plane and ratio of the shear distance to the Airy disk radius d/r_{Airy}. For the calculation, standardized reference focal lengths of the tube lenses L_t are used for infinity-focused objective lens, which are adopted by several microscope manufacturers.

$$d = \varepsilon \frac{L_t}{M} \quad (2.11)$$

In Eq. (2.11), M is the objective lens magnification. In particular, the reference focal length L_t is 180 mm for Olympus, 164.5 mm for Zeiss, and 200 mm for Nikon and Leica microscopes [9].

The radius of Airy disk r_{Airy} is determined by the following equation [9]:

$$r_{Airy} = 0.61 \frac{\lambda}{NA} \quad (2.12)$$

where NA is the objective lens numerical aperture.

The above data are shown for the green light ($\lambda = 532$ nm). The shear angle was about 4% less for the red light ($\lambda = 641$ nm).

We have measured parameters of the Olympus DIC prism U-DICTH, which was manufactured in previous years. Its shear angle is the same as the prism U-DICTHR. Mehta and Sheppard [23] found the angular shear of 74 μrad for the U-DICTS prism at wavelength 550 nm. This shear corresponds exactly to our observations for the U-DICT prism. Our other results indicate that shear angles of Nikon 60xI and Zeiss PA63x/1.40III DIC sliders are 76 and 71 μrad, respectively.

2.3 Bias Optimization

In polarization microscopy, the Bräce-Köhler compensator, an azimuthally rotatable birefringent plate with bias retardance up to $\lambda/10$, allows investigation of a specimen with small birefringence more precisely than the Senarmont compensator, which has a retardance of $\lambda/4$ [7,26]. For example, Swann and Mitchison [27], using the $\lambda/20$ Bräce-Köhler compensator, could detect a 0.028 nm retardation. In order to achieve high sensitivity with the LC-PolScope (a commercially available polarization microscope), we apply alternate bias retardance $\lambda/30$ [28,29]. The measured noise level using the five-frame algorithm was 0.036 nm. The bias $\lambda/5$, which was used for studying a sample with large retardance, produces much higher noise.

In conventional DIC microscopy, the situation is similar. Near extinction, the image field becomes dark gray and yields very high sensitivity, bringing out image regions with minute phase differences due to, for example, extremely shallow depressions or elevations or from isolated subresolution filaments. Use of bias less than 90° is advantageous in video-enhanced-DIC (VE-DIC) microscopy [30,31]. Holzwarth [12] studied a photon noise versus bias in PM-DIC microscopy. He showed theoretically and confirmed experimentally that the signal-to-noise ratio peaks when the bias equals the sample optical path difference.

The benefit of the optimized bias can be proved mathematically by using the following example. Let us consider a simplified sample with a binary gradient magnitude distribution such that half of the sample has a gradient magnitude of $+\gamma$ and the other half has a phase difference of $-\gamma$ (see Eq. (2.2)). The shear and gradient directions of the sample are parallel ($\theta = 0°$). We also take into account the depolarized light by using an extinction ratio $\xi = (I_c/\tilde{I})$.

We expect the best results to be achieved when the DIC images have the highest contrast C:

$$C = \frac{I_{max} - I_{min}}{I_{max} + I_{min}} \quad (2.13)$$

where I_{max} and I_{min} are the maximal and minimal intensities in the image.

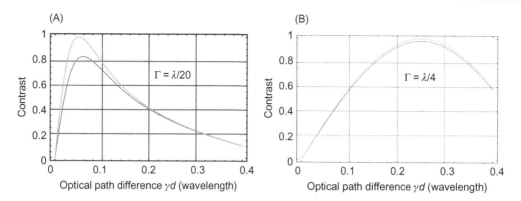

Figure 2.4
Contrast as a function of the sample optical path difference at biases 1/20th the wavelength (A) and quarter wavelength (B). The green color illustrates a case without depolarization (infinity extinction) and the red one corresponds to extinction 100. (For interpretation of the references to color in this figure legend, the reader is referred to the web version of this book.)

The maximal and minimal intensities can be determined using Eq. (2.2). For the specimen under consideration, the contrast of the DIC image is the following:

$$C = \frac{\sin((2\pi/\lambda)\Gamma)\sin((2\pi/\lambda)\gamma d)}{1 - \cos((2\pi/\lambda)\Gamma)\cos((2\pi/\lambda)\gamma d) + (2/\xi)} \quad (2.14)$$

Derivative $\partial C/\partial \Gamma$ of this equation is the next:

$$\frac{\partial C}{\partial \Gamma} = \sin\left(\frac{2\pi}{\lambda}\gamma d\right) \frac{(1+(2/\xi))\cos((2\pi/\lambda)\Gamma) - \cos((2\pi/\lambda)\gamma d)}{1 - \cos((2\pi/\lambda)\Gamma)\cos((2\pi/\lambda)\gamma d) + (2/\xi)} \quad (2.15)$$

The derivative is zero at $\Gamma \approx \gamma d$. Thus, the best optical contrast is achieved when bias equals optical path difference in the sample. This result agrees with data obtained by Salmon and Tran [31]. They found that for the edges of organelles and cells, the optical path difference corresponds to about $\lambda/10$th the wavelength or greater but for microtubules and tiny organelles in cell, optical path difference is very small, less than 1/100th the wavelength of green light. Salmon and Tran recommend using about 1/15th–1/20th the wavelength bias for observation of microtubules in order to have sufficient light at the camera. A similar result was found by Schnapp [32].

Contrast curves computed with formula (2.14) at different biases and extinctions are given in Figure 2.4. According to Figure 2.4, the contrast of microtubules and tiny organelles with a phase difference $\lambda/100$ equals 0.38 at bias $\lambda/20$ and infinity extinction, 0.27 at bias $\lambda/20$ and extinction 100, and 0.06 at bias $\lambda/4$ and both extinctions. The contrast at the small bias is greater than 6 times the contrast at the large bias. However, it is very important to have a high extinction in order to effectively use a small bias for studying tiny structures. The contrast of cell walls and organelle edges with a phase difference $\lambda/10$ is 0.79 at bias $\lambda/20$ and infinity extinction, 0.72 at bias $\lambda/20$ and extinction 100, and 0.59 at bias $\lambda/4$ and both extinctions.

In addition to reducing contrast, the low extinction decreases the dynamic range of measurement and can cause a diffraction anomaly in the Airy pattern. The lower extinction is a significant problem in microscopes equipped with high NA lenses. To reduce the beam depolarization, we can use a polarization rectifier [33,34].

Most current DIC microscopes employ one of the following methods of changing bias: (a) lateral shift of the DIC prism using a screw, (b) the Senarmont compensator, or (c) liquid crystal variable retarder. In principle, it is also possible to use other means for the bias adjustment. For example, one can employ the Babinet-Soleil compensator [26], the Ehringhaus compensator [26,35], the Berek compensator [26,36], or its analogue made of quartz [37], electro-optic, or piezo-optical modulators [38]. All of these methods could be readily calibrated for the quantitative bias variation.

The total optical path difference (bias) between two interfering beams is created by combination of two DIC prisms (see Figure 2.1). Each of the prisms introduces bias, which is not uniform and has a gradient. However, the prisms are oriented such that they mutually compensate the bias gradient. Thus, the total bias distribution becomes even across the objective back focal plane. The bias can be changed by a lateral shifting of one of the DIC prisms along the bias gradient direction. As it was shown in Section 2.1 (formula (2.4)), the unitless bias gradient equals the shear angle (in radians). The linear equation for the current bias $\Gamma(x)$ is written:

$$\Gamma(x) = \Gamma_0 + \varepsilon(x - x_0) \tag{2.16}$$

where x and x_0 are the current and initial positions of the prism and Γ_0 is the initial bias. The shear angle ε can be found in Table 2.1 or it can be measured, as described in the previous section.

Usually, the DIC prism is shifted by a translation screw. Then, the bias variation is determined using the pitch of a screw thread p and revolution number R in the following way:

$$\Gamma(R) = \Gamma_0 + \varepsilon p R \tag{2.17}$$

For example, the Olympus high-resolution DIC prism U-DICTHR has a translation screw with pitch of 2.5 mm. The screw allows a maximum of five rotations. According to Table 2.1, the prism shear angle is 40 μrad. Using Eq. (2.17), the bias variation $\Gamma = 100$ nm per 360° turn of the screw and the total range of the bias change $\Gamma_{tot} = 500$ nm is calculated. The General Olympus DIC prism U-DICT has 3 mm screw pitch and maximum

5 rotations. Consequently, the bias variation Γ is 225 nm per screw turn and the total bias range Γ_{tot} is 1125 nm.

As described in Section 2.1, the Senarmont technique produces bias, which is linearly proportional to the analyzer rotation. Using formulae (2.7) and (2.5) we can find dependence of the bias Γ on the analyzer orientation angle χ:

$$\Gamma = \frac{\chi}{180°} \lambda \qquad (2.18)$$

Rotation of the analyzer from $-90°$ to $90°$ introduces bias from $-\lambda/2$ to $\lambda/2$. The total bias range equals to one wavelength. The high bias accuracy is one of the advantages of the Senarmont method. For example, measuring the analyzer orientation with precision $0.1°$ will give the bias accuracy of 0.3 nm at wavelength 550 nm.

The liquid crystal variable retarder creates a bias, which depends on the applied voltage. However, this dependence is not linear. Therefore, it is necessary to have a calibration curve. The thinner liquid crystal variable retarders are faster and provide the bias change less than one wavelength. The thicker liquid crystals could give the bias change about two wavelengths. But they are considerably slower and more sensitive to the ambient temperature variation.

In order to create a large bias, it is possible to insert in the optical path an additional full-wave plate, which is also called the unit retardation plate or red plate [26]. This birefringent plate is cut of such a thickness as to have a retardation of one wavelength of yellowish green (550 nm), thus giving the sensitive first-order red between crossed polars in white light. Slight changes in the optical path difference by addition or subtraction of path differences by the specimen over which the plate is placed are very noticeable, and accordingly the plate is called a sensitive tint plate [39]. The color variation can be used for estimation of optical path difference of the specimen.

Figure 2.5 shows DIC images of 7 µm diameter glass rod immersed in Fisher Permount mounting medium (Fisher Scientific, http://www.fishersci.com). The refractive indices of the glass rod and the Permount at wavelength 546 nm were 1.554 and 1.524, respectively. The pictures were taken in white light using an Olympus BX-61 upright microscope equipped with the high-resolution DIC prism U-DICTHR, silicon oil objective lens UplanSApo30×/ 1.05Sil, photo eyepiece PE2.5× (Olympus America, http://www.olympusamerica.com), and Hamamatsu 3CCD Cooled Digital Color Camera ORCA-3CCD (Hamamatsu Photonics, http://www.hamamatsu.com). The shear plane is oriented in the northwest direction.

All images were captured with same exposure time, 0.1 s. Figure 2.5A represents a case when the bias is zero. Figure 2.5B was taken with bias of 100 nm (one full rotation of the translation screw). As one can see, the right bottom side of the rod is black. This means that

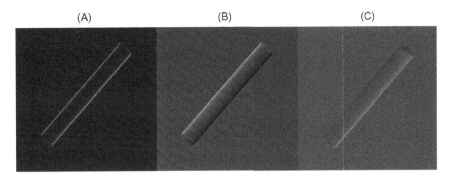

Figure 2.5

DIC images of 7 μm diameter glass rod captured in white light with zero bias (A), 100 nm bias (B), and 550 nm bias (C).

the optical path difference introduced by the rod side is subtracted by same amount of the bias (see formula (2.2)), and equals 100 nm. Figure 2.5C illustrates a case when bias is 550 nm. A full-wave plate U-TP530 was inserted in the optical path. Because the optical path difference is about one wavelength, the image shows Newton's interference colors. Then, the Michel-Levy chart is used to estimate the optical path difference [9,39]. The blue and orange colors of the rod sides correspond to 650 and 450 nm of optical path difference, respectively. After subtraction of the bias, the rod sides are shown to introduce about 100 nm of optical path difference. According to Table 2.1, the shear distance between the two interfering beams is 240 nm. Then, the optical path difference gradient of the rod side is 0.4 nm/nm. Hence, conventional DIC technique with calibrated bias can provide quantitative results.

2.4 A Quantitative OI-DIC Microscope with Fast Modulation of Bias and Shear Direction

OI-DIC technique records phase gradients within microscopic specimens quantitatively and independently of their orientation [6,20,21]. Unlike the other forms of phase and interference microscopes, this approach does not require a narrow illuminating cone. The new system is probably one of the few, if not the one and only, microscope system, which allows the generation of interference microscope images at truly high NAs.

A schematic of beam-shearing assembly with high speed switching shear direction and changing bias without any mechanical movement is shown in Figure 2.6 (center). The assembly consists of a linear polarizer P, first variable retarder (liquid crystal cell) LC1 with the principal axis orientated at 0°, a pair of DIC prisms (Nomarski or Wollaston) DIC1 and DIC2 with orthogonal shear directions, and second liquid crystal cell LC2 between the prisms. The second liquid crystal cell has the principal axis orientated at 45°.

32 Chapter 2

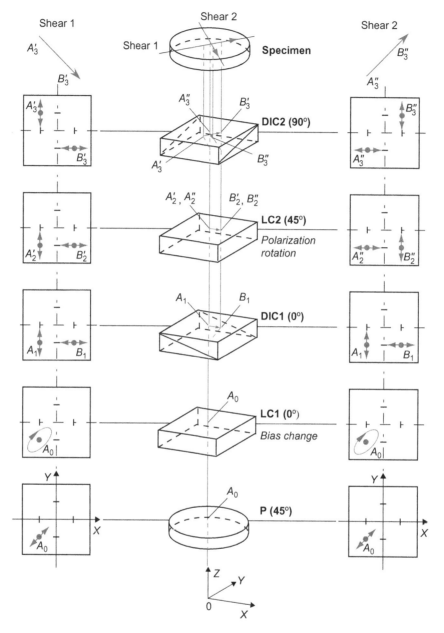

Figure 2.6
Schematic of switchable beam-shearing DIC assembly with liquid crystal variable retarders.

The shear direction of the first prism DIC1 is chosen as the initial X-direction to describe the orientation of the above-mentioned optical elements. The azimuth of the polarizer P is 45° in order to create equal intensities of polarization components that become spatially divided by the first prism. The first liquid crystal cell LC1 introduces bias Γ between the

X- and Y-polarization components, which depends on the applied voltage. The initial bias can be adjusted by sliding of the wedge component of one of the DIC prisms. In principle, the liquid crystal cell LC1 can be installed with the same orientation in any place in the assembly. The second liquid crystal cell LC2 switches the polarization by 90° when the plate retardation is a half wave ($\lambda/2$), and preserves the beam polarization if the plate retardance is 0 or full wave (λ).

The diagrams to the left and right in Figure 2.6 depict polarization transformations inside the assembly and the corresponding change of shear directions of the output beams. The left column illustrates a case when the beam polarizations between two prisms are preserved and the shear direction is 45° after the second prism DIC2. The right column describes a rotation of the output shear direction by 90° caused by switching the beam polarizations after the first prism DIC1.

Let us examine in more detail how the beam-shearing assembly works. For example, the initial ray A_0 falling on the first prism DIC1 has coordinates $(-1,-1)$. Here, the first number means a coordinate along the X-axis, and the second one is the coordinate along the Y-axis. The first DIC prism does not change the position of the ray with Y-polarization and deflects the ray with X-polarization, which creates a shear distance of two units between the two components. Thus, the first output ray A_1 has Y-polarization and coordinates $(-1,-1)$. If the shear value equals 2 units, then the X-polarized second ray has B_1 coordinates $(1,-1)$.

The second liquid crystal cell LC2 preserves the linear polarized states of the beams A_1 and B_1 or turns each of the states by 90° without altering the ray positions. In the first case (left column), the output rays A_2 and B_2 are polarized along the Y- and X-axis, respectively. In the second case (right column), the rays A_2 and B_2 are polarized along the X- and Y-axis, respectively.

The second DIC prism is oriented orthogonal to the direction of the first prism. So, the shear direction of the second prism lies along the Y-axis. The prism DIC2 does not change the position of a ray with X-polarization, but it deflects a ray with Y-polarization. The shear value for this second prism is also 2 units. First, we consider a case, which is shown on the left side of Figure 2.6, where the liquid crystal cell LC2 has retardance 0°, and therefore the ray polarizations are not changed. Then, ray A_2 with Y-polarization and coordinates $(-1,-1)$ passes the second prism with displacement by 2 units in the Y-direction, and the output ray A_3' has coordinates $(-1,1)$. Meanwhile, the coordinates of ray B_2 are not changed, and the coordinates for output ray B_3' are $(1,-1)$. As a result, in the first case, the dual beam that falls on the specimen has shear direction $-45°$.

For the second case, after the liquid crystal cell LC2 rotates the polarization by 90°, the ray A_2 is linearly polarized along the X-axis. Therefore, its coordinates are not changed by the second DIC prism, and the output ray A_3'' has coordinates $(-1,-1)$. However, the second DIC prism moves the Y-polarized ray B_2 so that the second output ray B_3'' has coordinates $(1,1)$. Hence, the output beam has a shear direction $+45°$.

In both the cases, the shears are the same, which is equal to the shear introduced by a single DIC prism multiplied by $\sqrt{2}$. However, the shear directions are mutually orthogonal. The median axis of two beams passes through point $(0,0)$ in both the cases. Therefore, there is no misalignment between the images.

The intensity distribution in the image will be described by the transformed formula (2.2):

$$I_i(x,y) = \tilde{I} \sin^2\left\{\frac{\pi}{\lambda}\left[\Gamma + \sqrt{2}d\gamma(x,y)\cos\left(\theta(x,y) - (-1)^i\frac{\pi}{4}\right)\right]\right\} + I_c(x,y) \tag{2.19}$$

where $i = 1, 2$ corresponds to the first or second state of shear direction, $-45°$ or $+45°$.

In order to find the two-dimensional distribution of magnitude and azimuth γ and θ, we capture two sets of raw DIC images at shear directions $-45°$ and $+45°$ with negative, zero, and biases: $-\Gamma$, 0, and $+\Gamma$ [6,20]. The following group of equations represents these six DIC images:

$$I_{ij}(x,y) = \tilde{I} \sin^2\left\{\frac{\pi}{\lambda}\left[j\Gamma + \sqrt{2}d\gamma(x,y)\cos\left(\theta(x,y) - (-1)^i\frac{\pi}{4}\right)\right]\right\} + I_c(x,y) \tag{2.20}$$

where $j = -1, 0, 1$.

Initially two terms are computed ($i = 1, 2$):

$$A_i(x,y) = \frac{I_{i,1}(x,y) - I_{i,-1}(x,y)}{I_{i,1}(x,y) + I_{i,-1}(x,y) - 2I_{i,0}(x,y)} \tan\left(\frac{\pi\Gamma}{\lambda}\right) \tag{2.21}$$

Using Eq. (2.20), we can show that

$$\begin{aligned} A_1(x,y) &= \tan\left(\frac{2\sqrt{2}\pi}{\lambda}d\gamma(x,y)\cos\left(\theta(x,y) + \frac{\pi}{4}\right)\right) \\ A_2(x,y) &= \tan\left(\frac{2\sqrt{2}\pi}{\lambda}d\gamma(x,y)\sin\left(\theta(x,y) + \frac{\pi}{4}\right)\right) \end{aligned} \tag{2.22}$$

Using the obtained terms, we can calculate the quantitative two-dimensional distributions of the gradient magnitude and azimuth of optical paths in the specimen as:

$$\begin{aligned} \gamma(x,y) &= \frac{\lambda}{2\sqrt{2}\pi d}\sqrt{(\arctan A_1(x,y))^2 + (\arctan A_2(x,y))^2} \\ \theta(x,y) &= \arctan\left(\frac{\arctan A_2(x,y)}{\arctan A_1(x,y)}\right) - \frac{\pi}{4} \end{aligned} \tag{2.23}$$

The gradient magnitude represents the increment of the optical path difference, which is in nanometers, along the lateral coordinate, which is also in nanometers. Thus, the gradient magnitude is unitless. The shear amount d can be measured as it is described in Section 2.1 or found in Table 2.1.

Note that the algorithm considered earlier employs ratios between intensities of light that have interacted with the specimen. Therefore, it suppresses contributions of absorption by the specimen or from nonuniformity of illumination, which can otherwise deteriorate a DIC image.

Also, after computing the optical path gradient distribution, enhanced regular DIC images $I_{enh}(x,y)$ can be restored with any shear direction σ, different bias Γ, and another shear amount d using the next formula:

$$I_{enh}(x, y) = \sin^2\left\{\frac{\pi}{\lambda}[\Gamma + d\gamma(x, y)\cos(\theta(x, y) - \sigma)]\right\} \quad (2.24)$$

The enhanced image provides a calculated image for any desired shear direction and bias without the requirement to directly collect an image for that shear direction and bias. Moreover, the enhanced image will have less noise than a regular DIC image, and it suppresses deterioration of the image due to specimen absorption and illumination nonuniformity.

Optical path difference shows the dry mass distribution of a specimen and can be obtained by computing a line integral [6,40]. Also, other techniques for phase computation can be used, for instance, iterative computation [41], noniterative Fourier phase integration [15], or nonlinear optimization with hierarchical representation [42]. Biggs has developed an iterative deconvolution approach for computation of phase images, based on the same principles as deconvolution techniques normally used to remove out-of-focus haze [43–45].

In principle, a regular research grade microscope equipped with DIC optics can be modified for obtaining OI-DIC images. The setup was implemented on an upright microscope Olympus BX-61 (Olympus, Center Valley, PA, http://www.olympus.com). We built two custom beam-shearing DIC assemblies, one placed in the illumination path and the another in the imaging path (Figure 2.7). Using MATLAB (The MathWorks Inc., Natick, MA, http://www.mathworks.com/), we developed software for setup control and image processing.

Figure 2.8 shows an example of the phase OI-DIC image of a live crane fly spermatocyte during metaphase of meiosis I taken with an Olympus UPlanFl 100×/1.30 oil immersion objective lens and 546/30 nm interference filter. Image size is 68 µm × 68 µm. Here, the image brightness is linearly proportional to the phase (refractive index) distribution. The image acquisition and processing took about 1 s each. The three autosomal bivalent

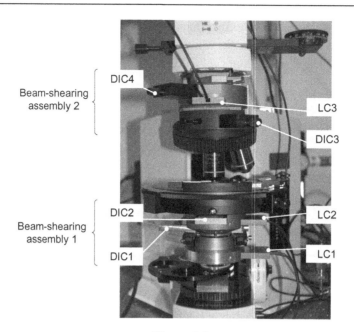

Figure 2.7
OI-DIC microscope setup with two switchable beam-shearing DIC assemblies. The first assembly consists of DIC prisms DIC1 and DIC2 and two liquid crystal variable retarders LC1 and LC2. The second beam-shearing DIC uses prisms DIC3 and DIC4 and liquid crystal variable retarders LC3.

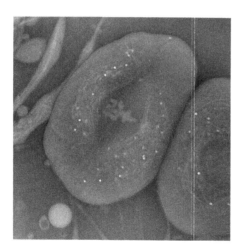

Figure 2.8
Phase OI-DIC image of live crane fly spermatocyte at metaphase of meiosis I.

chromosomes are in sharp focus at the spindle equator, along with one of the $X-Y$ sex univalents, which is located on the right. The tubular distribution of mitochondria surrounding the spindle is clearly evident. Both polar flagella in the lower centrosome are in focus, appearing as a letter "L" lying on its side. The experiment was done together with

Prof. James LaFountain (State University of New York at Buffalo, Buffalo, NY). The phase image was computed by Dr. David Biggs (KB Imaging Solutions, Waterford, NY) by employing the iterative deconvolution approach mentioned earlier.

The following are the comments made on this picture by MBL Distinguished Scientist Shinya Inoué:

> *The image is a real WOW! That is so striking; I have never seen such a view of a dividing cell, ever! Absolutely; have a huge blow up of this image for your poster at ASCB. The crane fly spermatocyte meiosis image is just mind blowing. You see the many thin-thread-shaped mitochondria surrounding the spindle, the chromosomes themselves and even the spindle fibers, all in striking 3-D. Also the scattered dyctiosomes show as prominent bright spots. But even more, the image shows some cytoplasmic structures that I had never seen in my life. Those must relate to Keith Porter's endoplasmic reticulum (membrane-related structures?) but in a different form. Their contrast is low but they definitely appear to be indented where astral rays would be expected. I don't think anyone has shown such structural differentiation of the cytoplasm until now, using any mode of microscopy whether in live or fixed and stained cells! In a nutshell this image shows the unusual capability of your orientation independent DIC system exceptionally well.*

2.5 Combination of OI-DIC and Orientation-Independent Polarization Imaging

DIC microscopy produces images of optical phase gradients in a transparent specimen, which is caused by variation of refractive index within thin optical section. Polarized light microscopy reveals structural or internal anisotropy due to form birefringence, intrinsic birefringence, stress birefringence, and other factors. An image in DIC microscope is determined by the optical phase distribution in the specimen, while an image in a polarized light microscope is produced by the polarization splitting of the optical phase caused by the specimen anisotropy. Thus, polarization microscopy data and DIC results are complementary. Both methods, however, have the same shortcomings: they require the proper orientation of a specimen in relation to the optical system in order to achieve high quality results, and the images are not quantitative. Hence, it would be beneficial to combine the OI-DIC microscope with an orientation-independent differential polarization (OI-Pol) system [45].

The differential OI-Pol microscope captures several images of the specimen with slightly different polarization settings of illumination or imaging beam [29,45–47]. Then, the images are subtracted one from another and further processed in order to obtain the quantitative picture of retardance and fast axis azimuth distribution.

The combined system is unique, providing complementary phase images of thin optical sections of the specimen that display a distribution of refractive index gradient and

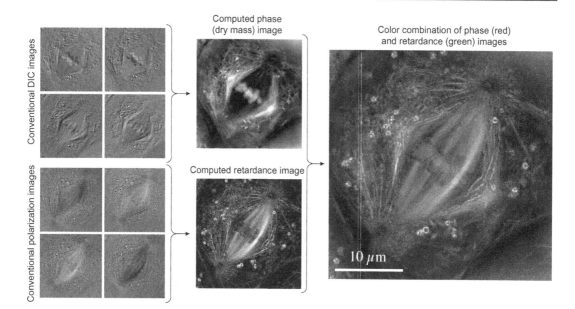

Figure 2.9
Metaphase of meiosis I in a crane fly spermatocyte. Color combination of phase (red) and retardance (green) images. (For interpretation of the references to color in this figure legend, the reader is referred to the web version of this book.)

distribution of birefringence due to structural or internal anisotropy of the cell structure. For instance, in a live dividing cell, the OI-DIC image clearly shows detailed shape of the chromosomes while the polarization image quantitatively depicts the distribution of the birefringent microtubules in the spindle, both without any need for staining or other modifications of the cell.

Figure 2.9 presents combined pseudo-color OI-DIC and OI-Pol image of spermatocytes from the crane fly, *Nephrotoma suturalis*, during metaphase of meiosis I. The experiment was done together with LaFountain. The phase image was computed by Biggs. The set-up was a Nikon Microphot-SA microscope equipped 60×/1.4NA oil immersion at wavelength 546 nm [45]. The changing of bias and rotating the shear direction during DIC image acquisition were done manually. Also, in order to switch between polarization and DIC imaging modes, the pair of liquid crystal waveplates was replaced with DIC prisms. At present, these mechanical manipulations take significant time relative to the temporal resolution of the test. A microscope which rapidly switches between the OI-DIC and the OI-Pol modes without any mechanical movement is theoretically possible and a prototype is in development.

The figure shows a group of 4 conventional DIC images (left top) and a group of 4 conventional polarization images (left bottom). These raw images were employed for computing phase (dry mass) and retardance images (middle top and bottom, accordingly). On the far right is a color combination of the phase (dry mass) and retardance modes, in which red and green colors correspond to dry mass distribution and retardance, respectively. Morphological structures, such as chromosomes, are especially prominent in the phase mode image. The birefringent spindle fibers (actually bundles of microtubules) exhibit much better contrast in the retardance mode. The combined picture provides clear evidence of our notion that the proposed technique can reveal architecture (morphology) of live cells without fluorescent labeling using 1.4NA optics.

2.6 Combination of OI-DIC and Fluorescence Imaging

In recent years, advances in fluorescent biosensors have made fluorescence imaging of living cells a key tool for cell biologists. However, quantitative data such as refractive indices and birefringences of whole specimens provide important information about protein concentration, density, and structural organization inside cells, but cannot be measured using fluorescence imaging. Although OI-DIC microscopy provides an extraordinary level of detail, the technique is more informative of cellular structures than of specific molecules. Fluorescence microscopy, on the other hand, provides detailed information on the distribution patterns of specific molecules and ions, but this information is often hard to interpret in the absence of additional structural information.

Combining fluorescence microscopy with OI-DIC imaging is an ideal technology that provides specificity of fluorescence and quantification of OI-DIC. For example, the type of organelles in a live specimen could be scored using fluorescence markers, and subsequent organelle development followed during a long time series using the OI-DIC with minimal phototoxicity.

Two schematics of the OI-DIC in combination with fluorescence microscopy were built: (1) simultaneous OI-DIC and fluorescence with two CCD cameras and (2) sequential OI-DIC and fluorescence with a switchable beamsplitter cube. Both approaches have been used successfully.

Figure 2.10 shows simultaneous images of chromosomes in a live *Spisula* oocyte treated with Hoechst 33342 (DNA-specific fluorescent dye) obtained by OI-DIC microscopy and wide field fluorescence microscopy. The micrographs were taken with a new Olympus 30×/1.05NA silicon oil objective lens UPLSAPO 30XS and photo eyepiece PE2.5×. The fluorescence image contains four selected square areas with chromosomes. These areas are enhanced in the computed quantitative gradient and phase images. The white level is 0.1 nm/nm in the enhanced areas and 0.3 nm/nm in other areas of the gradient picture. Brighter OI-DIC signal in the phase image indicates higher optical density, which corresponds to higher dry mass or concentration of the objects. The dry mass image of

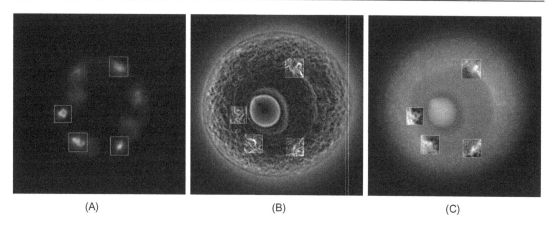

Figure 2.10
Simultaneous observation of live *Spisula* oocyte treated with DNA-specific fluorescent marker (Hoechst 33342) by (A) fluorescence and (B and C) OI-DIC microscopy techniques—(B) gradient and (C) phase (dry mass). Image size is 64 μm × 64 μm.

chromosomes is somewhat larger than the fluorescence image because it shows both proteins and DNA, while the fluorescence depicts stained DNA only.

This result indicates that OI-DIC can be used to visualize chromosomes in living cells and provides sufficient detail to recognize chromosomes by their dynamic individual three-dimensional shapes. OI-DIC imaging assessed the distribution of DNA in healthy living cells without using fluorescent DNA markers and irradiating with excess excitation light. Time-lapse imaging of DNA distribution during the mitotic phase of the cell cycle by this noninvasive methodology is an important tool. Moreover, as OI-DIC provides a two-dimensional distribution of the amount of DNA within an extremely thin optical section, OI-DIC enables reconstruction of the three-dimensional distribution of DNA in the nucleus.

Acknowledgment

This work was supported by NIH Grant Number R01-EB005710.

References

[1] F.H. Smith, Interference microscope, US Patent 2601175, (5 August 1947).
[2] F.H. Smith, Microscopic interferometry, Research (London) 8 (1955) 385.
[3] A.A. Lebedeff, Polarization interferometer and its applications, Rev. Opt. 9 (1930) 385.
[4] G. Nomarski, Interferential polarizing device for study of phase object, US Patent 2924142 (14 May 1952).
[5] R.D. Allen, G.B. David, G. Nomarski, The Zeiss–Nomarski differential equipment for transmitted light microscopy, Zeitschrift für Wissenschaftliche Mikroskopie und Mickroskopische Technik 69(4) (1969) 193.

[6] M. Shribak, S. Inoué, Orientation-independent differential interference contrast microscopy, Appl. Opt. 45 (2006) 460.

[7] M. Pluta, Advanced Light Microscopy. Vol. 2: Specialized Methods, Elsevier Science Publishing Co., Inc., New York, NY, 1989.

[8] S. Inoué, K.R. Spring, Video Microscopy: The Fundamentals, second ed., Plenum Press, New York, NY, 1989.

[9] R. Oldenbourg, M. Shribak, Microscopes, in: M. Bass (Ed.), Handbook of Optics, third ed., volume I: Geometrical and Physical Optics, Polarized Light, Components and Instruments, McGraw-Hill Professional, New York, NY, 2009 (chapter 28).

[10] S. Inoué, Ultrathin optical sectioning and dynamic volume investigation with conventional light microscopy, in: J. Stevens, L. Mills, J. Trogadis (Eds.), Three-Dimensional Confocal Microscopy: Volume Investigation of Biological Systems, Academic Press, San Diego, CA, 1994, pp. 397–419.

[11] G.M. Holzwarth, S.C. Webb, D.J. Kubinski, N.S. Allen, Improving DIC microcopy with polarization modulation, J. Microsc. 188(Pt 3) (1997) 249.

[12] G.M. Holzwarth, D.B. Hill, E.B. McLaughlin, Polarization-modulated differential-interference contrast microscopy with a variable retarder, Appl. Opt. 39 (2000) 6288.

[13] H. Ooki, Y. Iwasaki, J. Iwasaki, Differential interference contrast microscope with differential detection for optimizing image contrast, Appl. Opt. 35 (1996) 2230.

[14] P. Hariharan, M. Roy, Achromatic phase-shifting for two-wavelength phase-stepping interferometry, Opt. Commun. 126 (1996) 220.

[15] M.R. Arnison, K.G. Larkin, C.J.R. Sheppard, N.I. Smith, C.J. Cogswell, Linear phase imaging using differential interference contrast microscopy, J. Microsc. 214(Pt 1) (2004) 7.

[16] S.V. King, A.R. Libertun, C. Preza, C.J. Cogswell, Calibration of a phase-shifting DIC microscope for quantitative phase imaging, Proc. SPIE 6443 (2007) 64430M.

[17] H. Ishiwata, M. Itoh, T. Yatagai, A new method of three-dimensional measurement by differential interference contrast microscope, Opt. Commun. 260 (2006) 117.

[18] H. Ishiwata, M. Itoh, T. Yatagai, A new analysis for extending the measurement range of the retardation-modulated differential interference contrast (RM-DIC) microscope, Opt. Commun. 281 (2008) 1412.

[19] A. Noguchi, H. Ishiwata, M. Itoh, T. Yatagai, Optical sectioning in differential interference contrast microscopy, Opt. Commun. 282 (2009) 3223.

[20] M. Shribak, Orientation-independent differential interference contrast microscopy technique and device, US Patent 7233434, (17 December 2003).

[21] M. Shribak, Orientation-independent differential interference contrast microscopy technique and device, US Patent 7564618 (17 December 2003).

[22] E.B. van Munster, L.J. van Vliet, J.A. Aten, Reconstruction of optical path length distributions from images obtained by a wide-field differential interference contrast microscope, J. Microsc. 188 (1997) 149.

[23] S.B. Mehta, C.J.R. Sheppard, Sample-less calibration of the differential interference contrast microscope, Appl. Opt. 49 (2010) 2954.

[24] C.B. Müller, K. Weiß, W. Richtering, A. Loman, J. Enderlein, Calibrating differential interference contrast microscopy with dual-focus fluorescence correlation spectroscopy, Opt. Express 16 (2008) 4322.

[25] D.D. Duncan, D.G. Fischer, A. Dayton, S.A. Prahl, Quantitative Carré differential interference contrast microscopy to assess phase and amplitude, J. Opt. Soc. Am. A 28 (2011) 1297.

[26] N.H. Hartshorne, A. Stuart, Crystal and the polarizing microscope, fourth ed., Edward Arnold, London, 1970.

[27] M.M. Swann, J.M. Mitchison, Refinements in polarized light microscopy, J. Exp. Biol. 27 (1950) 226.

[28] M. Shribak, R. Oldenbourg, Sensitive measurements of two-dimensional birefringence distributions using near-circularly polarized beam, in Polarization Analysis, Measurement, and Remote Sensing V, D.H. Goldstein and D.B. Chenault (Eds.), Proc. SPIE, 4819 (2002) 56.

[29] M. Shribak, R. Oldenbourg, Technique for fast and sensitive measurements of two-dimensional birefringence distribution, Appl. Opt. 42 (2003) 3009.

[30] E.D. Salmon, VE-DIC light microscopy and the discovery of kinesin, Trends Cell Biol. 5 (1995) 154.

[31] E.D. Salmon, P. Tran, High resolution video-enhanced differential-interference contrast (VE-DIC) light microscopy, Methods Cell Biol. 56 (1998) 153.

[32] B.J. Schnapp, View single microtubules by video light microscopy, Methods Enzymol. 134 (1986) 561.

[33] M. Shribak, S. Inoué, R. Oldenbourg, Rectifiers for suppressing depolarization caused by differential transmission and phase shift in high NA lenses, in Polarization Analysis, Measurement, and Remote Sensing IV, D.H. Goldstein and D.B. Chenault (Eds.), Proc. SPIE 4481 (2001) 163.

[34] M. Shribak, S. Inoué, R. Oldenbourg, Polarization aberrations caused by differential transmission and phase shift in high NA lenses: theory, measurement and rectification, Opt. Eng. 41 (2002) 943.

[35] D.A. Holmes, Wave optics theory of rotary compensators, J. Opt. Soc. Am. A 54 (1964) 1340.

[36] F. Rinne, M. Berek, Anleitung zu Optischen Untersuclhungen mit dem Polarizationsmikroskop, Schweizerbart'sche Verlagsbuchhandlung, Stuttgart, Germany, 1953.

[37] M. Shribak, Use of gyrotropic birefringent plate as quarter-wave plate, Sov. J. Opt. Technol. 53 (1986) 443.

[38] A. Yariv, P. Yeh, Optical Waves in Crystals: Propagation and Control of Laser Radiation, John Wiley & Sons, New York, NY, 1984.

[39] F.E. Wright, The Methods of Petrographic-Microscopic Research, Carnegie Inst., Washington, DC, 1911.

[40] B. Heise, A. Sonnleitner, E.P. Klement, DIC image reconstruction on large cell scans, Microsc. Res. Tech. 66 (2005) 312.

[41] C. Preza, Rotational-diversity phase estimation from differential-interference-contrast microscopy images, J. Opt. Soc. Am. A 17 (2000) 415.

[42] F. Kagalwala, T Kanade, Reconstructing specimens using DIC microscope images, IEEE Trans. Syst. Man Cybern. Part B Cybern. 33 (2003) 728.

[43] T.J. Holmes, S. Bhattacharyya, J.A. Cooper, D. Hanzel, V. Krishnamurthi, W. Lin, et al., Light microscopic images reconstructed by maximum likelihood deconvolution, in: J.B. Pawley (Ed.), Handbook of Biological Confocal Microscopy, Plenum Press, New York, NY, 1995, pp. 389–402.

[44] D. Biggs, M. Andrews, Acceleration of iterative image restoration algorithms, Appl. Opt. 36 (1997) 1766–1775.

[45] M. Shribak, J. LaFountain, D. Biggs, S. Inoué, Orientation-independent differential interference contrast microscopy and its combination with an orientation-independent polarization system, J. Biomed. Opt. 13 (2008) 014011.

[46] R. Oldenbourg, G. Mei, New polarized light microscope with precision universal compensator, J. Microsc. 180 (1995) 140.

[47] M. Shribak, Complete polarization state generator with one variable retarder and its application for fast and sensitive measuring of two-dimensional birefringence distribution, J. Opt. Soc. Am. A 28 (2011) 410.

CHAPTER 3

Long-Term Recordings of Live Human Cells Using Phase Contrast Microscopy

Yumi Uetake, Greenfield Sluder
Department of Cell Biology, University of Massachusetts Medical School, Worcester, MA

Editor: Lisa L. Satterwhite

3.1 Introduction

In this chapter, we describe a method based on phase contrast microscopy to conduct long-term (~5−6 days) continuous time-lapse observations of individual untransformed human cells and later conduct correlative immunofluorescence characterization of cells previously followed in vivo. From an instrumentation standpoint, there are multiple ways to prepare and observe cells; here, we describe what has worked well for us in a convenient and cost-effective fashion. We start with some general considerations of the optical systems suitable for live cell time-lapse analysis, and then we will cover the specifics of our methodology.

Phase contrast and differential interference contrast (DIC) microscopy, described in details in Chapters 1 and 2 respectively, are the most effective and commonly used methods for long-term live cell time-lapse observations. Fluorescence microscopy, used with fluorescent protein tagged cellular structures, has also been used to good effect for specific and generally shorter-term applications. However, this contrast mode suffers from the greatly increased likelihood of photodamage to the live cells. Cells are stressed by blue light used to excite green fluorescent protein (GFP) and we have found that green excitation light for the red fluorescent proteins is only slightly less damaging (Galdeen and Sluder, unpublished). Over time, photodamage leads to cell cycle arrest and eventually cell death. Short of cell death, a pernicious consequence of photodamage is compromised cell viability that can influence experimental outcome and lead to false conclusions.

We use phase contrast microscopy for long-term observations because it provides good contrast in a direct independent fashion and minimizes the amount of light that impinges upon the cells. DIC is a polarized light-based contrast mode and when used in a high extinction mode, relatively high light levels at the specimen are required to produce images of sufficient intensity to allow reasonable exposure times. Thus, for multiple day runs with untransformed cells which are more likely to undergo apoptosis (cell death) from photo-damage, that is, demonstrate a robust p53 response, phase contrast microscopy is a more reliable choice.

3.2 The Microscope

The comments that follow apply to both upright and inverted microscope stands. The microscopes of all major manufacturers, be they the older 160 mm tube length systems or the current infinity corrected systems, offer good quality optics for phase contrast imaging. Given that live cell imaging should be conducted with monochromatic light (typically green light), there is no need to buy expensive apochromatic objectives; cost-effective achromats will work just as well at equivalent numerical apertures.

Focus stability is what sets one microscope apart from another in terms of suitability for time-lapse imaging. Passive or intrinsic focus stability is sufficient for low magnification/low numerical aperture imaging. Passive focus stability also works for high magnification/high numerical aperture applications, but it is imperative to empirically test the focus stability of the stand under constant temperature conditions. We emphasize constant temperature conditions because all stands, regardless of design, can show significant X, Y, and Z movements at the specimen with temperature changes. We find that some of the 30–40 year old stands with gear-driven focus blocks do a fine job. Some of the modern motorized focus drives are also focus stable. Stands with recirculating ball focus drives are sometimes not focus stable. We note that all major manufacturers offer for their newest generation stands optical feedback systems that use an infrared beam to monitor the position of the coverslip–water interface and drive the motorized focus to maintain a set focus position with high fidelity. These systems, though costly, work well and can be important for high magnification applications. We do not recommend the use of software-based auto focus routines because they can establish an undesired plane of focus within or outside the cell, and they take time which exposes the cells to more light.

Time-lapse imaging using a microscope with a conventional stage limits observations to a single field of cells for the duration of the film run. To increase the number of cells observed during a filming session, one can equip the microscope with a motorized stage and a motorized focus mechanism, both under computer control. In this way, one can concurrently film a number of fields, each indexed for X, Y, and Z positions. It is important to set up the acquisition routine so that image sequences from each field are put into separate files.

3.2.1 Illumination

Attention must also be paid to the illumination parameters. Phase contrast is not light intensive so tungsten illuminators with voltage stable power supplies work well. For our systems without built-in power supplies, we have replaced the simple rheostat-type power supplies with laboratory grade voltage and current stabilized power supplies. Since most of the output of tungsten lamps are in the infrared, one should use a good quality "heat cut" or long wavelength cutoff filter. Digital cameras are to some extent sensitive to infrared wavelengths and if they do not have a built-in infrared cutoff filter, there is the possibility of having a poorly focused and poorly corrected infrared image superimposed on the visible light image. For wavelength selection, we use good quality interference filters, not gelatin filters that pass a wide range of wavelengths. For our computer-driven microscopes, we shutter the transmitted light path; this allows us to raise the lamp intensity to allow relatively short exposures for multiple field acquisition while minimizing the dosage of light given to the cells. For our simple microscope systems with which we follow only a single field, we use continuous illumination, cut back the intensity with neutral density filters, and use long exposures integrated on the camera chip. We normally use green (546 nm) light; for cells that are extremely photosensitive one could use red light (~620 nm) which is substantially less phototoxic (Galdeen and Sluder, unpublished). Lastly, it is of great importance to properly align the microscope illumination pathway for Köhler illumination to ensure even illumination of the field. Image acquisition parameters set for high gain and contrast stretch accentuate even subtle variations in illumination intensity across the field of view.

3.2.2 Temperature Control

Live cell imaging of mammalian somatic cells requires precise 37°C temperature control of the specimen with little temperature fluctuations. If the temperature rises too high, even transiently, the cells become stressed and viability is compromised. Since stress can be persistent and additive (see Ref. [1]), temperature overshoot can sensitize cells to other environmental or experimental parameters that alone would not be a problem. If the temperature is even a few degrees too low the timing of events will not be normal and may not reflect the true kinetics of any given process [2]. Some processes may not be linear with temperature. For example, shifting HeLa cells from 37 to 33°C doubles their generation time and further reductions in temperature can differentially affect the kinetics of different cell cycle stages [2,3].

There are a number of ways to control the specimen temperature, but here we discuss the strategy we have found to be effective, safe, and reliable. The web pages listed at the end of this chapter provide a few sources of information that lay out other control strategies. To start, the entire microscope is covered with a box leaving the illuminator and camera outside the enclosure. For our simple systems, we use a corrugated cardboard box cut to provide doors for access to the stage areas (Figure 3.1A).

For our more complicated systems with motorized stages our machine shop built, Plexiglas enclosures with slider doors high on the front to allow access to the stage and slider doors

 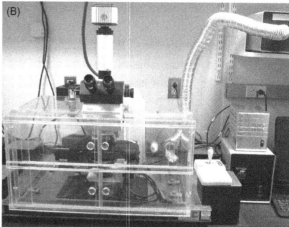

Figure 3.1
(A) Microscope enclosure made from a cardboard box. The box has cuts to provide "doors" to access the stage area, focus knobs, and the photochanger headpiece of the microscope. Eye pieces and the camera are not enclosed inside the box. For microscopes with a tungsten illuminator being run at moderate voltages, the illuminator can be left inside the box. (B) Plexiglas enclosure for an upright microscope with slider doors which allow access to the stage, switches, and other controls. The round "buttons" on the front of the enclosure are the handles for the slider doors. The microscope and the enclosure are on a vibration isolation table that has a "bread board" surface with tapped holes for screws. The enclosure is made in three pieces (front and two sides) that are held down with metal tabs that are held with bolts screwed into the top of the anti-vibration table. Also visible in the upper right corner of the image is the heater unit with an aluminum air duct going to the enclosure.

lower on the front to give access to the microscope switches and other controls (Figure 3.1B). For those who do not have the ability to have enclosures made, all major microscope manufacturers and independent suppliers offer Plexiglas enclosures that cover the upper portions of their microscopes. We cover the entire microscope because this keeps the stand at a constant temperature thereby obviating the otherwise inevitable changes in specimen position and focus if the room temperature changes. In addition, we can readily switch between lower and higher magnification immersion objectives without the bother of using objective heaters and changing the wiring from one lens heater to the other. We run the temperature control systems continuously; this allows us to put cell preparations on the microscope on short notice without having to wait for the microscope to warm up and become dimensionally stable. Also, the lack of temperature cycling reduces the chances that the glass in the condenser and objective lenses will become strained. Strain in the optics degrades the extinction, and hence contrast, for polarizing and DIC microscopes.

To heat the microscope enclosure one can buy heater/control systems integrated into a single box (e.g., the "Air-Therm ATX" from World Precision Instruments Inc., Sarasota,

FL). Alternatively, one can use separate control and heating/fan units (e.g., see products from Farnam Incorporated, Arden, NC). We build our temperature control units from parts obtained from Omega Engineering (Stamford, CT). We have had good results with proportional drivers for the heating elements that clip the waveform of the AC current driving the heating element and relay-driven "on-off" power sources. For the latter, we recommend against control units with mechanical relays because they eventually fail. Instead, we use solid-state relays that are sized to more than adequately handle the maximum wattage of the heating elements. We also recommend against using high thermal mass temperature sensors, such as fluid-filled bulbs that lead to a mechanical relay through a metal tube. These have slow response times and consequently produce long on-off cycles that lead to temperature cycles. This, in our experience, can result in dimensional changes in the microscope and cyclical movement of the specimen.

In selecting an air heater, care must be taken to ensure safety and durability. If the microscope is to be left unattended for significant periods of time (e.g., overnight and weekends), we recommend against consumer hair dryers rewired so that the heating coil is driven by the control box. These devices are noisy and the fans will eventually fail with prolonged usage. When the fan fails, the box cools and the heater is driven constantly without air flow thereby risking a fire. Commercial grade heat guns designed for heat shrink tubing applications also work well but should be replaced after approximately a year of continuous use to avoid fan failure. We use custom heaters built by our machine shop that consist of a metal box containing a steel tube enclosing a 250 W heater element. At one end we mount a ball bearing Muffin style fan and at the other we have a snout over which we slip a 2–3 inch aluminum heater hose. In addition, we mount a cut out thermostat (Omega Engineering, Stamford, CT) on the steel tube that cuts power to the heating element should the fan fail and the tube becomes overly hot. Lastly, we use a double pole single throw switch to control power so that the fan and power feed to the heater are coordinately turned on or off. That way the heater cannot be energized unless the fan is on. Powering the heating element without the fan running poses obvious fire dangers.

We favor using lightweight flexible aluminum air ducts because they can be bent to the desired shape and they will hold the shape without support. Also, we have found that sufficiently hot air will extract oils from plastic hoses and deposit them on the microscope.

3.2.3 Other Considerations

For the use of sealed preparations (chambers described later, mineral oil capped dishes, or Rose chambers), a single warmed air feed to the microscope enclosure works well. For open dish preparations, one must control CO_2 and humidity. To do so we configured the heater unit with inlet and outlet snouts to allow for delivery and recovery air ducts. Recirculating the warmed air maintains the humidity and less CO_2 is expended. At the heater box, we tapped in a tube to deliver CO_2 that has come through a flow meter. We monitor CO_2 levels with a separate portable CO_2 meter and run the system at steady state. In principle, one could alternatively set up an integrated system with a CO_2 monitor that controls gas flow into the air system. Humidity can be maintained by putting into the end of the warm air feed at the enclosure a small reservoir for fluid with a paper towel wick. We suggest using Phosphate buffered saline (PBS) instead of distilled water to prevent osmotic loading of water into the dish of cells.

3.2.4 Observation Chambers

We routinely use sealed chambers that allow for uninterrupted ∼100 h film runs without loss of cell viability or slowing of the cell cycle at later times for untransformed and transformed human somatic cells. Film runs are typically terminated when the cultures become confluent (Figure 3.2).

Details of chamber construction are described in Sluder et al. [4]. The only change we have made in the assembly of the chambers is that we no longer introduce fluorocarbon oil into the preparations.

Briefly, the chambers consist of a 1 × 3 inch aluminum slide 3 mm thick, with one or two square cutouts (14 mm × 14 mm) smaller than a 22 mm × 22 mm coverslip for ample overlap of the coverslip over the metal support to ensure a tight and durable seal (Figure 3.3).

We use only #1.5 (0.17 mm thick) coverslips for which all common objectives are corrected. Use of any other thickness of coverslip produces spherical aberration that substantially degrades image contrast and intensity. The thickness of the slide is intended to

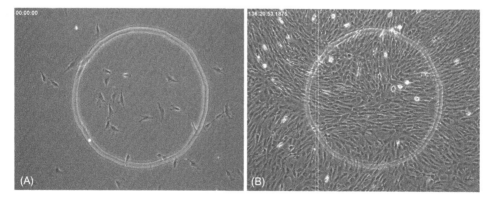

Figure 3.2
Example of RPE1 cell propagation at the start and end of a 136 h filming run. A and B are images of the same field marked by the scribed circle. In this particular case, the medium was not changed during the film run. *Source*: Preparation and images courtesy of Dr. Anna Krzywicka-Racka.

Figure 3.3
A two "window" cell preparation for filming. (A) Preparation seen from the side to show the raised ridges of both sides at the ends of the aluminum support slide. (B) Preparation as seen from the top. The left "window" is an assembled cell preparation for which the top and bottom coverslips are held on only with silicone grease. The right "window" is an assembled preparation for which a thin bead of VALAP is used to strengthen the seating of the cell bearing coverslip. Also the coverslip is marked with a felt tipped pen to indicate the orientation of the preparation for its removal and replacement on the stage of the microscope. (C) Enlarged image of the portion of the preparation enclosed by the dotted line in panel B. This enlargement provides better views of the VALAP bead at the margin of the coverslip and the pen line inked on the coverslip. These two "window" preparations are well suited for use with a dry objective with sufficient working distance to clear the VALAP bead mounted on a microscope with a motorized stage to allow filming of multiple fields in both windows. With two preparations on the same slide, one can image cells subjected to two experimental conditions or control plus experimental samples. For immersion objectives with short working distances, one should use only one "window" of the two or use a support slide with only one "window."

coordinately maximize the volume of the medium while maintaining optical properties that allow Köhler illumination with standard condensers. The ridges going across the end of the slide on both the top and the bottom surfaces provide clearance for the top and the bottom coverslip so that they do not drag on the stage. These ridges are put on both top and bottom surfaces so that the preparation can be moved between an upright and an inverted microscope without having to be rebuilt.

3.2.5 Preparing Chambers

1. Coverslips should be cleaned to remove oils and contaminants carried over from manufacturing and handling. We soak the coverslips in dilute Liquinox (Alconox, White Plains, NY) detergent in warm distilled water and sonicate the beaker containing them for 10 min. After extensive washing and sonication in distilled water, we soak the coverslips in 1 M HCl at 50–60°C for 4–6 h. Thereafter, we remove them with forceps and store them in 100% ethanol until use.
2. At the time of use, individual coverslips are removed with forceps from the ethanol and blotted on a paper towel. We then briefly pass each coverslip through the flame of an alcohol lamp to burn off all the ethanol. Then place the coverslip in a 3.5 cm dish (or a 6-well plate) and add cells suspended in culture media. We use media appropriate for each cell type supplemented with 12.5 mM HEPES buffer, 10% fetal calf serum and antibiotics (75 U penicillin and 75 µg streptomycin per milliliter; Invitrogen # 15140-122). It is important to constitutively culture cells with HEPES so that they remain acclimatized to this buffer.
3. Wipe the aluminum support slide with ethanol and air dry the residual ethanol. Use a small spatula to apply a thin rim of silicone grease (e.g., high vacuum grease from DOW Corning, Midland, MI) around the top and bottom margins of the opening. Flame a cleaned blank coverslip and attach the coverslip to the bottom of the support slide. Use a pair of forceps to gently tamp the area of the coverslip that overlaps the aluminum slide. Then, place a few layers of Kimwipes on the coverslip and press the overlap area gently using fingertips to ensure a tight seal and remove any excess grease. Place the prepared slide on the filter paper placed in a 10 cm plastic culture dish and expose it to Ultraviolet (UV) light for 10 min in a tissue culture hood. After putting a lid on the 10 cm plate to keep the observation chamber sterile, place the plate in an incubator to equilibrate the chamber to 37°C.
4. Place 3–5 ml of culture media in a small culture dish and keep in a CO_2 incubator for a few hours before filming begins.
5. In the tissue culture hood, fill the chamber with ∼0.8 ml of warm, buffered culture media. Take a coverslip with cells from the culture dish (protocol #2) and quickly place the coverslip, cell side down, on the chamber. Tamp the coverslip down onto the silicone grease with forceps, aspirate off excess media, and press using Kimwipes and fingertips to ensure a tight seal and remove excess grease. Wash the top of the preparation with water first, and then ethanol using a cotton swab to remove any residual media that could upon drying leave deposits on the coverslip.

6. Silicone grease provides a tight seal which will eliminate evaporation of the media for many days, if the coverslip is initially well-bedded down. If desired, a small bead of melted VALAP (1:1:1 mixture of vaseline, lanolin, and paraffin) can be applied around the edges of the top coverslip to mechanically stabilize the seal (see Figure 3.3). We use this extra sealing method when we need to change media during filming (see protocol #8). Keep the amount of VALAP used to a minimum to reduce the chance that any of the VALAP will deposit on the front element of the objective when the preparation is in use. If this happens, use a small amount of Naphtha (lighter fluid) and a cotton swab to clean the objectives.
7. Warm the assembled chamber to 37°C, place on the microscope stage, and start filming.
8. Although these chambers support cell viability for more than 100 h, we often change the media every few days to be safe. To do this, we stop the film run and take the preparation off the microscope being careful not to disturb the XY position of the stage. We sometimes mark one edge of the cell bearing coverslip with a felt tipped pen to indicate the polarity of the preparation so that it can be replaced on the microscope stage with proper orientation (see Figure 3.3). In the tissue culture hood, the bottom coverslip (the one without cells) and surrounding area are cleaned with a cotton swab soaked in ethanol. Sharp forceps are used to gently move the coverslip in both the X- and the Y-axes to loosen the attachment (2 mm in each direction). Then, slide the coverslip off and remove it prior to aspirating the medium and replacing it with fresh warm buffered media (protocol #4). Ahead of time, flame a cleaned blank coverslip, apply a thin rim of silicone grease around the edges of the coverslip, and expose it to UV light for 10 min in a tissue culture hood. We then reassemble the preparation as described in protocol #5, find the same field and restart the image sequence acquisition. This procedure is also used to add or withdraw drugs from the cells. We use 6–8 washes (total 5–6 ml of fresh media) to wash drugs out of the cells.

3.3 Correlative Live Cell–Fixed Cell Observations

Often studies require long-term time-lapse observation followed by fixation and immunostaining of cells previously followed in vivo. The methodology described below has enabled us to reliably mark and later relocate fixed cells for which we had acquired long-term time-lapse sequences.

1. To mark the field of interest, we use a diamond scribe mounted in one of the positions in the objective lens (e.g., Leitz Wetzlar, Germany) turret on the nosepiece of the microscope. These scribes have the overall appearance of an objective lens with a retractable diamond point instead of a glass front element. The scribe physically scores a circle around the area of interest on the surface of coverslip. Lower the microscope

stage to make enough distance for the scribe, and rotate the nosepiece so that the scribe is centered over the field of interest. Select the diameter of the circle by adjusting the rotatable ring on the barrel of the scribe. While watching from the side, carefully raise the stage until the coverslip lightly contacts the diamond tip of the scribe. Gently rotate the knurled ring on the scribe to rotate the diamond tip. Light contact is sufficient to make an easily visible scribe groove without cracking the coverslip. After scribing, lower the stage away from the tip of the scribe and rotate in an objective. Before taking the preparation off the microscope it is important to take an image of the cells in relation to the slightly out of focus scribed circle (Figure 3.2). This image can be printed and will serve as a map to identify and relocate cells of interest after fixation and immunostaining.

2. Once filming is terminated, the cell bearing coverslip is carefully removed (see protocol #8) and immediately fixed in whatever manner one is using.

References

[1] Y. Uetake, J. Loncarek, J.J. Nordberg, C.N. English, S. Laterra, A. Khodjakov, et al., Cell cycle progression and de novo centriole assembly after centrosomal removal in untransformed human cells, J. Cell Biol. 176 (2007) 173–182.

[2] C.L. Rieder, R.W. Cole, Cold shock and the mammalian cell cycle, Cell Cycle 1 (2002) 169–175.

[3] P.N. Rao, J. Engelberg, Hela cells: effects of temperature on the life cycle, Science 148 (1965) 1092–1093.

[4] G. Sluder, J.J. Nordberg, F.J. Miller, E.H. Hinchcliffe, A sealed preparation for long-term observations of cultured cells, in: D.L. Spector, R.D. Goldman (Eds.), Live Cell Imaging: A Laboratory Manual, Cold Spring Harbor Laboratory Press, Cold Spring Harbor, 2005, pp. 345–349.

Further Reading

http://micro.magnet.fsu.edu/primer/techniques/livecellimaging/index.html—Introduction to Live-Cell Imaging Techniques by Florida State University.

http://www.microscopyu.com/articles/livecellimaging/index.html—Introduction to Live-Cell Imaging Techniques by Nikon

http://www.olympusmicro.com/primer/resources/livecells.html—Live-Cell Imaging Resources by Olympus

J.C. Waters, Live cell fluorescence imaging, in: G. Sluder, D.E. Wolf (Eds.). Digital Microscopy, third ed., Methods in Cell Biology 81 (2007) 115–140, Elsevier, Academic Press, London.

CHAPTER 4

Phase Imaging in Plant Cells and Tissues

Vassilios Sarafis

Botany School University of Melbourne, Vic, Australia
Mathematics and Physics School, University of Queensland, Brisbane St Lucia, QLD, Australia
Biomedical Science and Engineering School, University of Adelaide, SA, Australia
Birla Institute of Technology and Science, Hyderabad, Andhra Pradesh, India

Editor: Zeev Zalevsky

Phase imaging in plant cells and tissues depends on the boundaries between places where concentration of water is different or there is a boundary between water and oil containing compartments. This chapter has listed 10 microscopy methods to make the boundaries to be visible to the human eye so that they can be well studied. These methods can be applied for transparent specimens of different characteristics.

The out of focus imaging is the first and the earliest method that produces contrast at the boundary of the specimen using the out of focus approach.

The dark field microscopy creates dark background around the specimen using the elimination of the unscattered beam from the image of the specimen.

The phase contrast techniques, presented Chapter 1, allow a living specimen to be studied due to its exclusion of staining showing the boundaries where refractive index changes.

There are also other types of contrast microscopy. One is to use Becke line test to create contrast between the boundaries of specimens. Differential contrast microscopy (DIC), presented in Chapter 2, is another type of contrast microscopy. It uses the interference of dual polarizations to form visible images of specimens due to the phase difference. Also, the Hoffman modulation contrast (HMC) microscope is able to vary the contrast of different regions within a specimen using its phase gradients.

The adaptive optics microscopy, on the other hand, uses adaptive optics elements such as deformable membrane mirror (DMM) or spatial light modulators to optimize the image quality of the specimen, usually in fluorescence.

The interferometric microscopy, also uses phase contrast technique with the addition of a spatial light interference microscopy (SLIM) module to capture the image of a specimen.

The second harmonic imaging microscopy (SHIM) obtains the half wavelength signal which is due to the nonlinear optical effect of the specimen for imaging.

The final part of this chapter shows the flowing image in the plants using the magnetic resonance imaging (MRI) as well as static water and lipids.

4.1 Out of Focus Imaging

The earliest method for observing phase variations is imaging at various focal positions. This earliest method is known as out of focus contrast. Let us consider an object in focus. It is utterly transparent but has a boundary separating it from water with the same refractive indices in and out of the boundary. If in true focus we shall see nothing in a transparent unstained specimen. Just out of focus it will have contrast which reverses as we go through true focus to the other side off the true focus. The contrast is due to partial coherence of the light waves originating from parts of the object prior to and after the object being observed [1]. Using Adobe Photoshop, it is possible to increase the contrast of such images considerably.

Such a method has been used by some Indian botanist scientists to observe enhanced contrast in pollen grains mounted in water. In that case, there is a difference in concentration of materials in the pollen grain and the outside which is pure water and this enhances the contrast due to refraction at the boundaries as well.

4.2 Dark Field

An excellent method for visualizing boundaries is dark field. In this method, direct light from the microscope condenser is excluded from entering the objective [2,3]. A high aperture may be used for illumination up to 1.42 in commercial systems and the diffracted light only admitted into the objective, which can then be with an aperture of up to 1.4. Central dark field can also be used when the direct light is occluded in the objective itself and the diffracted light gathered. The limit for this technique comes when the light enters the specimen horizontally and then only diffracted light enters the objective. This method much advocated by Siedentopf [4] can reveal any particle however small provided it has a different refractive index than the medium it is in. Variations of this methodology can be introduced by allowing some direct light or by using annular illumination of varying aperture as through the Heine condenser made for many years in Leitz.

4.3 Phase Contrast Techniques

Zernike made a significant discovery in making it possible without staining to make a refractive index change visible under the microscope under one condition [5,6] which is very important. This condition is that the refractive index varies rapidly between the object being examined in the specimen and its surrounds; ideally a sharp boundary is the most desired. This method was to make predominantly a ring illuminator in the condenser and then introduce a ring in the objective with a cavity or increased height introducing thus a phase delay or advances in the objective to the light entering it and also causing the phase delayed light or accelerated light wave to have a different absorption in this region. The theory is shown graphically in Figure 4.1.

Images that are produced using such methodology are shown in Figure 4.2. Note the halo around each boundary. There is also a further anomaly in the image. The outer part of the image has different contrast than the regions in the center and there is a further problem in

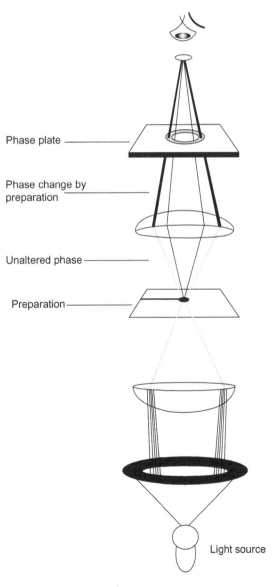

Figure 4.1
Formation of phase contrast image.

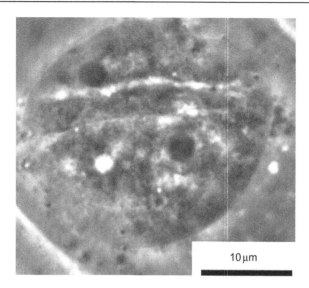

Figure 4.2
Inner epidermal nucleus from *Allium cepa* bulb scale.

that the image is only ideal for one wavelength and is suitable only for thin objects. The annular illumination introduces a large depth of focus to the images which can be problematic in thick specimens.

This methodology was discovered by Zernike in 1930s [5–9]. It is much used in optical microscopy and also it has had applications in X-ray and electron microscopy. The method basically relies on illuminating the specimen with an annular beam of light although earlier methods used axial illumination and also a double annular illumination as in the interphako Zeiss microscope now obsolete. The usual method is a dark contrast system where the specimen has a darker image than the surrounds. The method works best for thin specimens and those of small extent.

The main disadvantages of this approach are related to the artifacts that are caused by decay of contrast from the edges and a halo caused by absorption of light in this method [1]. The depth of focus is greater than in classical microscopy and results in disturbing fringes causing problems when imaging thick specimens.

A new variation of through-focus imaging introduced by Nugent and colleagues and also by PhaseView in France works well for small angles and works excellently for specimens where the aperture of illumination is low [10]. The images are good for high apertures too but the accuracy declines dramatically with increased aperture. It is manufactured by Ultima in Australia and PhaseView in France.

The phase ring used in classical phase microscopy is usually designed for a single wavelength but there have been versions which were achromatic and even apochromatic corrected for three colors.

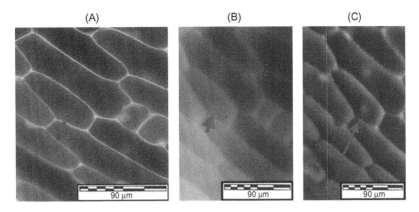

Figure 4.3
Onion inner epidermis from bulb scale in brightfield (A), low pass filter (B), and high pass filter (C).

Mitigation of the halos surrounding the phase delays are currently utilized by Nikon as apodized phase microscopy and reduce the halos considerably.

Let us consider a specimen to be imaged by annular illumination with a phase ring in the objective which matches it.

Figure 4.2 is of an onion inner epidermis nucleus showing contrast variations in black and white caused by phase differences due to different materials and different concentrations.

The alternative method of phase contrast imaging uses the transport of energy formulation. Results are published by Barone-Nugent et al. [11] for fossils with relatively opaque specimens, cells with absorbing structures such as algae like Spirogyra and for phase view which has a similar development of algorithms was also done in a French company (Figure 4.3).

4.4 Becke Line in Optical Microscopy

The method developed by Becke gives an easy determination of the refractive index of the desired object [12]. This is often used in mineral microscopy for looking at objects with a refractive index which one wishes to estimate and is applicable to botanical specimens. It is one of the earliest contrasting methods made between specimens and their surroundings and is due to their phase differences.

Figure 4.4A explains how the Becke lines appear between transparent materials of different refractive indices. This figure demonstrates the case where a specimen that is denser than its surroundings (i.e., higher refractive index) is placed on the stage of the microscope. When elevating the objective or lowering the stage of the microscope, the Becke line will

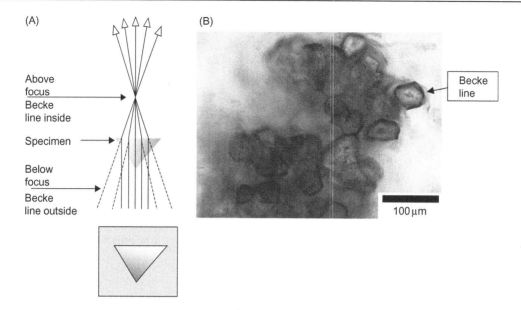

Figure 4.4
(A) Becke line principle (top) and its formation at the inside of specimen (bottom). (B) Image of Becke lines in pear stone cells.

be formed inside the specimen at the above focus point. The formation of the Becke lines inside the specimen is illustrated in Figure 4.4A.

A microscopical image that shows the formation of Becke lines inside the pear stone cells is shown in Figure 4.4B. The stone cells are surrounded by the line in black around each cell. This causes refractive index of cellulose encrusted with lignin and the depth of each stone cell creating a large phase difference between the cell and its surroundings. The surrounding medium is water which has a lower refractive index.

4.5 DIC Microscopy

Since the transparent specimen is not able to be visualized by the human eye, the contrasting method plays an important role to overcome this problem. Of current microscopical methodologies for optical microscopy, the DIC is the most widely used and has fewest artifacts in imaging thick and absorbing specimens [13]. This kind of microscope works by illuminating the transparent specimen with dual polarized light sources that are split up from a single polarized light source via a Nomarski prism, such that they become slightly incoherent (i.e., the offset is illustrated by the red and blue dash lines that indicate the wavefront of each polarized light source) and perpendicular to each other. The wavelength of the incident light changes as it passes through the transparent specimen due to the refraction effect (usually becomes shorter due to the denser medium of the transparent specimen), hence resulting in the changes of phase. While still not able to be

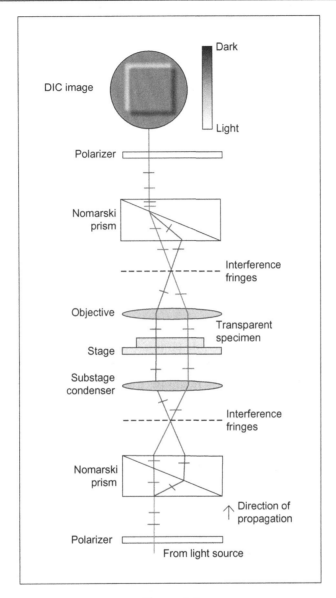

Figure 4.5
DIC light microscopy.

visualized by the human eye, the image of the transparent specimen can be produced using the phase information of two polarizations. The separation between the two beams is less than the resolution of the object. By introducing the interference through the recombination of these dual polarized lights using the second Nomarski prism, the relative phase difference will be formed, hence enabling visualization for the detector. Figure 4.5 illustrates the process of how the image of transparent specimen can be produced.

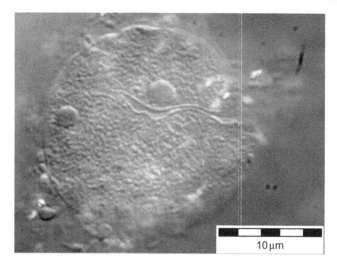

Figure 4.6
An illustration of the process of image production in a DIC microscope.

Nomarski in 1955 [14] developed this methodology using polarized light but versions using unpolarized light also existed in the Interphako microscope. However, the highest apertures and the best focal depths were given by Nomarski's methodology and it can also be extended to imaging in three dimensions by transfer of specimens in the visual field from one end to the other of the field of view thus creating two distinct illumination angles yielding a combined 3D effect as in Figure 4.6 (Sarafis, 1984 with the agreement also of Nomarski) [15].

It is possible to make a lower resolution image by DIC from data such as shown in the bright-field image in (Figure 4.7) on phase contrast imaging using software but they are not as good as direct DIC microscopy.

4.6 Hoffmann Modulation Contrast

HMC is a method that makes phase gradients to be displayed at different levels of intensity relative to the background, such that the transparent object can be visualized on a microscope [1]. It can be retrofitted to any microscope and gives relief contrast, high resolving power in the images, and a small depth of focus [16].

The HMC microscope works by placing a modulator at the back focal plane of the objective. This modulator consists of three regions that have different transmission levels (usually 1%, 15%, and 100%). Also, a slit has to be placed at the front focal plane of the substage condenser and calibrated to allow the incident light passing through the region of the modulator that has 15% transmission level, such that the background of the image

Figure 4.7
DIC image produced by software from data shown in Figure 4.3.

becomes uniformly gray. Upon placing the specimen on the stage of HMC microscope, some of the incident light will be refracted and passed through either the 100% or the 1% region of the modulator. Hence, the image of the specimen, as visualized by the slit, will appear to be bright or dark, respectively. The incident light that is not refracted will still be passing through the 15% region of the modulator. While the HMC microscope functions like a DIC, it is immune to birefringence that can be caused by some specimens. Note that the specimen must be rotated, allowing the asymmetries and symmetries in the specimen to be differentiated. The structure of an HMC microscope and its operating concept is shown in Figure 4.8.

It seems that this method has now lapsed as a patent and is still marketed under names such as Varel by Zeiss [17] and by Nikon as NAMC [18].

The author is unaware of its use for plant cells and tissue imaging thus far but it does have strong potential in his field of endeavor.

4.7 Adaptive Optics Microscopy

Derived from astronomy primarily, it has so far only been used for fluorescence although it should find use also in transmission and reflection imaging with quasicoherent light. Recently, Zeiss has had precommercial instruments offering this technology.

A typical example of applying adaptive optics in the area of microscopy is demonstrated by Poland et al. [19]. The experiment of this demonstration utilized a laser-scanning confocal microscope (later reconfigured as a multiphoton fluorescent microscope) with DMM as its

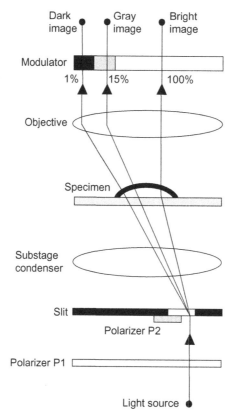

Figure 4.8
HMC microscope and its operating concept.

adaptive optics element. The optimization of the image quality of the sample is based on the shape calibration of DMM using an algorithm to mitigate aberration issue. This experiment produced results that illustrated the capability of single aberration correction in enhancing the image quality over the conventional wide-field fluorescence microscopy. Figure 4.9 shows an example of using adaptive optics with a single plane illumination microscope on *Arabidopsis* root for a fluorescent case from work led by John Girkin of Durham University. Such microscopy method is identical to the one performed in Ref. [20].

4.8 Interferometric Microscopy

The principle of the SLIM in conjunction with a Zeiss manufactured "Axio Observer Z1" phase contrast microscope is shown in Figure 4.10 while imaging neurons grown in vitro (taken from Ref. [21]). A halogen lamp is used as the light source of the microscope. The specimen on the stage can be pinpointed by the incident lights with the aid of the

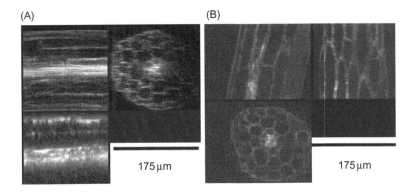

Figure 4.9
Adaptive optics in *Arabidopsis* root sections (A) as produced from fluorescence microscope (B) after image processing.

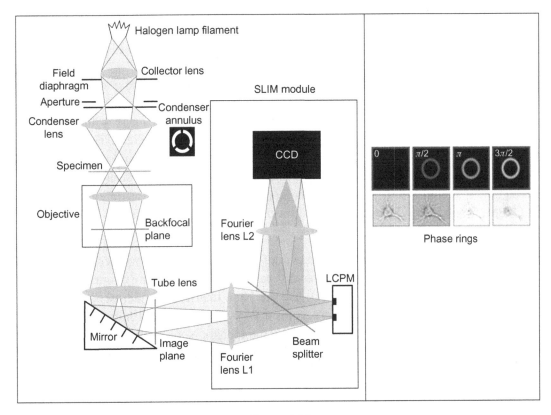

Figure 4.10
Configuration of SLIM microscope (left). The images of variation of phase rings as captured by the CCD (right).

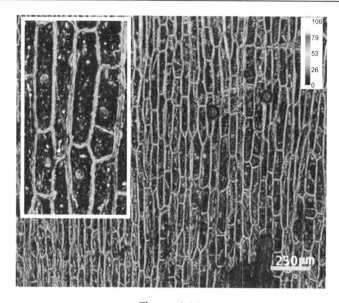

Figure 4.11
Image acquired by SLIM from the outer onion epidermis. Source: *Courtesy of Mustafa Mir and Gabriel Popescu, Quantitative Light Imaging Laboratory, University of Illinois at Urbana-Champaign.*

annulus condenser of the microscope. The phase ring that is fitted on the objective of the conventional phase contrast microscope works as a phase-shifter and attenuator. The post-processing procedure can be performed by the SLIM module on the image of the specimen, which is redirected from the microscope to the image plane via a tube lens. The Fourier lens L1 serves as a relay between the back focal plane of the objective and the surface of the liquid crystal phase modulator (LCPM). The phase delay can be precisely modulated between the scattered and the unscattered components to visualize the variation of masks on the LCPM. Consequently, the image of the specimen will be captured by the CCD produced by the Fourier lens L2 in the SLIM module.

Unlike, in the DIC microscopy of Nomarski double beam microscopy with complete separation of the two partially coherent beams was in vogue from many commercial sources. Polskie Zakłady Optyczne (PZO) had the Pluta Interference microscope, Leitz had a double microscope for this purpose, Zeiss from West Germany had a Jamin Lebedeff interference microscope, and Zeiss Jena had the Interphako. None of these are any longer in production. Recently, Popescu from Illinois University has developed a commercial microscope called SLIM. Figure 4.11 shows the outer epidermis of an onion scale.

Image of a cell layer from white onion skin was acquired using SLIM. The image covers an area of 1939×1453 μm^2 created by stitching 5×5 images together. At each of the 25 locations, a total of 34 z-sections were acquired spaced 4.5 μm apart and the maximum value from each slice was projected to create the image shown. The inset shows a

250×400 μm² detail of an area from this image. The yellow scale bar is 250 μm and the color bar shows optical path length in nanometers.

SLIM microscopy has been extended recently to super resolving microscopy by [22].

4.9 Second Harmonic Imaging Microscopy

SHIM was originated with confocal microscopy in transmission. Double photon absorption by orientated structures such as cell walls, starch grains, and chloroplasts is remitted in a shorter wavelength. The fluorescence that can result from this is always a longer wavelength than the remitted light which is at a shorter wavelength than the irradiation with pulsed wavelength.

This kind of microscope obtains the half wavelength of the light source using the nonlinear optical effect (i.e., second-harmonic generation) caused by the noncentrosymmetric structure of the specimen as a contrasting method to the image of the specimen. Such method is equivalent to the way that a conventional optical microscope obtaining contrast of specimen by detecting variations in optical density, path length, and/or the refractive index of the specimen. Campagnola and Leow [22] specify the advantages of this microscopy as unsusceptible to phototoxicity or photo-bleaching and also molecules with exogenous probes not necessarily to be labeled since many of the biological structures can produce strong second-harmonic generation signals, and hence will not affect the functioning of these biological systems.

This is commercially available from Till photonics and Leica. In the latter, only the light emitted forward is imaged, but in the Till microscope one collects both directions and these give different information. A conventional laser-scanning double-photon microscope can also be simply upgraded to achieve the SHIM's capability with minor modification [23].

The paper by Reshak et al. [24] shows the case in chloroplasts of a moss leaf.

4.10 Magnetic Resonance Imaging of Plants Showing the Flow Component

After early work with Sir Paul Callaghan and Xia, MRI was used by Sarafis and Campbell (25) to demonstrate flow by imaging the xylem and phloem of plants.

Figures 4.12 and 4.13 show the power of this technology which examines the flow of the fluids in these channels and is unlike Doppler methods in optical microscopy [26].

The Doppler methodologies are not commercial whereas the MRI is done on commercial machines which are not commercial; these latter depend on the carrying of particulates in the flow streams, but the MRI imaging is utterly independent of these components.

Field of view (FOV): 12 × 12 mm
Matrix size: 128 × 128
Rapid acquisition with refocused echoes (RARE) factors: 8
Repetition time (TR): 1250 ms
Echo time (TE): 5 ms.

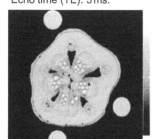

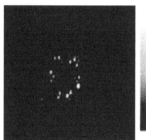 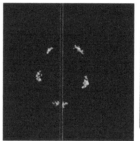

Figure 4.12
The cross-section of *Cucurbita* by MRI from intact plant showing distribution of water and flow in the xylem (left) and phloem (right).

Field of view (FOV): 12.5 × 12.5 nm
Diameter: 7.6 mm

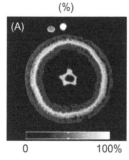 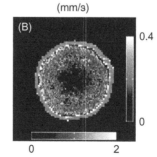

Figure 4.13
Poplar stem sections imaged by MRI showing water distribution (A) and flow in the xylem and phloem (B).

Newer emerging methods involving the hyperpolarization of water will change the imaging further and make it less demanding in time and increase the signal-to-noise component considerably by at least one and probably more orders of magnitude.

Windt et al. [26,27] originally at Wageningen University Center for MRI showed an example of flow imaging in plants in Figures 4.12 and 4.13.

MRI microscopy can also be used to show different phases such as water, oil, and essential components [28,29].

Acknowledgments

Grateful thanks to:

> Brendan Allman for providing the phase images (Figures 4.3 and 4.7), Melbourne University.
>
> John Girkin for adaptive optics (Figure 4.9), Durham University.
>
> Gabriel Popescu for interference image by SLIM (Figures 4.1 and 4.11), University of Illionis.
>
> Henk Van As for providing MRI images (Figures 4.12 and 4.13), University of Wageningen.
>
> Rabia Ghaffar for acquiring phase contrast, DIC, and bright-field images (Figures 4.1, 4.2, 4.4, and 4.6) and in compiling the article, University of Vienna.
>
> Chow Yii Pui for assistance in the final compilation of this book chapter and the images (Figures 4.5, 4.8, and 4.10), University of Adelaide.

With sincere regrets we inform that Prof. Vassilious Sarafis has passed away just before his chapter was finalized for publication. On the behalf of the editors we express our condolences to his family and dedicate his chapter to a memory of an extraordinary and multidisciplinary scientist, a colorful and unique person and a dear colleague.

References

[1] R. Wayne, Light and Video Microscopy, Elsevier, New York, NY, 2009, pp. 80–81
[2] S.H. Gage, Modern dark-field microscopy and the history of its development, Am. Microsc. Soc. (39) (1920) 95–141.
[3] S.H. Gage, The Microscope. Dark-Field Edition, Comstock Publishing Co., Ithaca, NY, 1925.
[4] H. Siedentopf, On the rendering visible of ultra-microscopic particles and of ultra-microscopic bacteria, J. R. Microsc. Soc. 23(573–578) (1903) 5.
[5] F. Zernike, Phase contrast, a new method of microscopic observation of transparent objects, Physica II(9) (1942) 947–986.
[6] F. Zernike, Phase-contrast, a new method for microscopic observation of transparent objects, Physica I(9) (1942) 686–698.
[7] F. Zernike, Phase-contrast, a new method for microscopic observation of transparent object, in: A. Bouwers (Ed.), Achievements in Optics, Elsevier, New York, NY, 1946, pp. 116–135.
[8] F. Zernike, Observing the phase of light waves, Science 107 (1948) 463.
[9] G.L.E. Turner, Frits Zernike, 1888–1966, R. Microsc. Soc. 17 (1982) 100–101.
[10] D. Paganin, K.A. Nugent, Noninterferometric phase imaging with partially coherent light, Phys. Rev. Lett. 80(12) (1998) 2586–2589.

[11] E.D. Barone-Nugent, A. Barty, K.A. Nugent, Quantitative phase-amplitude microscopy I: optical microscopy, J. Microsc. 206(3) (2002) 194–203.

[12] A. Allaby, M. Allaby, Becke Line Test, A Dictionary of Earth Sciences, January 1999 (Online). Available from: <http://www.encyclopedia.com/doc/1O13-Beckelinetest.html/> (accessed 16.03.2012).

[13] G. Nomarski, Interferential polarizing device for study of phase objects, US Patent 2,924,142, 9 February 1960.

[14] G. Nomarski, Microinterféromètre différentiel à ondes polarisées, J. Phys. Radium, Paris 16 (1955) 9S–11S.

[15] V. Sarafis, X-ray microscopy as a possible tool for the investigation of plant cells, Springer Series in Optical Sciences, X-ray microscopy, 1984.

[16] Modulation Optics, Making Transparent Specimens Clear and Vividly Detailed, Modulation Optics, 2012 (Online). Available from: <http://www.modulationoptics.com/> (accessed 16.03.2012).

[17] B. Hohman, E. Keller, Varel: A New Contrasting Method for Microscopy, Carl Zeiss Inc., Thornwood, NY, 2001.

[18] Laboratorytalk, NAMC Enhances Image Sharpness and Definition, Laboratorytalk, 9 September 2009 (Online). Available from: <http://www.laboratorytalk.com/news/nik/nik223.html/> (accessed 16.03.2012).

[19] S.P. Poland, A.J. Wright, S. Cobb, J.C. Vijverberg, J.M. Girkin, A demonstration of the effectiveness of a single aberration correction per optical slice in beam scanned optically sectioning microscopes, Micron 42(4) (2011) 318–323.

[20] J.M. Taylor, C.D. Saunter, G.D. Love, J.M. Girkin, Heart synchronization for SPIM microscopy of living zebra fish, in: SPIE 7904, San Francisco, 2011.

[21] M. Mir, Z. Wanga, Z. Shen, M. Bednarzd, R. Bashira, I. Golding, et al., Optical measurement of cycle dependent cell growth, Proc. Natl. Acad. Sci. U.S.A. 108(32) (2011) 1–6.

[22] S.D. Babacan, Z. Wanga, M. Do, G. Popescu, Cell imaging beyond the diffraction limit using sparse deconvolution spatial light interference microscopy, Biomed. Opt. Express 2(7) (2011) 1815–1857.

[23] P.J. Campagnola, L.M. Loew, Second-harmonic imaging microscopy for visualizing biomolecular arrays in cells, tissues and organisms, Nat. Biotechnol. 21(11) (2003) 1356–1360.

[24] A.H. Reshak, V. Sarafis, R. Heintzmann, Second harmonic imaging of chloroplasts using the two-photon laser scanning microscope, Micron 40(3) (2009) 378–385.

[25] Y. Xia, V. Sarafis, E.O. Campbell, P.T. Callaghan, Noninvasive imaging of water flow in plants by NMR microscopy, Protoplasma 173(3–4) (1993) 170–176.

[26] C.W. Windt, Nuclear Magnetic Resonance Imaging of Sap Flow in Plants, University of Wageningen, Wageningen, 2007.

[27] C.W. Windt, F.J. Vergeldt, P.A. DeJager, H.V. As, MRI of long-distance water transport: a comparison of the phloem and xylem flow characteristics and dynamics in poplar, castor bean, tomato and tobacco, Plant Cell Environ. 29(9) (2006) 1715–1729.

[28] V. Sarafis, H. Rumel, J. Pope, W. Kuhn, Non-invasive histochemistry of plant materials by magnetic resonance microscopy, Protoplasma 159(1) (1990) 70–73.

[29] J. Pope, V. Sarafis, NMR microscopy, Chem. Aust. 57(7) (1990) 221–224.

PART 2
Digital Holographic Phase Microscopy

CHAPTER 5

Digital Holographic Microscopy for Measuring Biophysical Parameters of Living Cells

Benjamin Rappaz[1], Christian Depeursinge[2], Pierre Marquet[1,3]

[1]*Laboratory of Neuroenergetics and Cellular Dynamics, Brain and Mind Institute, Ecole Polytechnique Fédérale de Lausanne (EPFL), Lausanne, Switzerland* [2]*Microvision and Microdiagnostic Group, STI, Ecole Polytechnique Fédérale de Lausanne (EPFL), Lausanne, Switzerland* [3]*Département de Psychiatrie-CHUV, Centre des Neurosciences Psychiatriques, Site de Cery, Prilly, Switzerland*

Editor: Natan T. Shaked

5.1 Introduction

Recording noninvasive high-resolution 3D images of living cells in real time remains a technical challenge. Most biological cells are transparent, i.e., they differ only slightly from their surroundings in terms of optical properties (including absorbance, reflectance, etc.) which precludes the generation of adequate contrast required to obtain high-resolution video microscopy images.

Consequently, developing methods to access the phase information for imaging purposes has been investigated for a long time. In practice, phase contrast (PhC), initially proposed by Zernike and presented in detail in Chapter 1, and Nomarski's differential interference contrast (DIC), discussed in detail in Chapter 2, are two widely used PhC-generating techniques available for high-resolution light microscopy. In contrast to fluorescence technique, PhC and DIC are noninvasive and allow the visualization of transparent specimens, particularly the fine subcellular structural organization, without using any staining contrast agent. However, despite their high sensitivity, PhC and DIC do not allow the direct and quantitative measurement of phase shift or optical path length. Consequently, the interpretation of minute DIC or PhC signal variations in terms of quantitative modification of specific biophysical cell parameters including cell volume remains difficult. More recently, efforts have been made to employ DIC as well as PhC as a quantitative imaging technique [1–3]. In contrast, interference microscopy has the capacity to provide a direct measurement of the optical path length based on the interference between the light wave passing the specimen, called the object wave, and a reference wave. Although quantitative phase measurements with interference microscopy applied to cell imaging were already known in 1950s, from the seminal work of Barer [4], only a few attempts [5] have been reported to dynamically image live cells in biology. Indeed, phase shifts are very sensitive to experimental artifacts, including lens defects, and noise originating from vibrations or thermal drift. Therefore, temporal phase shifting interferometry requires demanding and costly opto-mechanical designs which hinder wider applications in biology (Figure 5.1).

In parallel, holography techniques were developed by Gabor in 1948 [7] who demonstrated its lensless imaging capabilities, thanks to the reconstruction of an exact replica of the full wave front (amplitude and phase), emanating from the observed specimen (object wave). Due to costly opto-mechanical designs, few applications were developed at this time. However, scientific advances lowering the cost of lasers and data acquisition equipment and reducing the processing time to reconstruct digitally recorded holograms opened new avenues to holographic research. The digital treatment, contrary to the classical approach, considers the wave front as a combination of amplitude and phase, leading to the development of various quantitative phase microscopy (QPM) approaches through holography using different interferometric configurations. This resulted in a considerably simpler implementation than classical interference microscopy while providing a reliable and quantitative phase mapping of living specimens [8–52].

The development of easily accessible detection arrays made the distinction between holography and interferometry difficult to appreciate. In addition to the widespread use of laser sources having long temporal and spatial coherence properties, holography now commonly employs interferometric set-ups to produce holograms. The two research fields are presently widely interconnected. Similar to holography, several QPM techniques related to interferometric approaches have been recently developed to explore cell imaging [26,46,53–76].

Furthermore, a QPM technique based on transport-intensity equations has been developed. Through the acquisition of several images taken at different depths along the optical axis around the focus position, quantitative phase can be recovered [77–79]. Methods based on wave front characterization, such as Shack–Hartmann detectors, were recently applied to phase imaging, thanks to the development of high-resolution sensors [80]. Finally, phase-retrieval algorithmic methods were also developed [81,82]. However, due to the purely computational treatment of the signal, accuracy in phase recovery is usually lower, making a quantitative interpretation more difficult.

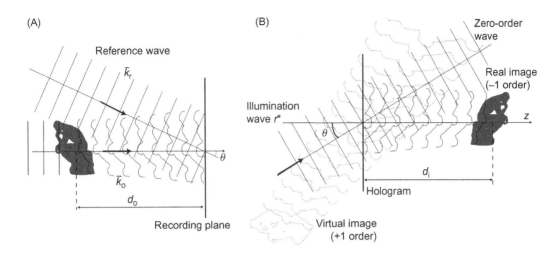

Figure 5.1
Principle of off-axis (A) recording and (B) reconstruction in DHM. A hologram resulting from the interference of the reference and object waves is recorded on the camera (in the recording plane) with a small incident angle, θ. For the reconstruction process, the hologram is digitally re-illuminated and the zero order is filtered to obtain a complete representation of the wave front (both in terms of intensity and phase) originally emitted by the object. Source: *Modified from Ref. [6]*.

The observations of quasi-transparent specimens, including living cells, are usually performed in a transmission implementation providing a quantitative phase signal through the specimen proportional to the integrated refractive index along each optical path length. Thus, quantitative phase signals, depending on both the thickness and the refractive index, provide unique information about the morphology and content of the observed specimen that can be regarded as a powerful endogenous contrast agent. On the other hand, using quantitative phase signals can be difficult to interpret underlying biological processes due to its dual dependence.

In Section 5.2, we present the principle of classical holography as well as the recent developments of digital holography as a quantitative phase technique aiming at exploring cell structure and dynamics. In addition, we address the issues of interpreting the quantitative phase signal. We present some recent technical developments that allow us to measure the refractive index and the morphology of living cells. In Section 5.3, we present a series of biological applications to illustrate how specific cell biophysical parameters, such as absolute volume, morphology shape, transmembrane water permeability, and membrane fluctuations at the nanometric scale are derived from the quantitative phase signal and applied to obtain a better understanding of cell dynamics.

5.2 Technical Introduction

5.2.1 Classical Holography

Holography techniques were developed by Gabor in 1948 with the aim of improving the detection of spatial resolution in the X-ray wavelength by exploiting its lensless imaging capabilities. This resulted in the possibility of generating, during the illumination of the recorded hologram (reconstruction process), an exact replica, with a specific magnification, of the full object wave front created by the observed specimen [7,83]. However, as already identified by Gabor, the imaging possibilities of holography are greatly reduced in quality because of the presence of different diffraction orders in the propagation of the diffracted wave front by the hologram when illuminated during the reconstruction process. This was resolved by Leith and Upatneiks [84,85], who proposed the use of a reference wave of a slightly different propagation direction than the object wave. This method, referred to as off-axis geometry, has been analyzed from a computational point of view [86] with a formalism based on diffraction, and is used in the first quantitative phase measurements [87].

However, the first developments in off-axis configuration were performed through a common-path configuration due to the coherence limitation of the light sources. Moreover, the emergence of laser light sources, enabling very long coherence lengths with high power, takes advantage of the versatility of "interferometric configuration." Note that short coherence length sources have been investigated recently in various cases [88–90]. Shorter coherence lengths have the capability to improve the lateral resolution as well as to decrease the coherence noise which could limit the quality of the reconstructed image, particularly the phase sensitivity. However, these low coherence implementations require more complicated arrangements where the coherence zone (in both spatial and temporal domains) must be adapted to ensure optimal interference [91].

5.2.2 From Classical to Digital Holography

The use of digital means in holography gradually occurred at the end of 1960s, when Goodman and Lawrence [92] used a vidicon detector to encode a hologram, which could be reconstructed on a computer. However, the interest in digital holography rose with the availability of cheaper digital detectors and charge-coupled device (CCD) cameras. The use of CCD cameras for holographic applications was validated in the mid 1990s, in the case of reflection macroscopic holograms [93] and microscopic holograms in endoscopic applications [94].

Another approach for hologram reconstruction was developed by taking advantage of the capability of digital detectors to rapidly record multiple frames, through the use of a phase-shifting technique [95] as developed first for interferometry [96,97].

Up to this point, holography was essentially considered an imaging technique, enabling the possibility of lensless imaging or the capability of focusing images recorded out of focus through the recovery of the full wave front. The digital treatment, contrary to the classical approach, considers the wave front as a combination of amplitude and phase, which led to the development of quantitative phase imaging through holography [98].

5.2.3 Digital Holography Methods

The two main approaches to recovering the object wave are namely temporal decoding, i.e., phase shifting, and spatial decoding, i.e., off-axis methods.

Phase shifting reconstruction methods are based on the combination of several frames, and enable the suppression of the zero order and the twin image through temporal sampling [99,100]. The most well-known phase-shifting algorithm, proposed by Yamaguchi and Zhang [95] as previously mentioned, is based on the recording of four frames separated by a phase shift of a quarter of a wavelength. Various combinations of frames derived from interferometry have been considered [96,99] and many different waves have been developed to produce the phase shift including high-precision piezo-electric transducers which move a mirror in the reference wave or acousto-optics modulators which use the light frequency shift, etc.

One of the main issues in the phase shifting method is the requirement of several frames for reconstruction in interferometric set-ups, which are commonly very sensitive to vibrations, so that it could be difficult to ensure stable phase shifts and an invariant sample state during acquisition. In addition, the requirements on the accuracy of the phase shifts are rather high with regard to displacements in the order of magnitude of hundreds of nanometers, implying the use of high-precision transducers. Consequently, several attempts were made to either reduce the required number of frames for reconstruction, which led to two-frame reconstruction [101,102] or to enable the recoding of the various phase shift frames simultaneously, by employing, for example, multiplexing methods [103]. On the other hand, more refined algorithms were developed in order to loosen the accuracy requirements of phase shift methods [104–106].

The second main approach to recovering the object wave is based on off-axis configuration, so that the different diffraction terms encoded in the hologram (zero-order wave, real image, and virtual image, cf. Figure 5.1) are propagating in different directions, enabling their separation for reconstruction. This configuration was the one employed for the first demonstration of a fully numerical recording and reconstruction holography [93,94].

In practice, reconstruction methods based on off-axis configuration usually rely on Fourier methods to filter one of the diffraction terms. This concept was first proposed by Takeda et al. [107] in the context of interferometric topography. The method was later extended for smooth topographic measurements for phase recovery [108] and generalized for use in digital holographic microscopy (DHM) with amplitude and phase recovery [98]. The main characteristic of this approach is its capability of recovering the complex object wave through only one acquisition, thus greatly reducing the influence of vibrations. However, as the diffraction terms are spatially encoded in the hologram, this one shot capability comes potentially at the cost of usable bandwidth. In addition, the frequency modulation, induced by the angle between the reference and the object waves, has to guarantee the separability of the information contained in the different diffraction terms that are encoded in the hologram while carrying a frequency compatible with the sampling capacity of digital detectors. Despite the fact that digital detectors have a sampling capability significantly lower than photosensitive plates, their use in off-axis digital holography microscopy does not represent a drawback. Rather, the microscope objective (MO) introduced in the interferometer allows the wavefield to adapt to the sampling capacity of the camera: the lateral components of the wave vector $k_{x \text{ or } y}$ are divided by the magnification factor M of the MO, therefore permitting the wavefield to be adequately sampled by the electronic camera and to obtain a reconstructed quantitative phase image with a diffraction-limited lateral resolution [34,98].

As mentioned earlier, holographic measurements often rely on recording the object wave in a nonimaging plane such as in lensless holography. Consequently, it is necessary to propagate the recovered object wave resulting from the filtered spectrum of the hologram, to retrieve a focus image. Depending on the distance between the recording position and the focus plane, different approximations are commonly used for numerical implementation of the object wave propagation [98,109–111]. The digital propagation can be considered as one of the major advantages provided by DH, as it enables off-line autofocusing [30], as well as extends the depth of focus, enabling the reconstruction of different focal planes from a single hologram [112]. In addition, the demodulation of the filtered spectrum, resulting from the off-axis geometry, as well as aberration compensation can be efficiently associated with the numerical calculation of the object wave propagation [98,113–115]. A priori, any type of aberration, including apparent spherical aberration resulting from the mismatch between the reference and the object waves, can be numerically compensated enabling optimization of imaging capabilities and making the use of simplified interferometric set-ups possible. In summary, the development of digital means has greatly changed the research field in holography leading to the development of various simplified interferometric configurations and numerical reconstruction procedures well suited to explore many different fields, whose scope extends beyond that of this chapter.

5.2.4 Cell Imaging and Quantitative Phase Signal Interpretation

In the field of cell imaging, transparent cells are probed by quantitatively measuring the phase retardation that they induce on the transmitted wave front. This phase signal is related to the biophysical cell parameter:

$$\Phi = \frac{2\pi}{\lambda}(\bar{n}_c - n_m)d \tag{5.1}$$

where d is the cellular thickness, $\bar{n}_c$ is the intracellular refractive index averaged over the optical path length through the specimen, and n_m is the refractive index of the surrounding medium. Consequently, the information concerning the intracellular content related to $\bar{n}_c$ is intrinsically mixed with the morphological information related to the thickness d. Due to this dual dependence, the phase signal remains difficult to interpret. As an illustration, a simple hypotonic shock induces a phase signal decrease [41] difficult to interpret as cellular swelling but consistent with an $\bar{n}_c$ decrease from a dilution of the intracellular content by an osmotic water influx. Accordingly, some strategies have been developed to separately measure the morphology and $\bar{n}_c$. Kemper et al. [20] and Lue et al. [116] measured the intracellular refractive index by trapping cells between two cover glasses separated by a known distance. These approaches, preventing cell movements, preclude the possibility of exploring cell dynamic processes. We have developed another approach to separately measure the parameters $\bar{n}_c$ and d from the phase signal Φ, based on a modification of the extracellular refractive index n_m. Basically, this method consists of performing a slight alteration of the extracellular refractive index n_m and recording two holograms corresponding to the two different values of n_m, allowing the reconstruction of two quantitative phase images (Φ_1, Φ_2) described by the following system of two equations for each pixel i:

$$\Phi_{1,i} = \frac{2\pi}{\lambda_1}(\bar{n}_{c,i} - n_{m,1})d_i \tag{5.2}$$

$$\Phi_{2,i} = \frac{2\pi}{\lambda_2}(\bar{n}_{c,i} - n_{m,2})d_i \tag{5.3}$$

where (λ_1, λ_2) are the wavelengths of the light source.

By solving this system of two equations, we obtain $\bar{n}_{c,i}$ and d_i for each pixel i. We have considered two different approaches to modify n_m: the first approach requires sequentially perfusing a standard cell perfusion solution, and a second solution with a different refractive index but with the same osmolarity (to avoid cell volume variation) to record the two corresponding holograms, at a single wavelength, ($\lambda_1 = \lambda_2$). The refractive index of the second solution is increased by replacing mannitol (a hydrophilic sugar present in the standard perfusion solution) with equal molarity of the hydrophilic molecule Nycodenz, a small molecule known for its high capability to modify the refractive index [41]. Nevertheless, this approach, due to the solution exchange time, precludes the possibility of monitoring dynamic changes of cell morphology and $\bar{n}_c$ occurring during fast biological processes. To overcome these drawbacks, we have developed a dual-wavelength ($\lambda_1 \neq \lambda_2$) DHM [117], which exploits the dispersion of the extracellular medium, enhanced by the utilization of an extracellular dye ($n_{m,1} = n_m$

$(\lambda_1) \neq n_m (\lambda_2) = n_{m,2})$, to achieve separate measurements of the intracellular refractive index and the cell morphology in real time [40].

5.3 Biological Applications

DHM provides new ways for noninvasive contactless and label-free quantitative PhC recording to enable the visualization of transparent living cells. In combination with the unique possibilities presented by digital means (real-time imaging, extended depth of focus, etc.) several original applications have been performed in the field of cell imaging. However, although providing unique information about cell morphology and intracellular content, the quantitative phase signal, due to its dual origin, could remain difficult to interpret in terms of biological processes. Consequently, in this part we will present a selection of relevant new applications of digital holography in the field of cell biology with a special emphasis on the interpretation of the phase signal. For specific applications, a description of which important biophysical parameters can be extracted from the phase signal will be provided as well as how they can contribute to bring a new understanding of specific biological mechanisms.

Straightforward applications with DHM exploit the ability of the technique to generate contrast from transparent specimens in order to image them and hence offer a quantitative alternative to PhC or DIC. For instance, Mölder et al. [36] compared a DHM to the conventional manual cell counting method using a hemocytometer and showed that digital holography can be used in noninvasive automatic cell counting as precisely as conventional manual cell counting, thereby offering a faster and easier alternative. Applications in reproduction research include the development of a holographic set-up to improve the visualization and detection of cow spermatozoa [118] or to compare the quantitative phase signal of human sperm heads in normozoospermia (normal sperm) and oligoasthenozoospermia (reduced sperm motility and low count) [119]. Hence, proposing that quantitative evaluation of the phase shift recorded by DHM could provide new information on the exact structure and composition of the sperm head could be useful for clinical practice. Other recent developments involved combining digital holography with a total internal reflection set-up [120], allowing QPM of cell−substrate interfaces and quantifying focal adhesions at the cellular membrane. Similarly, Pache et al. [121] studied the cellular modifications of the actin cytoskeleton in microgravity using a DHM. Such results were previously obtainable only with the use of labeled probes in conjunction with conventional fluorescence microscopy, with all the classically described limitations in terms of bias, bleaching, and temporal resolution.

5.3.1 Extended Depth of Focus and 3D Tracking

In addition to visualization applications, other groups have taken advantage of the ability of digital propagation to apply autofocusing [30,122,123] as well as extended depth of focus [112,124,125], hence opening the ability to track particles and cells in 3D [126,127] and offering an alternative to the shallow depth of field of conventional microscopy which hampers 3D tracking of cells in their environment with a high temporal resolution. Sheng et al. [128] used DHM to track predatory dinoflagellates in 3D and reveal prey-induced changes in their swimming behavior in dense suspensions. Using this technique, they were also able to analyze their response to various stimuli and toxins [129].

Three-dimensional tracking with DHM was also extensively used in hematology where the measurement of blood flow with high spatial and temporal resolutions in a volume is a challenge in biomedical research fields. Choi and Lee [14], with a cinematographic holography technique, continuously tracked individual red blood cells (RBCs) in a microtube flow to characterize their trajectories as well as their 3D velocity profiles. Similarly, Sun et al. [45] studied the rapid movement of RBCs in the blood stream of live *Xenopus* tadpoles. Both studies took advantage of the extended depth of focusing offered by holography to track RBCs in 3D after offline reconstruction of multiple focal planes with high spatial and temporal resolution.

5.3.2 Dry Mass and Cell Cycle

In addition to its noninvasive capability for visualizing and 3D tracking, DHM offers other advantages when considering the direct interpretation (without requiring any decoupling procedure) of the phase signal for biological applications. The measured quantitative phase shift induced by the observed cell on the transmitted light wave front is proportional to the intracellular refractive index, which largely depends on protein content. Therefore, it can be directly interpreted to monitor protein production, thanks to a relation established more than 50 years ago by Barer [4,130] within the framework of interference microscopy. The phase shift induced by a cell is related to its dry mass (DM) by the following equation (converted to the International System of Units):

$$\mathrm{DM} = \frac{10\lambda}{2\pi\alpha} \int_{S_c} \Delta\varphi \, ds = \frac{10\lambda}{2\pi\alpha} \Delta\overline{\varphi} S_c \tag{5.4}$$

where $\Delta\overline{\varphi}$ is the mean phase shift induced by the whole cell, λ is the wavelength of the illuminating light source, S_c is the projected cell surface and α is a constant known as the specific refraction increment (in cubic meters per kilogram) related to the intracellular content. α can be approximated by $1.8-2.1 \times 10^{-3}$ m^3/kg, when considering a mixture of all the components of a typical cell [4]. A direct application of this relationship is to study the DM production rate and the dry mass surface density (DMSD) in the cell cycle. This interpretation had been first proposed by Barer in 1950s, but due to the difficulty of manipulating an interferometric set-up it did not show any biological applications until recently when many groups using QPM techniques started to exploit this phase—DM relationship [19,39,131–134]. Some other applications in the quantitative monitoring of the cell division will be discussed in Chapter 5.

Cells need to double their content during each cell cycle to stay a constant size. Hence, a good indicator of the biomass is the DM (nonaqueous material), defined as the weight of the cell when water has evaporated and which mainly depends on protein concentration. Consequently, monitoring the DM production provides a dynamic indicator of the real-time evolution of the cell cycle in a noninvasive manner, thus providing a label-free tool to screen for quantitative analysis of the cell.

Most cells reproduce by duplicating their content and then dividing to produce daughters of equal size. The ability to genetically manipulate yeast cells, coupled with their rapid growth rate, has made them an attractive model to study cell growth and division. In a recent study, Rappaz et al. [39] illustrated this application by comparing the evolution of DM concentration during a complete cell cycle in wild-type and mutant fission yeast cells. Cells of the fission yeast *Schizosaccharomyces pombe* may be modeled as cylinders capped by hemispherical ends; they grow mainly by tip-elongation and divide by formation of a medially placed septum, which is cleaved to produce two daughter cells. In this study, we showed a marked difference between the DM production during the cell cycle of wild-type yeasts compared to a *cdc16-116* mutant. Cdc16p is required to limit the cell to a single septum per cell cycle; if it is inactivated, the cell synthesizes multiple additional septa, producing anucleate cell compartments [135] (Figure 5.2).

In this study, we observed that the DM production rates of the mutant and wild-type cells are significantly different. The wild-type DMSD shows a specific pattern resulting from the cytokinesis process, which is absent in the mutant. DM accumulation can be combined to other parameters easily measured by DHM, including cell volume and refractive index, to provide a better understanding of the cell cycle for DHM techniques can thus be further used for high-throughput, label-free screening, allowing a rapid quantitative characterization of the cell cycle, and the effects of specific pharmacological agents likely to affect cell cycle progression.

5.3.3 Biophysical Parameters in Hematology

In addition to the DM, the phase signal can be interpreted in many other ways, allowing to calculate clinically important RBC parameters including mean corpuscular volume (MCV),

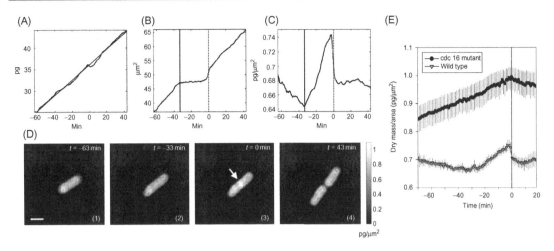

Figure 5.2
Time evolution of (A) dry mass (DM), (B) projected cell surface, and (C) DM concentration. Vertical lines: solid, end of surface growth; dashed, cytokinesis (cell cleavage). (D) Representative DM surface density images recorded (1) at the beginning, (2) at the end of cell growth, (3) just before cytokinesis (arrow: septum), and (4) at the end of the recording period. Scale bar = 5 µm. The two sister cells after cell division have been considered as a single cell in order to coherently appreciate the evolution of the measured parameters. (E) Comparison of the DM concentration for two phenotypes: wild-type (triangle) and cdc16 mutant (circle) fission yeast. Data expressed as mean ± SEM for five cells. The horizontal line denotes time of cytokinesis for wild type and the stage of maximal DM concentration for mutant. *Source: Modified from Ref. [39].*

mean corpuscular hemoglobin concentration (MCHC) as well as efficiently addressing the important issue of the erythrocyte flickering. DHM permits online tracking of changes in individual cells during osmotic fragility test or shear stress. It does not require the use of fluorescent probes and employing delicate and time-consuming deconvolution and image analysis procedures such as confocal laser scanning microscopy (CLSM). In addition, the intracellular hemoglobin content of individual cells, a parameter altered in various pathological states, can be directly estimated from the phase measurements. These tremendous possibilities have recently led to a large number of QPM publications in the field on hematology [12,14,48,136–146].

RBCs represent the main cell type in circulating blood and are responsible for the transport of oxygen. Parameters such as RBC shape, volume, refractive index, and hemoglobin content are important characteristics that can be used as good indicators of the body's physiological state. For instance, erythrocyte volume distribution is altered in patients with anemia, folate and vitamin B12 deficiency, and microcytic anemia [147]. Oxygen saturation also modulates the hemoglobin refractive index.

In addition, the duration of storage before transfusion may alter the RBC function and, therefore, influence the incidence of complications in patients. DHM was used to classify RBCs based on their duration of storage where it can replace traditional approaches through biomolecular assays that may be invasive, expensive, and time consuming. DHM allows for measurement of the RBCs' biconcave profile, resulting in a discriminative dataset that can be used by statistical clustering algorithms to discriminate the RBCs' populations. This approach was proven successful in classifying RBCs stored for 14 and 38 days [48].

The RBC MCV is a highly medically relevant parameter that can be easily calculated [141] using the decoupling procedure presented in Section 5.2. In addition, as the hemoglobin content is mainly responsible for the refractive index of the RBC, it can be used as a direct measure of the MCHC [141]. Barer [4] proposed a way to relate the phase signal to the mean corpuscular hemoglobin (MCH) (in a similar relationship than with the DM presented above in Eq. (5.4)):

$$\text{MCH} = \frac{10\varphi\lambda S_{\text{cell}}}{2\pi\alpha_{\text{Hb}}} \tag{5.5}$$

where φ is the mean phase shift induced by the whole cell, λ is the wavelength of the illumination light source, S_{cell} is the projected cell surface, and α_{Hb} is the hemoglobin refraction increment (1.96×10^{-3} dl/g at 663 nm). This formula allows to directly and noninvasively measure the MCH of single RBC. Knowledge of the MCV, obtained with the decoupling procedure presented in Section 5.3, allows to calculate the MCHC of individual red cells by dividing MCH by MCV.

In a previous paper, we showed that MCH and MCHC measurements obtained with DHM and those obtained with two other techniques, CLSM and an impedance volume analyzer (Sysmex KX-21), are in good agreement. The result of this comparison is presented in Table 5.1.

These RBCs' biophysical parameters, noninvasively monitored by DHM, are clinically relevant parameters that can be used as diagnostic tools (e.g., involved in the anemia classification). In addition, monitoring at a single cell level allows the investigation of changes occurring in the RBC subpopulation.

5.3.4 Cell Membrane Fluctuations

Throughout their lifespan of 100–120 days, RBCs are squeezed as they pass capillaries often smaller than the cell diameter. This ability can be attributed to the remarkable elastic properties of the membrane structure, which consists of a lipid bilayer that is coupled at specific binding sites to the underlying cytoskeleton. This structure exhibits a high resistance to stretching ensuring that no leakage through the lipid bilayer occurs, whereas

Table 5.1: Results of the Measurements of Normal Erythrocytes by Different Techniques

Technique	Volume (fl)	Surface (μm^2)	Diameter (μm)	Refractive Index	MCH (pg/cell)	MCHC (g/l)	N
DHM							
Mean	83.3	46.7	7.7	1.418	29.9	362	
STD	13.7	5.9	0.5	0.012	4.4	40	36
Dif_t	5.7	–	–	0.006	–	7	
CLSM							
Mean	90.7	46.9	7.7	–	–	–	
STD	16.7	6.8	0.6	–	–	–	34
Impedance volume analyzer							
Mean	81.8	–	–	–	28.6	349	
RDW–SD/CV	40.8/12.7	–	–	–	–	–	$\sim 3 \times 10^5$

DHM, digital holographic microscopy; CLSM, confocal laser scanning microscopy; impedance volume analyzer (Sysmex KX-21); N, number of cells; STD, standard deviation for cell population; Dif_t, averaged difference of individual cells for measurements taken at different times; RDW, red cell distribution width (expressed as coefficient of variation (CV, %) or standard deviation (SD, fl)).
Source: Modified from Ref. [141].

its low resistance to bending and shearing allows the cell to easily undergo morphological changes when passing through small capillaries. As a consequence of these elastic properties, RBCs show spontaneous cell membrane fluctuations (CMF), often called flickering, that can be observed under the microscope [140,142,148–151]. This phenomenon has been the subject of considerable scientific interest in the past. One reason is that CMF provide information on the elastic properties of the RBC membrane structure and might help in the diagnosis or understanding of diseases afflicting the RBC [151,152]. Understanding the origin of CMF may provide information about the functional status of erythrocyte under normal and pathological conditions.

DHM offers a powerful tool to noninvasively investigate the CMF of individual RBCs with a nanometric resolution [142]. The membrane fluctuations amplitude is derived from the deviation map (measured temporal deviation within the cell, in degrees), with the following equation:

$$\text{CMF [nm]}_{(x,y)} = \frac{\text{std}(\varphi_{\text{cell}})[°]_{(x,y)} \times \lambda}{2\pi(n_{\text{rbc}} - n_{\text{medium}})} \quad (5.6)$$

where $\text{std}(\varphi_{\text{cell}})[°]_{(x,y)}$ is the deviation map expressed in degrees, λ is the light source wavelength, and n_{rbc} and n_{medium} are the refractive indices of the RBC and medium, respectively. n_{rbc} can be measured using the decoupling procedure presented in Section 5.2. The morphological origin of the phase signal fluctuations was verified with the decoupling procedure.

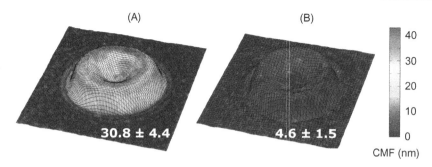

Figure 5.3
Normal CMF displayed as color code over the shape of the RBC for a representative healthy (A) and ethanol-fixed cell (B). Inset contains the mean ± SD (nm) of the CMF of the representative cell. Source: *Modified from Ref. [142]*.

As the measured CMF amplitude corresponds to fluctuations with an orientation parallel to the coverslip and not to the RBC membrane, further corrections should be applied to take into account the membrane slope and extrapolate the real fluctuations amplitude, normal to the membrane, described in Ref. [142] (Figure 5.3).

The local phase shift is proportional to the specimen thickness with an accuracy of 5–10 nm. As a noninvasive full-field technique, DHM is particularly well suited to assess and study membrane fluctuations of a large number of cells simultaneously. CMF have a mean fluctuation amplitude of 47 nm amplitudes and are heterogeneously distributed on the cellular surface and seem to correlate with the biconcave equilibrium shape of erythrocytes [141,153].

5.3.5 Neuroscience

Glutamate is the major excitatory neurotransmitter in the central nervous system and contributes to the synaptic transmission between neurons. One of its main actions is mediated by the glutamate N-Methyl-D-aspartate (NMDA) receptor which allows the flow of Na^+ and small amounts of Ca^{2+} ions into the cell and K^+ out of the cell in a voltage-dependent manner. The NMDA receptor is the predominant molecular device for controlling synaptic plasticity and memory function.

The passive diffusion or active transport of water through the plasma membrane is associated with several cellular processes. As far as the nervous system is concerned, the neuronal activity, i.e., mediated by the release of glutamate, has been shown to result in cell swelling linked to water diffusion. Consequently, neuronal volume changes could be used as an indirect marker of neuronal activity with relevance to functional imaging.

At the cellular level, the monitoring of absolute cell volume as well as the associated transmembrane flow of water can be achieved by DHM with recent studies having shown that phase signal changes mainly reflect intracellular refractive index variations [37,41]. Consequently, within this framework, the phase signal will be predominantly sensitive to any mechanism modifying the concentration of intracellular compounds, including transmembrane water movements, which accompany various ionic fluxes through the plasma membrane for osmotic reasons. Concretely, an entry of water will dilute the intracellular content, producing a decrease in the phase signal. In contrast, an outflow of water will concentrate the intracellular content, resulting, in this case, in an increase in the phase signal. This interpretation was confirmed by the decoupling procedure which effectively showed that a drop of phase was indeed related to a cellular swelling and vice versa [41].

This ion−water relationship was first illustrated using multimodality microscopy combining DHM with epifluorescence microscopy and allowing to simultaneously monitor the dynamics of intracellular ionic homeostasis and the DHM quantitative phase signal. The multimodal ability of this set-up was illustrated by imaging the early stage of neuronal response induced by glutamate, known to produce a cellular swelling [154]. Thanks to the decoupling procedure. It has been possible to measure simultaneously the dynamics of intracellular calcium, the concomitant transmembrane water movements and the corresponding neuronal volume changes, occurring at the early stage of the glutamate-mediated activity [37].

Following the demonstration that the phase signal can be used to monitor water movements linked to ionic fluxes, DHM showed that glutamate produces the following three distinct optical responses in mouse primary cortical neurons in culture, predominantly mediated by NMDA receptors: biphasic, reversible decrease (RD), and irreversible decrease (ID) responses. The shape and amplitude of the optical signal were not associated with a particular cellular phenotype but reflected the physiopathological status of neurons linked to the degree of NMDA activity [155]. Thus, the biphasic, RD, and ID responses indicated, respectively, a low level, a high level, and an "excitotoxic" level of NMDA activation. Moreover, furosemide and bumetanide, two inhibitors of sodium-coupled and/or potassium-coupled chloride movement, strongly modified the phase shift, suggesting an involvement of two neuronal cotransporters, NKCC1 (Na−K−Cl) and KCC2 (K−Cl), in the genesis of the optical signal. This observation is of particular interest since it shows that DHM is the first imaging technique able to monitor dynamically and in situ the activity of these cotransporters during physiological and/or pathological neuronal conditions (Figure 5.4).

5.3.6 Neuronal Cell Death

Glutamate, when present in excess, can provoke excitotoxic effects potentially triggering cell death mechanisms typically through calcium saturation. The cell then takes one of the many death pathways provoked by various mechanisms such as apoptosis or necrosis. In this context, it could be shown that glutamate-mediated excitotoxicity can induce various cell death pathways depending, for example, on the intracellular calcium uptake.

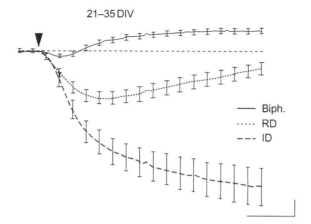

Figure 5.4
Application of glutamate triggers three main types of phase shifts. Averaged traces of the three characteristic optical signals induced by glutamate application (30 µM, 30 s, arrowhead) and recorded from 21 to 35 DIV (mature) neurons. These phase signals are classified into three categories: a biphasic response (Biph., $n = 137$), a reversible decrease of the phase signal (RD, $n = 80$), and an irreversible decrease of the phase signal (ID, $n = 35$). Scale bar: horizontal, 2 min; vertical, 5°. Source: *Modified from Ref. [155]*.

In a recent study, Pavillon et al. [37] correlated the calcium dynamics with phase measurements. Furthermore, we assessed the cell viability at the end of the experiment with propidium iodide (PI), a fluorescent dye relying on the loss of membrane permeability following necrotic mechanisms.

Most importantly, the neuronal cell volume regulation, accurately monitored using the high sensitivity of DHM quantitative phase signal, has permitted us to predict within a time frame of tens of minutes, whether a subsequent neuronal death would occur [156] (Figure 5.5).

5.3.7 Future Perspectives

As illustrated through these various applications, DHM allows developing a QPM specifically well adapted to study cell structure and dynamics. In practice, low energy levels and short exposure time are required to prevent any photo-damage even during long experiments. In addition, DHM, as a label-free technique, does not require any solution change or insertion of dye, providing efficient conditions for high-throughput screening.

Although the quantitative phase signal provides unique information about cell morphology and content with a high sensitivity, its interpretation in terms of biophysical parameters in analyzing specific biological mechanisms remains an issue. Within this framework,

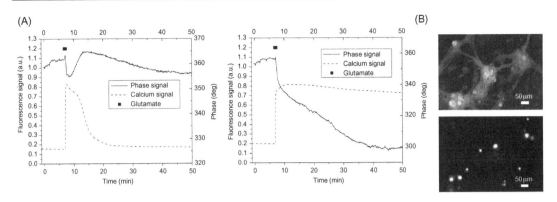

Figure 5.5
(A) Phase signal monitoring and calcium imaging through Fluo-4 measurement in the case of (left) a calcium regulation, linked with a rapid recovery of the phase signal and (right) a calcium deregulation linked with the strong phase signal decrease. (B) Image of neuron cells after a glutamate stimulation, shown in phase (top). The image can be linked with the fluorescence signal detected from propidium iodide (bottom), where a clear staining can be identified on several cells. Source: *Modified from Ref. [157].*

the development of real time and high-resolution DHM-based optical diffraction tomography, would avoid these potential interpretation difficulties of the phase signal by having direct access to the 3D map of refractive index [158–160]. Obtaining a high-resolution 3D map of the intracellular refractive index could provide invaluable information about cytoarchitecture and the emerging concept of the functional compartmentalization of cytoplasm, which plays a critical role in many fundamental cell mechanisms, including proteins synthesis. See Sections 4.1 and 4.2 of Chapter 4 for more information about the recent development of optical diffraction tomography.

On the other hand, promising future directions involve integrating DHM into a multimodal microscope able to provide various different types of information on the cell state, a necessary step to obtain a comprehensive understanding of cell dynamics. For instance, DHM will benefit from the specificity of fluorescent dyes to correlate cell morphology with specific protein or ion localization and dynamics [32,37,161]. Furthermore, information concerning transmembrane current obtained from electrophysiological recording combined with phase signal allows us to pave the way for studying the involvement of specific receptors or cotransporters in the physiological and/or pathological cellular processes [155].

Acknowledgments

The authors would like to thank the people of the Microvision and Microdiagnostic group at the EFPL and those at Lyncée Tec (www.lynceetec.com) for fruitful discussions during the writing of this chapter.

References

[1] M.R. Arnison, K.G. Larkin, C.J.R. Sheppard, N.I. Smith, C.J. Cogswell, Linear phase imaging using differential interference contrast microscopy, J. Microsc. 214(1) (2004) 7–12.

[2] M. Shribak, S. Inoué, Orientation-independent differential interference contrast microscopy, Appl. Opt. 45(3) (2006) 460–469.

[3] D. Fu, S. Oh, W. Choi, T. Yamauchi, A. Dorn, Z. Yaqoob, et al., Quantitative DIC microscopy using an off-axis self-interference approach, Opt. Lett. 35(14) (2010) 2370.

[4] R. Barer, Interference microscopy and mass determination, Nature (London) 169(4296) (1952) 366–367.

[5] G.A. Dunn, D. Zicha, P.E. Fraylich, Rapid, microtubule-dependent fluctuations of the cell margin, J. Cell. Sci. 110(24) (1997) 3091–3098.

[6] N. Pavillon, C. Depeursinge, P. Marquet, Cellular dynamics and three-dimensional refractive index distribution studied with quantitative phase imaging, in: LOA, EPFL, 2011.

[7] D. Gabor, A. New, Microscopic principle, Nature 161(4098) (1948) 777–778.

[8] A. Anand, V. Chhaniwal, B. Javidi, Quantitative cell imaging using single beam phase retrieval method, J. Biomed. Opt. 16(6) (2011) 060503 (3 pages).

[9] A. Anand, V.K. Chhaniwal, B. Javidi, Imaging embryonic stem cell dynamics using quantitative 3-D digital holographic microscopy, IEEE Photonics J. 3(3) (2011) 546–554.

[10] M. Antkowiak, M.L. Torres-Mapa, K. Dholakia, F.J. Gunn-Moore, Quantitative phase study of the dynamic cellular response in femtosecond laser photoporation, Biomed. Opt. Express 1(2) (2010) 414–424.

[11] A. Bauwens, M. Bielaszewska, B. Kemper, P. Langehanenberg, G. von Bally, R. Reichelt, et al., Differential cytotoxic actions of Shiga toxin 1 and Shiga toxin 2 on microvascular and macrovascular endothelial cells, Thromb. Haemost. 105(3) (2011) 515–528.

[12] I. Bernhardt, L. Ivanova, P. Langehanenberg, B. Kemper, G. von Bally, Application of digital holographic microscopy to investigate the sedimentation of intact red blood cells and their interaction with artificial surfaces, Bioelectrochemistry 73(2) (2008) 92–96.

[13] D. Carl, B. Kemper, G. Wernicke, G. von Bally, Parameter-optimized digital holographic microscope for high-resolution living-cell analysis, Appl. Opt. 43(36) (2004) 6536–6544.

[14] Y.S. Choi, S.J. Lee, Three-dimensional volumetric measurement of red blood cell motion using digital holographic microscopy, Appl. Opt. 48(16) (2009) 2983–2990.

[15] M. DaneshPanah, S. Zwick, F. Schaal, M. Warber, B. Javidi, W. Osten, 3D holographic imaging and trapping for non-invasive cell identification and tracking, J. Disp. Technol. 6(10) (2010) 490–499.

[16] C. Depeursinge, E. Cuche, P. Dahlgren, A. Marian, F. Montfort, T. Colomb, et al., Digital holographic microscopy applied to the study of topology and deformations of cells with sub-micron resolution: example of neurons in culture, in: Biomedical Topical Meeting, 2002.

[17] F. Dubois, C. Yourassowsky, O. Monnom, J.C. Legros, O. Debeir, P. Van Ham, et al., Digital holographic microscopy for the three-dimensional dynamic analysis of in vitro cancer cell migration, J. Biomed. Opt. 11(5) (2006) 54032–54035.

[18] Z. El-Schish, S. Mölder, L. Gisselsson, K. Alm, A. Gjörloff Wingren, Digital holographic microscopy—innovative and non-destructive analysis of living cells, Microsc. Sci. Technol. Appl. Edu. 2 (2010) 1055–1062.

[19] B. Kemper, A. Bauwens, A. Vollmer, S. Ketelhut, P. Langehanenberg, J. Muthing, et al., Label-free quantitative cell division monitoring of endothelial cells by digital holographic microscopy, J. Biomed. Opt. 15(3) (2010) 036009.

[20] B. Kemper, D. Carl, J. Schnekenburger, I. Bredebusch, M. Schäfer, W. Domschke, et al., Investigation of living pancreas tumor cells by digital holographic microscopy, J. Biomed. Opt. 11(3) (2006) 034005.

[21] B. Kemper, S. Kosmeier, P. Langehanenberg, G. von Bally, I. Bredebusch, W. Domschke, et al., Integral refractive index determination of living suspension cells by multifocus digital holographic phase contrast microscopy, J. Biomed. Opt. 12(5) (2007) 54009–5

[22] B. Kemper, P. Langehanenberg, A. Höink, G. von Bally, F. Wottowah, S. Schinkinger, et al., Monitoring of laser micromanipulated optically trapped cells by digital holographic microscopy, J. Biophoton. 3(7) (2010) 425–431.

[23] B. Kemper, A. Vollmer, C.E. Rommel, J. Schnekenburger, G. von Bally, Simplified approach for quantitative digital holographic phase contrast imaging of living cells, J. Biomed. Opt. 16(2) (2011) 026014.

[24] B. Kemper, G. von Bally, Digital holographic microscopy for live cell applications and technical inspection, Appl. Opt. 47(4) (2008) A52–A61.

[25] A. Khmaladze, M. Kim, C.M. Lo, Phase imaging of cells by simultaneous dual-wavelength reflection digital holography, Opt. Express 16(15) (2008) 10900–10911.

[26] M. Kim, Y. Choi, C. Fang-Yen, Y. Sung, R. Dasari, M. Feld, et al., High-speed synthetic aperture microscopy for live cell imaging, Opt. Lett. 36(2) (2011) 148–150.

[27] J. Klokkers, P. Langehanenberg, B. Kemper, S. Kosmeier, G. von Bally, C. Riethmuller, et al., Atrial natriuretic peptide and nitric oxide signaling antagonizes vasopressin-mediated water permeability in inner medullary collecting duct cells, Am. J. Physiol. Ren. Physiol. 297(3) (2009) F693–F703.

[28] P. Langehanenberg, G. von Bally, B. Kemper, Application of partially coherent light in live cell imaging with digital holographic microscopy, J. Mod. Opt. 57(9) (2010) 709–717.

[29] P. Langehanenberg, L. Ivanova, I. Bernhardt, S. Ketelhut, A. Vollmer, D. Dirksen, et al., Automated three-dimensional tracking of living cells by digital holographic microscopy, J. Biomed. Opt. 14(1) (2009) 14018-7

[30] P. Langehanenberg, B. Kemper, D. Dirksen, G. von Bally, Autofocusing in digital holographic phase contrast microscopy on pure phase objects for live cell imaging, Appl. Opt. 47(19) (2008) D176–D182.

[31] S. Liu, F. Pan, Z. Wang, F. Wang, L. Rong, P. Shang, et al., Long-term quantitative phase-contrast imaging of living cells by digital holographic microscopy, Laser Phys. 21(4) (2011) 740–745.

[32] C.J. Mann, P.R. Bingham, H.K. Lin, V.C. Paquit, S.S. Gleason, Dual modality live cell imaging with multiple-wavelength digital holography and epi-fluorescence, 3D Res. 2(1) (2010) p. 1–6.

[33] C.J. Mann, L. Yu, M.K. Kim, Movies of cellular and sub-cellular motion by digital holographic microscopy, Biomed. Eng. Online 5(21) (2006) p. 1–10.

[34] P. Marquet, B. Rappaz, P.J. Magistretti, E. Cuche, Y. Emery, T. Colomb, et al., Digital holographic microscopy: a noninvasive contrast imaging technique allowing quantitative visualization of living cells with subwavelength axial accuracy, Opt. Lett. 30(5) (2005) 468–470.

[35] M. Mihailescu, M. Scarlat, A. Gheorghiu, J. Costescu, M. Kusko, I.A. Paun, et al., Automated imaging, identification, and counting of similar cells from digital hologram reconstructions, Appl. Opt. 50(20) (2011) 3589–3597.

[36] A. Mölder, M. Sebesta, M. Gustafsson, L. Gisselson, A.G. Wingren, K. Alm, Non-invasive, label-free cell counting and quantitative analysis of adherent cells using digital holography, J. Microsc. 232(2) (2008) 240–247.

[37] N. Pavillon, A. Benke, D. Boss, C. Moratal, J. Kühn, P. Jourdain, et al., Cell morphology and intracellular ionic homeostasis explored with a multimodal approach combining epifluorescence and digital holographic microscopy, J. Biophotonics 3(7) (2010) 432–436.

[38] J. Persson, A. Mölder, S.-G. Pettersson, K. Alm, Cell motility studies using digital holographic microscopy, Microsc. Sci. Technol. Appl. Edu. 2 (2010) 1063–1072.

[39] B. Rappaz, E. Cano, T. Colomb, J. Kuhn, C. Depeursinge, V. Simanis, et al., Noninvasive characterization of the fission yeast cell cycle by monitoring dry mass with digital holographic microscopy, J. Biomed. Opt. 14(3) (2009) 034049 (5 pages).

[40] B. Rappaz, F. Charrière, C. Depeursinge, P.J. Magistretti, P. Marquet, Simultaneous cell morphometry and refractive index measurement with dual-wavelength digital holographic microscopy and dye-enhanced dispersion of perfusion medium, Opt. Lett. 33(7) (2008) 744–746.

[41] B. Rappaz, P. Marquet, E. Cuche, Y. Emery, C. Depeursinge, P.J. Magistretti, Measurement of the integral refractive index and dynamic cell morphometry of living cells with digital holographic microscopy, Opt. Express 13(23) (2005) 9361–9373.

[42] C. Remmersmann, S. Stürwald, B. Kemper, P. Langehanenberg, G. von Bally, Phase noise optimization in temporal phase-shifting digital holography with partial coherence light sources and its application in quantitative cell imaging, Appl. Opt. 48(8) (2009) 1463–1472.

[43] C.E. Rommel, C. Dierker, L. Schmidt, S. Przibilla, G. von Bally, B. Kemper, et al., Contrast-enhanced digital holographic imaging of cellular structures by manipulating the intracellular refractive index, J. Biomed. Opt. 15(4) (2010) 041509.

[44] N.T. Shaked, T.M. Newpher, M.D. Ehlers, A. Wax, Parallel on-axis holographic phase microscopy of biological cells and unicellular microorganism dynamics, Appl. Opt. 49(15) (2010) 2872–2878.

[45] H. Sun, B. Song, H. Dong, B. Reid, M.A. Player, J. Watson, et al., Visualization of fast-moving cells in vivo using digital holographic video microscopy, J. Biomed. Opt. 13(1) (2008) 14007–14009.

[46] M. Takagi, T. Kitabayashi, S. Ito, M. Fujiwara, A. Tokuda, Noninvasive measurement of three-dimensional morphology of adhered animal cells employing phase-shifting laser microscope, J. Biomed. Opt. 12(5) (2007) 54010–54015.

[47] N. Warnasooriya, F. Joud, P. Bun, G. Tessier, M. Coppey-Moisan, P. Desbiolles, et al., Imaging gold nanoparticles in living cell environments using heterodyne digital holographic microscopy, Opt. Express 18(4) (2010) 3264–3273.

[48] R. Liu, D.K. Dey, D. Boss, P. Marquet, B. Javidi, Recognition and classification of red blood cells using digital holographic microscopy and data clustering with discriminant analysis, J. Opt. Soc. Am. A: Opt. Image Sci. Vis. 28(6) (2011) 1204–1210.

[49] C. Hu, J. Zhong, J. Weng, Digital holographic microscopy by use of surface plasmon resonance for imaging of cell membranes, J. Biomed. Opt. 15(5) (2010) 056015 (8 pages).

[50] P. Klysubun, G. Indebetouw, A posteriori processing of spatiotemporal digital microholograms, J. Opt. Soc. Am. A: Opt. Image Sci. Vis. 18(2) (2001) 326–331.

[51] G. Indebetouw, P. Klysubun, Spatiotemporal digital microholography, J. Opt. Soc. Am. A: Opt. Image Sci. Vis. 18(2) (2001) 319–325.

[52] G. Indebetouw, Y. Tada, J. Leacock, Quantitative phase imaging with scanning holographic microscopy: an experimental assesment, Biomed. Eng. Online 5(63) (2006) p. 1–7.

[53] A.R. Brazhe, N.A. Brazhe, G.V. Maksimov, P.S. Ignatyev, A.B. Rubin, E. Mosekilde, et al., Phase-modulation laser interference microscopy: an advance in cell imaging and dynamics study, J. Biomed. Opt. 13(3) (2008) 034004.

[54] N.A. Brazhe, A.R. Brazhe, A.N. Pavlov, L.A. Erokhova, A.I. Yusipovich, G.V. Maksimov, et al., Unraveling cell processes: interference imaging interwoven with data analysis, J. Biol. Phys. 32 (2006) 191–208.

[55] N. Lue, W. Choi, K. Badizadegan, R.R. Dasari, M.S. Feld, G. Popescu, Confocal diffraction phase microscopy of live cells, Opt. Lett. 33(18) (2008) 2074–2076.

[56] N. Lue, W. Choi, G. Popescu, T. Ikeda, R.R. Dasari, K. Badizadegan, et al., Quantitative phase imaging of live cells using fast fourier phase microscopy, Appl. Opt. 46(10) (2007) 1836–1842.

[57] Y.K. Park, T. Yamauchi, W. Choi, R. Dasari, M.S. Feld, Spectroscopic phase microscopy for quantifying hemoglobin concentrations in intact red blood cells, Opt. Lett. 34(23) (2009) 3668.

[58] G. Popescu, T. Ikeda, R. Dasari, M. Feld, Diffraction phase microscopy for quantifying cell structure and dynamics, Opt. Lett. 31(6) (2006) 775–777.

[59] N.T. Shaked, J.D. Finan, F. Guilak, A. Wax, Quantitative phase microscopy of articular chondrocyte dynamics by wide-field digital interferometry, J. Biomed. Opt. 15(1) (2010) 010505 (3 pages).

[60] N.T. Shaked, M.T. Rinehart, A. Wax, Dual-interference-channel quantitative-phase microscopy of live cell dynamics, Opt. Lett. 34(6) (2009) 767–769.

[61] N.T. Shaked, L.L. Satterwhite, N. Bursac, A. Wax, Whole-cell-analysis of live cardiomyocytes using wide-field interferometric phase microscopy, Biomed. Opt. Express 1(2) (2010) 706–719.

[62] O.V. Sosnovtseva, A.N. Pavlov, N.A. Brazhe, A.R. Brazhe, L.A. Erokhova, G.V. Maksimov, et al., Interference microscopy under double-wavelet analysis: a new approach to studying cell dynamics, Phys. Rev. Lett. 94 (2005) 218103–218104.

[63] Y. Sung, W. Choi, C. Fang-Yen, K. Badizadegan, R.R. Dasari, M.S. Feld, Optical diffraction tomography for high resolution live cell imaging, Opt. Express 17(1) (2009) 266–277.
[64] V.P. Tychinsky, Coherent phase microscopy of intracellular processes, Phys. Usp. 44(6) (2001) 617–629.
[65] V.P. Tychinsky, Dynamic phase microscopy: is a "dialogue" with the cell possible? Phys. Usp. 50(5) (2007) 513–528.
[66] V.P. Tychinsky, A.V. Kretushev, I.V. Klemyashov, T.V. Vyshenskaya, A.A. Shtil, O.V. Zatsepina, Coherent phase microscopy, a novel approach to study the physiological state of the nucleolus, Dokl. Biochem. Biophys. 405(1-6) (2005) 432–436.
[67] V.P. Tychinsky, A.N. Tikhonov, Interference microscopy in cell biophysics. 1. Principles and methodological aspects of coherent phase microscopy, Cell Biochem. Biophys. 58(3) (2010) 107–116.
[68] V.P. Tychinsky, A.N. Tikhonov, Interference microscopy in cell biophysics. 2. Visualization of individual cells and energy-transducing organelles, Cell Biochem. Biophys. 58(3) (2010) 117–128.
[69] G.N. Vishnyakov, G.G. Levin, V.L. Minaev, V.V. Pickalov, A.V. Likhachev, Tomographic interference microscopy of living cells, Microsc. Anal. 18(1–6) (2004) 15–17.
[70] L. Xue, J. Lai, S. Wang, Z. Li, Single-shot slightly-off-axis interferometry based Hilbert phase microscopy of red blood cells, Biomed. Opt. Express 2(4) (2011) 987–995.
[71] T. Yamauchi, H. Iwai, M. Miwa, Y. Yamashita, Low-coherent quantitative phase microscope for nanometer-scale measurement of living cells morphology, Opt. Express 16(16) (2008) 12227–12238.
[72] T. Yamauchi, H. Iwai, Y. Yamashita, Label-free imaging of intracellular motility by low-coherent quantitative phase microscopy, Opt. Express 19(6) (2011) 5536–5550.
[73] C. Fang-Yen, S. Oh, Y. Park, W. Choi, S. Song, H.S. Seung, et al., Imaging voltage-dependent cell motions with heterodyne Mach-Zehnder phase microscopy, Opt. Lett. 32(11) (2007) 1572–1574.
[74] T. Ikeda, G. Popescu, R.R. Dasari, M.S. Feld, Hilbert phase microscopy for investigating fast dynamics in transparent systems, Opt. Lett. 30(10) (2005) 1165–1167.
[75] H. Iwai, C. Fang-Yen, G. Popescu, A. Wax, K. Badizadegan, R.R. Dasari, et al., Quantitative phase imaging using actively stabilized phase-shifting low-coherence interferometry, Opt. Lett. 29(20) (2004) 2399–2401.
[76] G. Popescu, L.P. Deflores, J.C. Vaughan, K. Badizadegan, H. Iwai, R.R. Dasari, et al., Fourier phase microscopy for investigation of biological structures and dynamics, Opt. Lett. 29(21) (2004) 2503–2505.
[77] A. Barty, K.A. Nugent, D. Paganin, A. Roberts, Quantitative optical phase microscopy, Opt. Lett. 23(11) (1998) 817–819.
[78] D. Paganin, K.A. Nugent, Noninterferometric phase imaging with partially coherent light, Phys. Rev. Lett. 80(12) (1998) 2586–2589.
[79] J.B. Tiller, A. Barty, D. Paganin, K.A. Nugent, The holographic twin image problem: a deterministic phase solution, Opt. Commun. 183(1–4) (2000) 7–14.
[80] P. Bon, G. Maucort, B. Wattellier, S. Monneret, Quadriwave lateral shearing interferometry for quantitative phase microscopy of living cells, Opt. Express 17(15) (2009) 13080–13094.
[81] R.W. Gerchberg, W.O. Saxton, A practical algorithm for the determination of phase from image and diffraction plane pictures, Optik (Stuttgart) 35(2) (1972) 237–250.
[82] J.R. Fienup, Phase retrieval algorithms: a comparison, Appl. Opt. 21(15) (1982) 2758–2769.
[83] D. Gabor, Microscopy by reconstructed wave-fronts, Proc. R. Soc. London, Ser. A 197(1051) (1949) 454–487.
[84] E.N. Leith, J. Upatnieks, Reconstructed wavefronts and communication theory, J. Opt. Soc. Am. 52(10) (1962) 1123–1130.
[85] E.N. Leith, J. Upatnieks, Wavefront reconstruction with diffused illumination and three-dimensional objects, J. Opt. Soc. Am. 54(11) (1964) 1295–1301.
[86] E. Wolf, J.R. Shewell, Diffraction theory of holography, J. Math. Phys. 11(8) (1970) 2254–2267.
[87] W.H. Carter, Computational reconstruction of scattering objects from holograms, J. Opt. Soc. Am. 60(3) (1970) 306–314.

[88] F. Dubois, N. Callens, C. Yourassowsky, M. Hoyos, P. Kurowski, O. Monnom, Digital holographic microscopy with reduced spatial coherence for three-dimensional particle flow analysis, Appl. Opt. 45(5) (2006) 864–871.

[89] B. Kemper, S. Stürwald, C. Remmersmann, P. Langehanenberg, G. von Bally, Characterisation of light emitting diodes (LEDs) for application in digital holographic microscopy for inspection of micro and nanostructured surfaces, Opt. Lasers Eng. 46(7) (2008) 499–507.

[90] Z. Yaqoob, T. Yamauchi, W. Choi, D. Fu, R.R. Dasari, M.S. Feld, Single-shot full-field reflection phase microscopy, Opt. Express 19(8) (2011) 7587–7589.

[91] Z. Ansari, Y. Gu, M. Tziraki, R. Jones, P.M.W. French, D.D. Nolte, et al., Elimination of beam walk-off in low-coherence off-axis photorefractive holography, Opt. Lett. 26(6) (2001) 334–336.

[92] J.W. Goodman, R.W. Lawrence, Digital image formation from electronically detected holograms, Appl. Phys. Lett. 11(3) (1967) 77–79.

[93] U. Schnars, W. Jüptner, Direct recording of holograms by a Ccd target and numerical reconstruction, Appl. Opt. 33(2) (1994) 179–181.

[94] O. Coquoz, R. Conde, F. Taleblou, C. Depeursinge, Performances of endoscopic holography with a multicore optical fiber, Appl. Opt. 34(31) (1995) 7186–7193.

[95] I. Yamaguchi, T. Zhang, Phase-shifting digital holography, Opt. Lett. 22(16) (1997) 1268–1270.

[96] P. Carré, Installation et utilisation du comparateur photoélectrique et interférentiel du bureau international des poids et mesures, Metrologia 2(1) (1966) 13–23.

[97] J.H. Bruning, D.R. Herriott, J.E. Gallagher, D.P. Rosenfeld, A.D. White, D.J. Brangaccio, Digital wavefront measuring interferometer for testing optical surfaces and lenses, Appl. Opt. 13(11) (1974) 2693–2703.

[98] E. Cuche, P. Marquet, C. Depeursinge, Simultaneous amplitude-contrast and quantitative phase-contrast microscopy by numerical reconstruction of Fresnel off-axis holograms, Appl. Opt. 38(34) (1999) 6994–7001.

[99] T. Kreis, Handbook of Holographic Interferometry, Wiley-VCH, Weinheim, 2005.

[100] P. Rastogi, Holographic Interferometry: Principles and Methods, Series in Optical Sciences, vol. 68, Springer-Verlag, NY, 1994 (Chapter 5).

[101] P. Guo, A.J. Devaney, Digital microscopy using phase-shifting digital holography with two reference waves, Opt. Lett. 29(8) (2004) 857–859.

[102] J.P. Liu, T.C. Poon, Two-step-only quadrature phase-shifting digital holography, Opt. Lett. 34(3) (2009) 250–252.

[103] Y. Awatsuji, M. Sasada, T. Kubota, Parallel quasi-phase-shifting digital holography, Appl. Phys. Lett. 85(6) (2004) 1069–1071.

[104] C.S. Guo, L. Zhang, H.T. Wang, J. Liao, Y.Y. Zhu, Phase-shifting error and its elimination in phase-shifting digital holography, Opt. Lett. 27(19) (2002) 1687–1689.

[105] Z. Wang, B. Han, Advanced iterative algorithm for phase extraction of randomly phase-shifted interferograms, Opt. Lett. 29(14) (2004) 1671–1673.

[106] X.F. Xu, L.Z. Cai, Y.R. Wang, X.F. Meng, W.J. Sun, H. Zhang, et al., Simple direct extraction of unknown phase shift and wavefront reconstruction in generalized phase-shifting interferometry: algorithm and experiments, Opt. Lett. 33(8) (2008) 776–778.

[107] M. Takeda, H. Ina, S. Kobayashi, Fourier-transform method of fringe-pattern analysis for computer-based topography and interferometry, J. Opt. Soc. Am. 72(1) (1982) 156.

[108] T. Kreis, Digital holographic interference-phase measurement using the Fourier-transform method, J. Opt. Soc. Am. A: Opt. Image Sci. Vis. 3(6) (1986) 847–855.

[109] M. Liebling, T. Blu, M. Unser, Complex-wave retrieval from a single off-axis hologram, J. Opt. Soc. Am. A: Opt. Image Sci. Vis. 21(3) (2004) 367–377.

[110] P. Picart, J. Leval, General theoretical formulation of image formation in digital Fresnel holography, J. Opt. Soc. Am. A: Opt. Image Sci. Vis. 25(7) (2008) 1744–1761.

[111] J.W. Goodman, Introduction to Fourier Optics, second ed., McGraw-Hill Company, Inc., New York, NY, 1996.

[112] T. Colomb, N. Pavillon, J. Kühn, E. Cuche, C. Depeursinge, Y. Emery, Extended depth-of-focus by digital holographic microscopy, Opt. Lett. 35(11) (2010) 1840–1842.

[113] P. Ferraro, S. De Nicola, A. Finizio, G. Coppola, S. Grilli, C. Magro, et al., Compensation of the inherent wave front curvature in digital holographic coherent microscopy for quantitative phase-contrast imaging, Appl. Opt. 42(11) (2003) 1938–1946.

[114] T. Colomb, E. Cuche, F. Charrière, J. Kühn, N. Aspert, F. Montfort, et al., Automatic procedure for aberration compensation in digital holographic microscopy and applications to specimen shape compensation, Appl. Opt. 45(5) (2006) 851–863.

[115] P. Ferraro, D. Alferi, S.D. Nicola, L.D. Petrocellis, A. Finizio, G. Pierattini, Quantitative phase-contrast microscopy by a lateral shear approach to digital holographic image reconstruction, Opt. Lett. 31(10) (2006) 1405–1407.

[116] N. Lue, G. Popescu, T. Ikeda, R.R. Dasari, K. Badizadegan, M.S. Feld, Live cell refractometry using microfluidic devices, Opt. Lett. 31(18) (2006) 2759–2761.

[117] J. Kuhn, T. Colomb, F. Montfort, F. Charriere, Y. Emery, E. Cuche, et al., Real-time dual-wavelength digital holographic microscopy with a single hologram acquisition, Opt. Express 15(12) (2007) 7231–7242.

[118] L. Miccio, A. Finizio, P. Memmolo, M. Paturzo, F. Merola, G. Coppola, et al., Detection and visualization improvement of spermatozoa cells by digital holography, in Molecular Imaging III, C. Lin and V. Ntziachristos, eds., Vol. 8089 of Proceedings of SPIE-OSA Biomedical Optics (Optical Society of America, 2011), paper 80890C. http://www.opticsinfobase.org/abstract.cfm?URI=ECBO-2011-80890C.

[119] I. Crga, J. Zakova, M. Huser, P. Ventruba, E. Lousova, M. Pohanka, Digital holographic microscopy in human sperm imaging, J. Assist. Reprod. Genet. (2011).

[120] W.M. Ash III, L. Krzewina, M.K. Kim, Quantitative imaging of cellular adhesion by total internal reflection holographic microscopy, Appl. Opt. 48(34) (2009) H144–H152.

[121] C. Pache, J. Kühn, K. Westphal, M.F. Toy, J. Parent, O. Büchi, et al., Digital holographic microscopy real-time monitoring of cytoarchitectural alterations during simulated microgravity, J. Biomed. Opt. 15(2) (2010) 026021 (9 pages).

[122] P. Langehanenberg, G. Bally, B. Kemper, Autofocusing in digital holographic microscopy, 3D Res. 2(1) (2011) pp. D176–D182.

[123] M. Liebling, M. Unser, Autofocus for digital fresnel holograms by use of a fresnelet sparsity criterion, J. Opt. Soc. Am. A: Opt. Image Sci. Vis. 21(12) (2004) 2424–2430.

[124] M. Antkowiak, N. Callens, C. Yourassowsky, F. Dubois, Extended focused imaging of a microparticle field with digital holographic microscopy, Opt. Lett. 33(14) (2008) 1626–1628.

[125] P. Ferraro, S. Grilli, D. Alfieri, S.D. Nicola, A. Finizio, G. Pierattini, et al., Extended focused image in microscopy by digital holography, Opt. Express 13(18) (2005) 6738–6749.

[126] I. Moon, M. Daneshpanah, A. Anand, B. Javidi, Cell identification with computational: 3-D holographic microscopy, Opt. Photonics News 22(6) (2011) 18–23.

[127] M. DaneshPanah, B. Javidi, Tracking biological microorganisms in sequence of 3D holographic microscopy images, Opt. Express 15(7) (2007) 10761–10766.

[128] J. Sheng, E. Malkiel, J. Katz, J. Adolf, R. Belas, A.R. Place, Digital holographic microscopy reveals prey-induced changes in swimming behavior of predatory dinoflagellates, Proc. Natl. Acad. Sci. U.S.A. 104(44) (2007) 17512–17517.

[129] J. Sheng, E. Malkiel, J. Katz, J.E. Adolf, A.R. Place, A dinoflagellate exploits toxins to immobilize prey prior to ingestion, Proc. Natl. Acad. Sci. U.S.A. 107(5) (2010) 2082–2087.

[130] R. Barer, Determination of dry mass, thickness, solid and water concentration in living cells, Nature (London) 172(4389) (1953) 1097–1098.

[131] D. Fu, W. Choi, Y. Sung, Z. Yaqoob, R.R. Dasari, M. Feld, Quantitative dispersion microscopy, Biomed. Opt. Express 1(2) (2010) 347−353.

[132] M. Mir, Z. Wang, Z. Shen, M. Bednarz, R. Bashir, I. Golding, et al., Optical measurement of cycle-dependent cell growth, Proc. Natl. Acad. Sci. U.S.A. 108(32) (2011) 13124−13129.

[133] G. Popescu, Y.K. Park, N. Lue, C. Best-Popescu, L. Deflores, R.R. Dasari, et al., Optical imaging of cell mass and growth dynamics, Am. J. Physiol. Cell Physiol. 295 (2008) 538−544.

[134] D. Zicha, G.A. Dunn, An image processing system for cell behaviour studies in subconfluent cultures, J. Microsc. 179(1) (1995) 11−21.

[135] M. Minet, P. Nurse, P. Thuriaux, J.M. Mitchison, Uncontrolled septation in a cell division cycle mutant of the fission yeast *Schizosaccharomyces pombe*, J. Bacteriol. 137(1) (1979) 440−446.

[136] M.S. Amin, Y.K. Park, N. Lue, R.R. Dasari, K. Badizadegan, M.S. Feld, et al., Microrheology of red blood cell membranes using dynamic scattering microscopy, Opt. Express 15(25) (2007) 17001−17009.

[137] Y. Park, C.A. Best, K. Badizadegan, R.R. Dasari, M.S. Feld, T. Kuriabova, et al., Measurement of red blood cell mechanics during morphological changes, Proc. Natl. Acad. Sci. U.S.A. 107(15) (2010) 6731−6736.

[138] Y.K. Park, M. Diez-Silva, G. Popescu, G. Lykotrafitis, W. Choi, M.S. Feld, et al., Refractive index maps and membrane dynamics of human red blood cells parasitized by *Plasmodium falciparum*, Proc. Natl. Acad. Sci. U.S.A. (2008) 13730−13735.

[139] G. Popescu, K. Badizadegan, R.R. Dasari, M.S. Feld, Observation of dynamic subdomains in red blood cells, J. Biomed. Opt. 11(4) (2006) 40503

[140] G. Popescu, Y. Park, W. Choi, R.R. Dasari, M.S. Feld, K. Badizadegan, Imaging red blood cell dynamics by quantitative phase microscopy, Blood Cells Mol. Dis. 41(1) (2008) 10−16.

[141] B. Rappaz, A. Barbul, Y. Emery, R. Korenstein, C. Depeursinge, P.J. Magistretti, et al., Comparative study of human erythrocytes by digital holographic microscopy, confocal microscopy, and impedance volume analyzer, Cytometry A 73a(10) (2008) 895−903.

[142] B. Rappaz, A. Barbul, A. Hoffmann, D. Boss, R. Korenstein, C. Depeursinge, et al., Spatial analysis of erythrocyte membrane fluctuations by digital holographic microscopy, Blood Cells Mol. Dis. 42(3) (2009) 228−232.

[143] N.T. Shaked, L.L. Satterwhite, M.J. Telen, G.A. Truskey, A. Wax, Quantitative microscopy and nanoscopy of sickle red blood cells performed by wide field digital interferometry, J. Biomed. Opt. 16(3) (2011) 030506 (3 pages).

[144] R. Wang, H. Ding, M. Mir, K. Tangella, G. Popescu, Effective 3D viscoelasticity of red blood cells measured by diffraction phase microscopy, Biomed. Opt. Express 2(3) (2011) 485−490.

[145] C.L. Curl, C.J. Bellair, P.J. Harris, B.E. Allman, A. Roberts, K.A. Nugent, et al., Single cell volume measurement by quantitative phase microscopy (QPM): a case study of erythrocyte morphology, Cell. Physiol. Biochem. 17(5−6) (2006) 193−200.

[146] T. Tishko, V. Titar, D. Tishko, K. Nosov, Digital holographic interference microscopy in the study of the 3D morphology and functionality of human blood erythrocytes, Laser Phys. 18(4) (2008) 486−490.

[147] J.D. Bessman, R.K. Johnson, Erythrocyte volume distribution in normal and abnormal subjects, Blood 46(3) (1975) 369−379.

[148] S. Levin, R. Korenstein, Membrane fluctuations in erythrocytes are linked to MgATP-dependent dynamic assembly of the membrane skeleton, Biophys. J. 60(3) (1991) 733−737.

[149] S. Tuvia, A. Almagor, A. Bitler, S. Levin, R. Korenstein, S. Yedgar, Cell membrane fluctuations are regulated by medium macroviscosity: evidence for a metabolic driving force, Proc. Natl. Acad. Sci. U.S.A. 94(10) (1997) 5045−5049.

[150] F. Brochard, J.F. Lennon, Frequency spectrum of the flicker phenomenon in erythrocytes, J. Phys. France 36(11) (1975) 1035−1047.

[151] Y.Z. Yoon, H. Hong, A. Brown, D.C. Kim, D.J. Kang, V.L. Lew, et al., Flickering analysis of erythrocyte mechanical properties: dependence on oxygenation level, cell shape, and hydration level, Biophys. J. 97(6) (2009) 1606−1615.

[152] M. Goldstein, I. Leibovitch, S. Levin, Y. Alster, A. Loewenstein, G. Malkin, et al., Red blood cell membrane mechanical fluctuations in non-proliferative and proliferate diabetic retinopathy, Graefes Arch. Klin. Exp. Ophthalmol. 242(11) (2004) 937–943.

[153] D. Boss, J. Kuehn, C. Depeursinge, P.J. Magistretti, P. Marquet, Exploring red blood cell membrane dynamics with digital holographic microscopy, in: Proc. SPIE 7715, 77153I (2010); http://dx.doi.org/10.1117/12.854717.

[154] M.M. Aarts, M. Tymianski, Molecular mechanisms underlying specificity of excitotoxic signaling in neurons, Curr. Mol. Med. 4(2) (2004) 137–147.

[155] P. Jourdain, N. Pavillon, C. Moratal, D. Boss, B. Rappaz, C. Depeursinge, et al., Determination of transmembrane water fluxes in neurons elicited glutamate ionotropic receptors and by the cotransporters KCC2 and NKCC1: a digital holographic study, J. Neurosci. 31(33) (2011) 11846–11854.

[156] N. Pavillon, J. Kuhn, C. Moratal, P. Jourdain, C. Depeursinge, P.J. Magistretti, et al., Early cell death detection with digital holographic microscopy, PLoS ONE 7(1) (2012) e30912.

[157] N. Pavillon, J. Kühn, P. Jourdain, C. Depeursinge, P.J. Magistretti, P. Marquet, Cell death detection and ionic homeostasis monitoring with digital holographic microscopy, in: Proc. SPIE 8090, 809004 (2011); http://dx.doi.org/10.1117/12.889246.

[158] W. Choi, C. Fang-Yen, K. Badizadegan, S. Oh, N. Lue, R.R. Dasari, et al., Tomographic phase microscopy, Nat. Methods 4(9) (2007) 717–719.

[159] F. Charrière, A. Marian, F. Montfort, J. Kühn, T. Colomb, E. Cuche, et al., Cell refractive index tomography by digital holographic microscopy, Opt. Lett. 31(2) (2006) 178–180.

[160] F. Charrière, E. Cuche, P. Marquet, C. Depeursinge, Biological cell (pollen grain) refractive index tomography with digital holographic microscopy—art. no. 609008, in: Three-Dimensional and Multidimensional Microscopy: Image Acquisition and Processing XIII, Ecole Polytech Fed Lausanne, Imaging and Application of Optical Institute, Lausanne, CH-1015 Switzerland. F. Charriere, Ec: SPIE-Int. Soc. Opt. Eng., 2006.

[161] S.I. Satake, T. Anraku, F. Tamoto, K. Sato, T. Kunugi, Three-dimensional simultaneous measurements of micro-fluorescent-particle position and temperature field via digital hologram, Microelectron. Eng. 88(8) (2011) 1875–1877.

CHAPTER 6

Holographic Microscopy Techniques for Multifocus Phase Imaging of Living Cells

Björn Kemper[1], Patrik Langehanenberg[1], Andreas Bauwens[2]

[1]Center for Biomedical Optics and Photonics, University of Muenster, Muenster, Germany
[2]Institute of Hygiene, University of Muenster, Muenster, Germany

Editor: Natan T. Shaked

6.1 Introduction

In cancer research and in the investigations of infection and inflammation processes, the analysis of dynamic processes such as temporal changes of the cell morphology and the concentration of intracellular substances is of particular interest. Here, the imaging of living single cells with light microscopy can give new insights into cell motility, biomechanical properties on cellular or subcellular level, and the response of cells to drugs and toxins. On a molecular level, live cell imaging aspects have been addressed widely by using a variety of fluorescence microscopy techniques (see Refs. [1–3] and references therein). The specificity of fluorescence signals is high due to a large number of available dyes, specific autofluorescence mechanisms, fluorescence life-time imaging, and a wide experience with the techniques. However, there still remain challenges in the application of these methods for live cell imaging as fluorescence staining is often toxic, restricted to specific subcellular structures, or limited to chemically fixated samples. Thus, in addition to research activities on the improvement of labeling techniques, label-free methods for quantitative cell imaging have been developed. To make use of the high accuracy of diffraction and interferometry-based metrology, the activities of a growing number of research groups focus on techniques for quantitative phase imaging. Quantitative phase imaging provides label-free data with low demands on light exposure and high data acquisition rates. In deference to the above-mentioned fluorescence techniques, such methods detect changes in the optical path length (OPL) that are caused by the specimen under investigation.

This chapter focuses on multifocus quantitative phase imaging of living cells with digital holographic microscopy (DHM). Holographic interferometric metrology is well established in industrial nondestructive testing and quality control [4–6]. In combination with microscopy, digital holography provides label-free, quantitative phase imaging [7–12]. The reconstruction of digitally captured holograms is performed numerically. Thus, in comparison to other phase contrast methods [13,14], related interferometry-based techniques [15–18], and optical coherence tomography or optical coherence microscopy [19–24], DHM provides quantitative phase contrast imaging with subsequent numerical refocusing (multifocus imaging) from a single recorded hologram.

In an overview, principles of DHM that have been found applicable for quantitative live cells imaging are presented. First, two experimental arrangements for the recording of digital off-axis and self-interference holograms are explained which are suitable for the modular integration of DHM into common research microscopes. Afterward, the numerical processing of digitally captured off-axis holograms by spatial phase shifting techniques and the further evaluation of the reconstructed quantitative phase images are explained. This includes a description of (subsequently) numerical autofocusing and the determination of the integral refractive index of suspended cells. Then, the application of DHM for quantitative cell imaging is illustrated by representative results. It is shown that DHM can be used for monitoring of toxin-mediated changes of the cell morphology and cell thickness. Furthermore, the analysis of vacuole formation and the imaging of apoptotic and necrotic effects are illustrated. Finally, the usage of DHM for quantitative monitoring of cell division and the tracking of cells in a three-dimensional (3D) environment are demonstrated. The results demonstrate that DHM represents a helpful tool to achieve novel insights in toxicity testing and cell dynamics.

6.2 DHM Setups for Live Cell Imaging

Various approaches for the implementation of DHM have been developed (for an overview, see Refs. [25,26] and references therein). For imaging of living cells in a biomedical laboratory environment, the combination of robust and flexible DHM setups with common research microscopy systems is advantageous. In this way, the usage of the method and the combination with other optical and nonoptical imaging methods for live cell imaging are simplified. Thus, in this section, two different principles for quantitative phase imaging with DHM are presented that are suitable for the integration into various inverted microscopes. In order to ensure a vibration insensitive data acquisition, off-axis arrangements in Mach–Zehnder and Michelson interferometer configuration are chosen, which enable a single-shot acquisition of digital holograms.

6.2.1 Fiber Optic Modular DHM

Figure 6.1 shows a fiber optic Mach–Zehnder interferometer concept for DHM, designed for the integration into common commercial research microscopes and for the investigations on transparent specimens such as living cells [27]. The light of a laser (e.g., a frequency-doubled Nd:YAG laser, $\lambda = 532$ nm with a large coherence length >1 m) is divided into object illumination wave (object wave) and reference wave. For variable light guidance polarization maintaining, single-mode optical fibers are used. The illumination of the sample with coherent laser light is performed in transmission by coupling the object wave into the microscope's condenser. Thus, an optimized illumination similar to the Koehler concept [28] is achieved.

For imaging, common microscope objectives are utilized. The reference wave is guided directly to an interferometric unit that is adapted to one of the microscope's camera ports. Off-axis holography is achieved by a beam splitter that tilts the reference wave front against

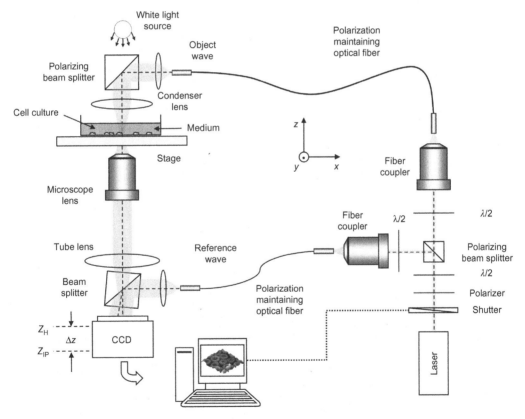

Figure 6.1
Schematic of the fiber optic modular integration of digital holography into an iMIC for live cell imaging. Source: *Modified from Ref. [27]*.

the wave front of the object wave. The interferogram that is formed by the superposition of object wave and reference wave is recorded by a charge-coupled device (CCD) camera and transferred to an image processing system for reconstruction and evaluation of the digitized holograms.

6.2.2 Self-Interference DHM

A drawback of many Mach–Zehnder-interferometer-based quantitative phase imaging arrangements [7,9–11,29] is the requirement for a separate reference wave, which results in a phase stability decrease and the demand for a precise adjustment of the intensity ratio between object and reference waves. To overcome these problems, several approaches were reported [17,30–34]. Here, in order to avoid a separately generated reference wave, a Michelson interferometer approach for DHM is presented [35].

Figure 6.2 shows a sketch of the Michelson interferometer-based DHM arrangement setup which can be attached to an inverted research microscope [35]. In analogy to the setup in

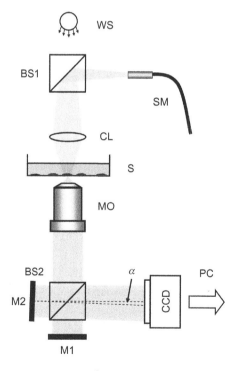

Figure 6.2
Schematic of an experimental setup for Michelson interferometer-based self-interference DHM. WS: white light source; BS1, BS2: beam splitter cubes; CL: condenser lens; SM: single-mode fiber; S: sample (here, Petri dish with adherent cells); MO: microscope lens; M1, M2: mirrors; CCD: charge-coupled device sensor; α: tilt angle; PC: computer [36].

Figure 6.1, the sample is illuminated by laser light via a single-mode optical fiber (SM) by inserting a nonpolarizing beam splitter cube (BS1) into the illumination path of the microscope's white light source (WS). Behind the microscope lens (MO), the light is coupled into a Michelson interferometer which is attached to a camera side port of the microscope. Mirror M2 is tilted by an angle α in such a way that an area of the sample that contains no object is superposed with the image of the specimen to create a suitable spatial carrier fringe pattern for off-axis holography (for illustration, see Figure 6.4). Note that due to the Michelson interferometer design in areas without specimen, two wave fronts with nearly identical curvatures are superimposed. This is even fulfilled for an imaging geometry with two slightly divergent waves that differ from the collimated arrangement, which is shown in Figure 6.2 to simplify the illustration of the measurement principle. The digital holograms are recorded by a CCD sensor. The numerical calculation of the quantitative DHM phase contrast images from the resulting self-interference digital holograms can be performed as described in Sections 6.3.2 and 6.3.3.

6.3 Recording and Numerical Evaluation of Digital Holograms

In this section, the recording and the numerical evaluation of digital off-axis holograms for multifocus quantitative phase imaging is described. First, the coding of the object wave in off-axis holograms is explained. Then, the spatial phase shifting evaluation for the retrieval of the object wave and the optional numerical propagation of the obtained wave fields to the image plane by a convolution approach of the Huygens–Fresnel principle are described. Afterward, the evaluation of off-axis and self-interference digital holograms for multifocus quantitative phase imaging and the principle of (subsequent) numerical autofocusing is illustrated.

6.3.1 Intensity Distribution in the Hologram Plane

The intensity distribution I_H of the interferogram in the hologram plane located at $z = z_H$ that is created with the experimental setups in Figures 6.1 and 6.2 by the interference of the object wave O and the reference wave R is [4]

$$\begin{aligned}
I_H(x, y, z_H) &= O(x, y, z_H)O^*(x, y, z_H) + R(x, y, z_H)R^*(x, y, z_H) \\
&+ O(x, y, z_H)R^*(x, y, z_H) + R(x, y, z_H)O^*(x, y, z_H) \\
&= I_O(x, y, z_H) + I_R(x, y, z_H) + 2\sqrt{I_O(x, y, z_H)I_R(x, y, z_H)}\cos\Delta\varphi_{HP}(x, y, z_H)
\end{aligned}$$
(6.1)

with $I_O = OO^* = |O|^2$ and $I_O = RR^* = |R|^2$ ($*$ represents the conjugate complex terms). The parameter $\Delta\varphi_H(x,y,z_H) = \varphi_R(x,y,z_H) - \varphi_O(x,y,z_H)$ is the phase difference between O and R at $z = z_H$. In the presence of a sample in the optical path of O, the phase distribution represents the sum $\varphi_O(x, y, z_H) = \varphi_{O_H}(x, y, z_H) + \Delta\varphi_S(x, y, z_H)$ with the pure object wave phase $\varphi_{O_H}(x, y, z_H)$ and the phase change $\Delta\varphi_S(x,y,z_H)$ that is effected by the sample. In the following section, it is explained how the sample-induced phase change $\Delta\varphi_s(x,y,z_{IP})$ in the image plane $z = z_{IP}$ is retrieved from the intensity distribution in Eq. (6.1).

6.3.2 Spatial Phase Shifting-Based Reconstruction of Digital Holograms

For the numerical reconstruction of temporal phase shifted and digital off-axis holograms, various methods have been developed (for an overview, see Refs. [12,26,37−41]). Here, the numerical reconstruction is performed in two steps. First, the complex object wave O is reconstructed in the hologram plane. This is performed by spatial phase shifting. Spatial phase shifting provides the retrieval of the object wave without the disturbing terms "twin image" and "zero-order intensity." In temporal phase shifting holography, several holograms for known (or unknown but equidistant) phase shifts of the reference wave (or alternatively of the object wave) are recorded to retrieve the object wave (for illustration, see Refs. [42−44]). In contrast, in spatial phase shifting holography, neighboring pixels of single off-axis holograms are evaluated. Therefore, the same numerical algorithms as for temporal phase shifting [45] can be used [46]. However, the interferogram equation (Eq. (6.1)) can also be solved pixel-wise within a squared area of pixels (in practice usually 5×5 pixels) around a given hologram pixel using the least squares principle. By definition of appropriate substitutions, the resulting nonlinear problem is transferred to a form that can be solved by linear algorithms [9,47]. The resulting robust reconstruction method for O was found particularly suitable for the application in DHM and has been reported to be applied successfully for the analysis of living cells in Refs. [9,29].

In case that objects have been recorded out of focus, numerical refocusing is required. Thus, in an optionally subsequent step, O is numerically propagated to the image plane. This is typically performed by the numerical implementation of the Fresnel−Huygens principle [38,48]. Here, the numerical wave propagation is illustrated by an approach of the convolution method in which the convolution theorem is applied after the Fresnel approximation [29,49]. The propagation of $O(x, y, z_H)$ to the image plane z_{IP} that is located at $z_{IP} = z_H + \Delta z$ in the distance Δz to the hologram plane (see Figure 6.1) is performed by using the following equation:

$$O(x, y, z_{IP} = z_H + \Delta z) = \mathbf{F}^{-1}\{\mathbf{F}\{O(x, y, z_H)\}\exp(i\pi\lambda\Delta z(\nu^2 + \mu^2))\} \quad (6.2)$$

In Eq. (6.2), λ is the wave length of the applied laser light, ν and μ are the coordinates infrequency domain, and $\mathbf{F}$ denotes a Fourier transformation. The advantage of this approach is that the size of the propagated wave field is preserved during the refocusing process. This is a particular advantage for numerical autofocusing as described in Section 6.3.4 because it simplifies the comparison of the image definition in different focal planes. Furthermore, in contrast to the propagation by digital Fresnel transformation, Eq. (6.2) allows the refocusing of only slightly defocused images of the sample near the hologram plane [26]. However, numerical refocusing also may be performed with other common numerical propagation methods, including in particular more general approaches of the convolution method [38] and the angular spectrum method [41,50]. During the propagation process, the parameter Δz in Eq. (6.2) is chosen so that the holographic amplitude image $|O|$ appears sharply, like, for example, a microscopic image under white light illumination. In the special case that the image of the sample is sharply focused in the hologram plane with $\Delta z = 0$ and thus $z_{IP} = z_H$, the reconstruction process can be accelerated because no propagation of O by Eq. (6.2) is required.

6.3.3 Quantitative Phase Imaging

In addition to the amplitude $|O(x, y, z_{IP})|$ that represents the image of the sample, from the numerically reconstructed and optionally propagated complex object wave $O(x, y, z_{IP})$ (see Eq. (6.2)) also the sample induced phase change $\Delta\varphi_S(x, y, z_{IP})$ is obtained [10,29]:

$$\Delta\varphi_S(x, y, z_{IP}) = \arctan \frac{\text{Im}\{O(x, y, z_{IP})\}}{\text{Re}\{O(x, y, z_{IP})\}} \ (\text{mod } 2\pi) \tag{6.3}$$

After removal of the 2π ambiguity by phase unwrapping [4], the data obtained by Eq. (6.3) can be used for quantitative phase imaging.

In *transmission mode* (see Figures 6.1 and 6.2), the induced phase change $\Delta\varphi_S(x,y,z_{IP})$ of a semitransparent sample is influenced by the sample thickness, the refractive index of the sample, and the refractive index of the medium that surrounds the investigated specimen. Thus, for quantitative cell imaging, information about the cellular refractive index is required. Different interferometric and holographic methods for the determination of the refractive index have been developed (see Refs. [12,51,52] and Section 6.4). For cells in cell culture medium with the refractive index n_{medium}, and the assumption of a known homogeneously distributed integral cellular refractive index n_{cell}, the cell thickness $d_{cell}(x, y, z_{IP})$ can be determined by measuring the OPL change $\Delta\varphi_{cell}$ of the cells to the surrounding medium [10,29]:

$$d_{cell}(x, y, z_{IP}) = \frac{\lambda \Delta\varphi_{cell}(x, y, z_{IP})}{2\pi} \cdot \frac{1}{n_{cell} - n_{medium}} \tag{6.4}$$

The parameter λ in Eq. (6.4) represents the wavelength of the applied laser light. For adherently grown cells, the parameter d_{cell} can be interpreted as the cell shape. Nevertheless, the results from Eq. (6.4) have to be handled critically, e.g., if toxically and osmotically induced reactions of cells [29,53] are analyzed that may cause dynamic changes in the cellular refractive index.

Figure 6.3 shows the numerical processing of a digital off-axis hologram by spatial phase shifting-based reconstruction.

Figure 6.3A shows a digital hologram of human red blood cells (RBCs) in phosphate buffered saline. The cells were recorded slightly defocused with a 63× microscope lens (numerical aperture (NA) = 0.75) with a setup as shown in Figure 6.1 by using a frequency-doubled Nd:YAG laser ($\lambda = 532$ nm) as light source. The enlarged area in Figure 6.3A shows a part of the carrier fringe pattern of the digital off-axis hologram that was used for holographic coding of the object wave. Figure 6.3B shows the unfocused reconstructed amplitude image of the RBCs in the CCD sensor plane. Figure 6.3C displays the sharply focused amplitude after numerical refocusing with Eq. (6.2). This image corresponds to a microscopic bright-field image under coherent illumination. The quantitative phase image after removal of the 2π ambiguity is shown in Figure 6.3D, while Figure 6.3E shows a corresponding pseudo-3D representation of the data. In Figure 6.3F, the cell thickness measurement along a cross-section through the phase data along the dashed line in Figure 6.3D is illustrated. The thickness values d_{cell} of the RBC were calculated from the phase contrast $\Delta\varphi_{cell}$ with Eq. (6.4) by estimating an integral cellular refractive index $n_{RBC} = 1.400$ [55]. The refractive index $n_{medium} = 1.337$ of the buffer solution was obtained by an Abbe refractometer. For completeness, Figure 6.3G shows a DHM differential phase contrast image (DHM DIC image) that has been obtained by the calculation of the

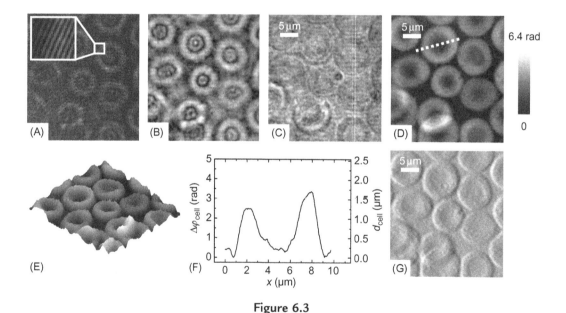

Figure 6.3
Evaluation of digital off-axis holograms by spatial phase shifting. (A) Slightly defocused recorded digital hologram of human red blood cells, (B) reconstructed amplitude in the hologram plane (defocused), (C) numerically refocused amplitude distribution, (D) unwrapped phase distribution coded to 256 gray levels, (D) pseudo-3D representation of the phase distribution in (E), (F) cross-section through a cell along the dashed line in (D), (G) first derivative of (E) in the x-direction. Source: *Modified from Ref. [54]*.

differential quotient in the *x*-direction of the data in Figure 6.3D. DHM DIC is comparable to Nomarski differential interference phase contrast [56] with the advantage of an adaptable sensitivity due to a variable digital shear [57].

The measurement principle for self-interference DHM as in Figure 6.2 is illustrated by white light images and by the evaluation of digital holograms that were obtained from three living human pancreatic tumor cells (PaTu 8988 T; for details, see Ref. [29] and references therein) by using a 40× microscope lens (NA = 0.6). For both white light imaging and hologram recording, the specimens were sharply focused onto the CCD sensor. Figure 6.4A–C shows white light images of the cells obtained from mirror M1 (optical path of mirror M2 blocked), from mirror M2 (optical path of mirror M1 blocked), and the superimposed image obtained by using both mirrors, M1 and M2. For hologram recording, the white light illumination was replaced by the highly coherent light of a frequency-doubled Nd:YAG laser (λ = 532 nm,

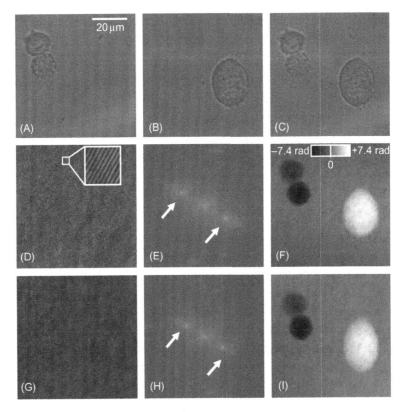

Figure 6.4
Evaluation of self-interference off-axis holograms. (A) White light image (Figure 6.2) of PaTu 8988 T cells from mirror M1 (M2 blocked); (B) white light image (Figure 6.2) from mirror M2 (M1 blocked); (C) white light image from M1 + M2; (D), (G) digital off-axis holograms with enlarged spatial carrier fringe pattern; (E), (H): 2D spatial frequency spectra of (D) and (G); (F), (I) quantitative phase images numerically reconstructed from (D) and (G) (coded to 256 gray levels) [36].

coherence length >1 m) or by a diode laser ($\lambda = 641$ nm, coherence length ≈ 0.75 mm). The angle α between the two mirrors M1 and M2 was chosen in such a way that a spatial carrier fringe frequency with a spatial phase gradient near 0.5π per pixel [58] for optimized numerical reconstruction was generated. The captured holograms are shown in Figure 6.4D (frequency-doubled Nd:YAG) and Figure 6.4G (diode laser). The resulting parallel carrier fringe pattern is depicted in the enlarged part of Figure 6.4D. Figure 6.4E and H shows the spatial frequency spectra of the holograms in Figure 6.4D and G that were calculated by a two-dimensional (2D) fast Fourier transformation. The narrow peaks that are marked in Figure 6.4E and H with arrows show the high linear degree of the carrier fringe patterns. Finally, Figure 6.4F and I presents the quantitative phase contrast images obtained by spatial phase shifting reconstruction as described in Section 6.3.2. Each of the two sheared wave fronts serves as a reference wave for the complementary wave. Thus, the cells appear in inverse phase contrast as they are imaged with different mirrors. For both light sources, the frequency-doubled Nd:YAG laser and the diode laser, phase contrast images with a similar quality were obtained.

A further characterization of the Michelson interferometer approach and comparative investigations to a modular DHM system based on a Mach–Zehnder interferometer with a fiber optic reference wave (see detailed description in Ref. [27]) were performed using the frequency-doubled Nd:YAG laser as light source. Due to the small OPL difference within the interferometer for the setup in Figure 6.2, an up to five times increased temporal phase stability comparison to the fiber optic Mach–Zehnder setup was achieved [35]. The increased interferometric stability also results in a significant reduction of reconstruction artifacts like unwrapping errors which is important for the automated evaluation of digital holograms obtained from time-lapse series.

However, the analysis of the spatial phase noise within the quantitative phase contrast images also shows that the accuracy for the detection of OPL changes for the Michelson interferometer-based DHM arrangement is lower than for a Mach–Zehnder setup [35]. This can be explained by the circumstance that in the Michelson interferometer setup both waves are affected by coherent disturbances due to scattering effects from the microscope imaging system and the medium in which the specimen is embedded. Nevertheless, in Refs. [35,36], it is demonstrated that even subcellular structures like the nuclear envelope and the nucleoli are resolved. Furthermore, the setup is particularly suitable for low-cost light sources with short coherence lengths $l_c < 1$ mm [36].

6.3.4 Subsequent Refocusing and Autofocusing

In digital holography, refocusing can be performed numerically by variation of the propagation distance Δz in Eq. (6.2). The combination of this feature with image sharpness quantification algorithms yields subsequent autofocusing (for an overview, see Ref. [54]

and references therein). Pure phase objects with negligible absorption such as technical reflective specimens or biological cells are sharply focused at the setting with the least contrasted contours in the amplitude distributions [59,60]. In contrast to the bright-field case, in digital holography this setting is of particular interest, as the amplitude and phase distributions are accessible simultaneously, and the focal setting with the least contrasted amplitude image corresponds to the best resolved structures in the quantitative phase contrast distribution. In Ref. [60], different numerical methods that calculate a scalar focus value for the quantification of the image sharpness were compared. Therefore, common criteria to evaluate the image sharpness in bright-field microscopy [61,62] were used. In agreement with previously reported results for microscopic imaging with white light illumination [63,64], in Refs. [60,65] the evaluation of the weighted and band-pass-filtered power spectra

$$\text{focus value} = \sum_{\mu,\nu} \log\{1 + [\mathbf{F}_F(|O(x,y)|)(\mu,\nu)]\} \tag{6.5}$$

was identified to be most suitable for a robust determination of the image sharpness in DHM in combination with the convolution method described in Section 6.3.2. In Eq. (6.5), (x, y) are the coordinates in spatial domain and (μ, ν) denote the corresponding spatial frequencies. The parameter $\mathbf{F}_F(|O|)(\mu, \nu)$ denotes the band-pass-filtered Fourier transform of the amplitude distribution $|O(x, y)|$. Logarithmic weighting is applied in order to consider also weak parts of the spatial frequency spectrum. The lower boundary of the band-pass filter is chosen in such a way that the constant background intensity is excluded. To minimize the computation time as maximum spatial frequency, the value given by the resolution of the optical imaging system is selected.

Figure 6.5 illustrates the principle of digital holographic autofocusing for the example of time-lapse imaging of the reactions of living human brain microvascular endothelia cells (HBMECs) co-incubated with a toxin. The experiments were performed with a DHM setup as shown in Figure 6.1 using a 40× microscope lens (NA = 0.6, λ = 532 nm).
Figure 6.5A–F shows representative amplitude and phase contrast images. At the beginning of the experiment ($t = 0$), the investigated HBMECs were imaged sharply in the hologram plane (Figure 6.5A and D). Due to an instability in the experimental setup, after $t = 31.5$ h, without numerical focus correction the cells appeared unfocused in the reconstructed amplitude and phase distributions (Figure 6.5B and E). Figure 6.5E shows further that defocusing-induced diffraction patterns can lead to phase singularities and thus may cause phase-unwrapping artifacts which inhibit a further data evaluation. The resulting images after numerical refocusing are shown in Figure 6.5C and F. The cells appear sharply and subcellular structures are clearly resolved without phase-unwrapping errors. Furthermore, toxin-induced effects become visible. In Figure 6.5G, exemplary focus value curves in dependence of the propagation distance Δz are plotted which were calculated by Eq. (6.5)

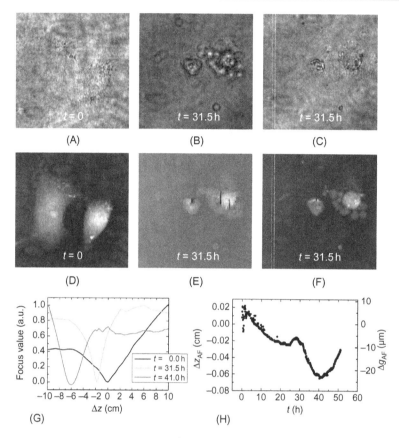

Figure 6.5
Subsequent numerical refocusing during time-lapse investigations. (A)–(C) Reconstructed amplitude and (D)–(F) unwrapped phase distributions obtained from long-term investigations on living HBMECs; (A), (D): $t = 0$; (B), (E): $t = 31.5$ h with fixed mechanical focus; (C), (F): $t = 31.5$ h with holographic autofocusing; (G): exemplary focus values in dependence of the propagation distance Δz, calculated by Eq. (6.5) for digital holograms recorded at $t = 0$, 31.5, and 41 h; (H): temporal dependency of autofocus position z_{AF} and corresponding change of the axial object position Δg_{AF}. Source: *Modified from Ref. [65]*.

from digital holograms recorded at $t = 0$, 31.5, and 41 h. The global minima of the curves indicate the propagation distances for which the amplitude images of the phase specimen appear sharply in focus with minimized contrast in the amplitude distributions. Finally, Figure 6.5H shows the temporal dependency of the focal position that was detected by numerical autofocusing for the whole measurement period ($t_{max} = 52$ h). The nonlinear drift also illustrates the need for a permanent focus control during long-term investigations. Furthermore, quantitative measurement data for the axial object position are provided (see also Section 6.7) that may be used to improve the stability of the experimental setup or to identify the sources of instability.

6.4 Refractive Index Determination of Cells in Suspension

The cellular refractive index represents an important parameter in quantitative phase imaging of living cells. The cellular refractive index influences the visibility of cells and subcellular structures in the quantitative phase images and limits of the accuracy for the determination of the thickness of transparent samples [29]. In addition, the determination of the cellular refractive indices are useful for the utilization with optical tweezers and related optical manipulation systems as the individual cellular refractive index values influence the resulting optical forces [66,67] and to analyze intracellular (protein) concentrations [68]. Although in many cases a high measurement accuracy can be achieved, the refractive index determination of adherent cells with quantitative phase imaging methods, as described, for example, in Refs. [29,69–71], can be time consuming or it requires special experimental equipment. In contrast, quantitative phase imaging methods for the determination of the integral refractive index of suspended cells can be carried out without further sample preparation [52,72,73]. In addition, the multifocus feature of DHM enables an increased data acquisition as several suspended cells that may be located laterally separated in different focal planes can be recorded simultaneously. Here, the determination of the integral refractive index of suspended cells is illustrated by a method in which a sphere model is fitted line by line to the DHM phase data. For sharply focused spherical cells in suspension, located at $x = x_0$, $y = y_0$, and with radius R, the cell thickness $d_{cell}(x,y)$ is as follows:

$$d_{cell}(x, y) = \begin{cases} 2 \cdot \sqrt{R^2 - (x-x_0)^2 - (y-y_0)^2} & \text{for } (x-x_0)^2 + (y-y_0)^2 \leq R^2 \\ 0 & \text{for } (x-x_0)^2 + (y-y_0)^2 > R^2 \end{cases} \quad (6.6)$$

Insertion of Eq. (6.4) in Eq. (6.6) gives

$$\Delta\varphi_{cell}(x,y) = \begin{cases} \frac{4\pi}{\lambda} \cdot \sqrt{R^2 - (x-x_0)^2 - (y-y_0)^2} \cdot (n_{cell} - n_{medium}) & \text{for } (x-x_0)^2 + (y-y_0)^2 \leq R^2 \\ 0 & \text{for } (x-x_0)^2 + (y-y_0)^2 > R^2 \end{cases}$$
$$(6.7)$$

with the unknown parameters n_{cell}, R, x_0, and y_0. In order to obtain the parameters n_{cell}, R, x_0, and y_0, Eq. (6.7) can be fitted iteratively to the measured phase data of spherical cells in suspension, for example, with the Gauss–Newton method [52,74].

Figure 6.6A shows the phase contrast image of a suspended cell, coded to 256 gray levels. The data $\Delta\varphi_{cell}(x,y)$ for the fitting process is selected by a threshold value that specifies the phase noise in the area around the cell. In Figure 6.6B, the fit of Eq. (6.7) to the phase data along dashed line in Figure 6.6A is depicted. Figure 6.6C and D show in comparison pseudo-3D plots of the phase distribution in Figure 6.6A and the result of line-by-line fitting of Eq. (6.7) to the measurement data. The mean value of the cell refractive index,

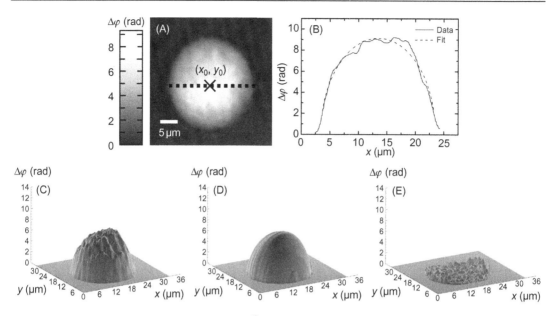

Figure 6.6

Refractive index determination of a suspended cell. (A) Quantitative digital holographic phase contrast image of a spherical trypsinized pancreatic tumor cell located at x_0, y_0; (B) phase data $\Delta\varphi$ along the cross-section marked by the dashed line in (A) and the fitted data corresponding to Eq. (6.7); (C) rendered pseudo-3D plot of the phase distribution in (A) that is used for the determination of the integral refractive index; (D) rendered pseudo-3D plot of data that is obtained by line-wise fitting of Eq. (6.7); (E) rendered pseudo-3D plot of the phase difference data between (C) and (D) [52].

here $n_{cell} = 1.372 \pm 0.002$, is used for further analysis. The uncertainty for n_{cell} can be estimated by the standard deviation obtained from all line fits. The average cell radius is obtained by the determination of $R = R_x(y = y_0) \equiv R_{fit,max}$ from all fitted lines (for Figure 6.6: $R_{fit,max} = 10.2 \pm 0.1$ μm). The uncertainty for R is obtained from the standard deviation of ± 5 neighboring lines to the central line at $R = R_{fit,max}$. Figure 6.6E shows a pseudo-3D plot of the absolute values of the phase difference between the measured phase contrast data in Figure 6.6C and the fitted data in Figure 6.6D (mean value = 0.3 rad) which indicates a homogeneous distribution of the fitting errors.

The application of the method for the refractive index determination of living cells is illustrated by the analysis of different human pancreatic tumor cell types (PaTu 8988 T ($N = 28$), PaTu 8988 S ($N = 15$), and PaTu 8988 T pLXIN E-cadherin ($N = 20$)). The cells were investigated in cell culture medium (Dulbecco's modified Eagle's medium) in comparison to a human liver tumor cell line (HepG2, $N = 55$; for details, see Ref. [52]). The cells were trypsinized, and for each cell line the parameters n_{cell} and $R_{fit,max}$ were determined with $n_{medium} = 1.337 \pm 0.001$ as described above. For data evaluation, only cells

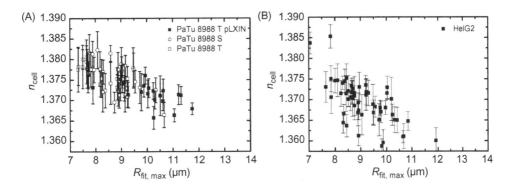

Figure 6.7
Refractive index of human pancreatic tumor cells (PaTu 8988 XX) cells (A) and human liver tumor cells (HepG2) cells (B) versus the cell radius $R_{\mathrm{fit,max}}$. Source: *Modified from Ref. [52]*.

with spherical shape were selected which were observed in the main fraction of the recorded holographic phase contrast images for all investigated cell lines. Figure 6.7A and B shows the refractive index n_{cell} in dependence of $R_{\mathrm{fit,max}}$ for pancreas and liver tumor cells. Within the range of uncertainty, no significant differences between the refractive index of the three different types of pancreas tumor cells are observed (Figure 6.7A). On the other hand, Figure 6.7A shows that the cellular refractive index of the cells decreases with increasing $R_{\mathrm{fit,max}}$. The data from the liver tumor cells (Figure 6.7B) show the same behavior. In the case of tumor cells of pancreas, a mean refractive index of $\overline{n_{\mathrm{cell}}} = 1.375 \pm 0.004$ for a mean cell radius $\overline{R_{\mathrm{fit,max}}} = 9.1 \pm 1.1$ is determined, while for the tumor cells of liver, $\overline{n_{\mathrm{cell}}} = 1.369 \pm 0.005$ with $\overline{R_{\mathrm{fit,max}}} = 9.1 \pm 0.9$ is obtained. The decrease in the refractive index with increasing cell radius, and thus the cell volume, can be explained by the cellular water content [16]. Figure 6.8A and B shows the histogram plots of the refractive index for both cell lines. The maximum numbers of cells lie near the mean value of the refractive index.

The exemplary results in Figures 6.7 and 6.8 demonstrate that the described method represents an efficient and reliable way to determine the integral refractive index of living cells. In Ref. [72], it is shown that similar results are obtained if the refractive index is retrieved by a 2D fitting process. The advantage of two-dimensional fitting of Eq. (6.7) to the measurement data of spherical cells is that a higher accuracy for the cell radius can be achieved.

6.5 DHM Analysis of Morphology and Thickness Changes

In this section, it is shown by representative examples how quantitative DHM phase contrast imaging can be used for the analysis of toxin-mediated changes of the cell

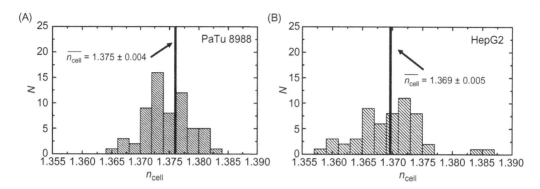

Figure 6.8
Histogram of the refractive index data for PaTu 8988 XX cells (A) and HepG2 cells (B) [52].

morphology. First, results from the temporal analysis of phase contrast and thickness changes of endothelial cells due to toxin exposure are presented. Then, the impact of vacuole formation in cells is illustrated. Finally, taking pancreatic tumor cells as an example, quantitative phase images of apoptotic and necrotic processes are compared.

6.5.1 Cell Reaction on Chemical Substances

The analysis of cell reactions to chemical substances on individual cells is demonstrated by a comparative study of HBMECs and umbilical vein endothelial cell-derived EA.hy 926 cells which were exposed to a toxin (Shiga toxin 1: Stx1; for details, see Ref. [75]). Therefore, an iMIC (Till Photonics, Gräfelfing, Germany) with attached DHM module based on a fiber optic Mach–Zehnder interferometer as shown in Figure 6.1 (for illustration see Ref. [76]) was used. An incubator was used for stabilized temperature. The coherent light source for the recording of digital holograms was a frequency-doubled Nd:YAG laser ($\lambda = 532$ nm). The cells were investigated in Petri dishes and exposed to 500 ng/ml of Stx1 at 37 °C. Digital off-axis holograms of single cells were recorded continuously every 3 min. The reconstruction of the quantitative phase images from the digitally captured holograms was performed as described in Section 6.3.

Figure 6.9 shows different time courses of Stx1-induced cell death of a single HBMEC and an EA.hy cell by dynamic changes of the cells' shape and thickness after exposure to Stx1. The results shown are representative data of the three independent single-cell analyses which were performed for each cell type. Figure 6.9A shows false color-coded DHM phase contrast images of cells at indicated time after Stx1 addition. In Figure 6.9B, cross-sections through the quantitative DHM phase images along the lines in panel (A) are depicted. The parameters $\Delta\varphi$ and d denote the phase contrast in radian and corresponding cell thickness calculated by Eq. 6.4, respectively, for a cellular refractive index of $n_{cell} = 1.37$

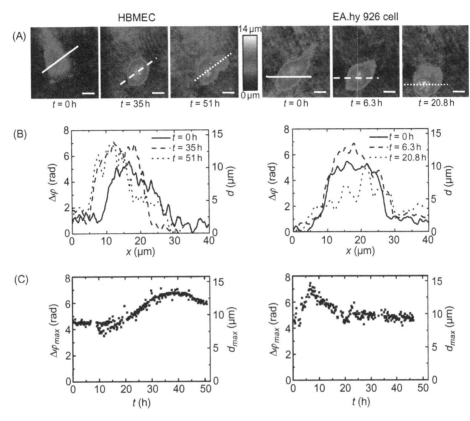

Figure 6.9

DHM investigation of Stx1-induced cell death of a single HBMEC and an EA.hy cell. (A) False color-coded digital holographic phase contrast images of cells at indicated time after Stx1 addition. The magnification bars correspond to 10 μm; (B) cross-sections through the quantitative DHM phase images whereby the x-axis matches the lines in panel (A). The parameters $\Delta\varphi$ and d denote phase contrast in radian and corresponding cell thickness; (C) temporal dependence of the maximum OPL $\Delta\varphi_{max}$ and the corresponding maximum cell thickness d_{max}.
Source: *Modified from Ref. [75]*.

($n_{medium} = 1.337$) that was determined on suspended cells as described in Section 6.4. Figure 6.9C shows the temporal dependence of the maximum OPL $\Delta\varphi_{max}$ and the corresponding maximum cell thickness d_{max}.

For the HBMEC (Figure 6.9, left panels), the first observed reaction was cell rounding after 20 h followed by an increase in the cell thickness by up to 50% after 35 h (Figure 6.9C, left panel), with simultaneously decreased intracellular fluctuations. After 45 h, the phase contrast started to decrease as a sign of progressive cell disintegration. Finally, the cell death, accompanied by an extensive cell leakage, occurred after <51 h of Stx1 treatment (Figure 6.9A, left panel). In contrast, initial Stx1-mediated cell rounding of the EA.hy 926 cell occurred already after 3 h and the cellular impairment progressed rapidly (Figure 6.9, right panels). After a peak in the cell thickness at 8 h, a necrotic death with cell burst and cytoplasm release was observed after 20 h (Figure 6.9A, right panel). These findings are in accordance with the results of scanning electron microscopy investigations and apotosis/necrosis assays [75].

In summary, the results in Figure 6.9 show that Stx1 induces cell swelling followed by necrotic cell death, which occurred earlier in macrovascular than in microvascular cells. In a comparative study on the exposure of the same cells to a different toxin (Stx2) that is described in detail in Ref. [75], it is demonstrated by similar results from DHM measurements that this toxin caused no necrotic effects but prevented cells from cell division.

6.5.2 Imaging of Vacuole Formation

The analysis of vacuole formation in cells is illustrated by the results of a DHM study in which HBMEC cells were treated by the toxin VD90 of EHEC-Vac (for details, see Ref. [77]). Therefore, single cells were observed from the time of toxin exposure until cell death. The same DHM experimental setup used to obtain the results in Figure 6.9 was applied.

Figure 6.10 shows typical stages in the destiny of vacuolated cells as white light images and quantitative DHM phase images. High-density cell components such as the nuclei and nucleoli are visible as bright spots in the DHM phase contrast, whereas the vacuoles appear as dark areas. This indicates a low refractive index of vacuoles near or equal to that of the cell culture medium and this can be interpreted as a low protein/nucleic acid concentration. The cell "A" which was fully vacuolated after 21.8 h of exposure to VD90 of EHEC-Vac (Figure 6.10A and B) rounded up during the next 24.5 h ($t = 46.3$ h after exposure) (Figure 6.10C), which led to a significant increase in the DHM phase contrast (Figure 6.10D). After an additional time of 25.4 h ($t = 71.7$ h), the cell poured out as shown in the white light image (Figure 6.10E) and evidenced by the decrease in the cell's DHM phase contrast (Figure 6.10F). A similar turnover from fully vacuolated status to necrosis was observed for another cell (cell "B," Figure 6.10C–F, respectively), during an even shorter time period (25.4 h). This experiment clearly demonstrated that the vacuolated cells undergo necrosis after treatment with VD_{90} of EHEC-Vac. However, probably due to the heterogeneity of unsynchronized HBMEC population, the time intervals between full vacuolization and necrosis differ among cells, resulting in a stepwise, rather than a sudden, necrotic death of the population [77].

6.5.3 Imaging of Apoptosis and Necrosis

Results from investigations on living pancreatic tumor cells (PaTu 8988 T) demonstrate in comparison the potentials of DHM for the label-free time-resolved visualization of

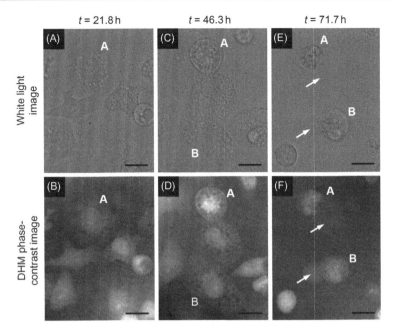

Figure 6.10

DHM analysis of vacuolated HBMECs. (A), (C), (E) White light images of HBMECs at different time points after exposure to VD90 of EHEC-Vac; (B), (D), (F) corresponding quantitative DHM phase contrast images. Cells designated "A" and "B" are examples of heavily vacuolated cells that underwent necrosis after different times of exposure to EHEC-Vac. Arrows in panels (E) and (F) indicate cellular leakage and the release of the cytoplasmic content from necrotic cells. The scale bars correspond to 10 μm. Source: *Modified from Ref. [77]*.

apoptotic and necrotic processes. In the first experiment, PaTu 8988 T cells were exposed to cell death inducing concentrations of taxol. Digital holograms of the cells were recorded continuously every 2 min in a temperature-stabilized environment (37 °C) for 12 h. Figure 6.11 shows exemplarily time-dependent results for the obtained unwrapped digital holographic phase distributions of two cells (denoted as A and B) after taxol addition in comparison to phase contrast images from a control measurement of two cells (denoted as C and D) without the addition of taxol. The phase contrast images in the upper row of Figure 6.11 show fast morphologic changes. The appearance of several small spherical cell fragments indicates apoptotic cell death. A second measurement without toxin addition was done. Here rather a necrotic process due to changes in pH value of the cell culture medium was observed at the end of the experiment in which the cells leak and release the cytoplasm to the cell culture medium as shown in the lower row of Figure 6.11. Although the images in Figure 6.11 clearly indicate apoptotic and necrotic effects, further investigations with apoptosis assays are necessary to validate the quantitative phase data. Nevertheless, it is illustrated that DHM has potentials to be a helpful label-free tool to study drug effects on cells' morphology quantitatively.

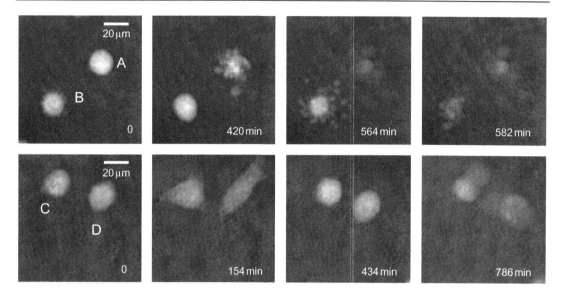

Figure 6.11
Label-free dynamic monitoring of cell death processes. Time-dependent observation of living pancreatic tumor cells (PaTu 8988 T) after addition of taxol. First row: representative gray level coded unwrapped phase distributions after toxin addition; second row: representative phase contrast images from a control measurement without taxol addition. Source: *Modified from Ref. [78]*.

6.6 Quantitative Cell Division Monitoring

In Refs. [79–81], it was shown that quantitative phase imaging can be used for the observation of cellular growth by the determination of changes in the density of the intracellular content. Here, the use of DHM for label-free quantitative dynamic monitoring of endothelial cell division [82] by cell tracking and cell thickness monitoring is demonstrated. For the experiments, HBMECs (see Section 6.5) were cultivated in a Petri dish. The cells were observed with an iMIC with attached DHM module as shown in Figure 6.1. An incubator was used for a stabilized temperature of 37 °C.

Figure 6.12 shows representative results for the obtained quantitative DHM phase contrast images (coded to 256 gray levels). After $t = 19.7$, 32.5, and 37.7 h the HBMECs denoted as A, B, and D, respectively, underwent a cell division. The corresponding daughter cells after the cell division process are designated as a_1, a_2, b_1, b_2, and d_1, d_2. Cell C did not divide during the experimental period. It was observed that the phase contrast increases significantly before mitosis. During the cell cycle phases in which the cells adhere on the substrate subcellular regions with a higher density like the nucleus, the nuclear membrane and the nucleoli become visible. In analogy, during metaphase, anaphase, and telophase, bright areas in the phase contrast images indicate the separation of the chromosomes, in

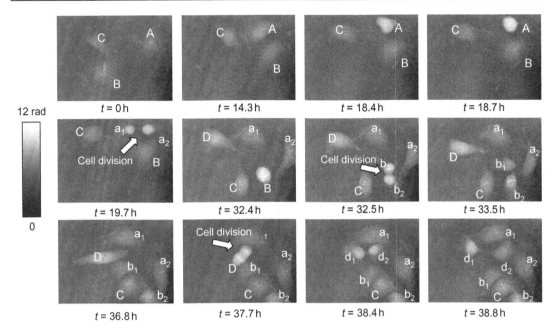

Figure 6.12
Time-dependent quantitative DHM phase contrast images of HBMECs (coded to 256 gray levels). The arrows indicate cell division after $t = 19.7$ h (cell A), $t = 35.5$ h (cell B), and $t = 37.7$ h (cell D). For cell C, no mitosis is observed during the experimental period. The corresponding daughter cells after the cell division process are denoted as a_1, a_2, b_1, b_2, d_1, d_2, respectively [82].

particular for cell D at $t = 37.7$ h. A gray level coded pseudo-3D representation of the quantitative contrast images of Figure 6.12 is shown in Figure 6.13. For cell cycle phases in which the cells adhere on the substrate, the quantitative phase contrast images correspond to the cell shape (see Eq. (6.4)), while during cell rounding, the projection of the cell thickness is measured (see Section 6.4 and Ref. [52]).

For further evaluation of the DHM phase contrast images, the maximum phase contrast $\Delta\varphi_{cell,max}$, the maximum cell thickness $d_{cell,max}$, and the cell migration trajectories of all cells were determined. Therefore, in a first step, the phase distributions were low-pass-filtered 2 times, with a box average filter of 5×5 pixels. In this way, substructures of the specimen in the phase distributions and noise—e.g., due to parasitic interferences and coherent noise—were reduced. Afterward, within a region of interest (ROI) in which the analyzed cell was located, the pixel coordinates of the maximum phase contrast were determined. The automated tracking of dynamic displacements of time-lapse sequences was performed by successive recentering of the ROI to the coordinates of the preceding maximal phase value [83]. During the evaluation procedure, the trajectories of the remaining daughter cells that were not detected by the automated cell tracking algorithm were started by manual selection after the cell division.

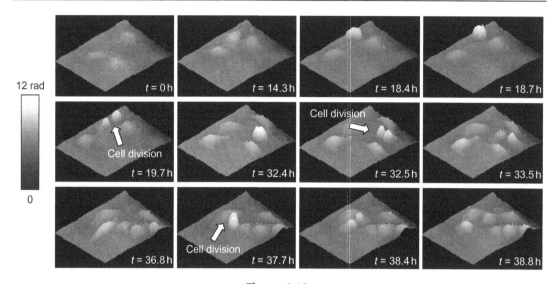

Figure 6.13
Gray level coded pseudo-3D representation of the quantitative contrast images of living HBMECs shown in Figure 6.12 [82].

Figure 6.14 shows for each cell the temporal dependence of the maximum phase $\Delta\varphi_{cell,max}$ as well as the corresponding cell thickness $d_{cell,max}$ that is obtained from Eq. (6.4) by using a value $n_{cell} = 1.373$ for the integral cellular refractive index that was retrieved from DHM measurements on suspended HBMECs as described in Section 6.4. Prior to the cell division, the cell morphology is changed to a spherical shape which causes a significant increase in the maximum phase contrast and the cell thickness. The resulting peaks are marked with arrows in Figure 6.14. After the cell division, the daughter cells adhere to the Petri dish. This is accompanied with a decrease in $\Delta\varphi_{cell,max}$. For cell C that undergoes no cell division, only few fluctuations of phase contrast and cell thickness are observed.

Figure 6.15 shows the automated obtained cell migration x,y trajectories for all evaluated cells and demonstrates the reliability of the applied cell tracking algorithm.

Figures 6.12–6.14 show the ability of DHM for dynamic cell division monitoring by simultaneous cell thickness measurement and 2D cell tracking. The mitosis is clearly detected by the phase contrast images and by the temporal dependency of phase contrast and cell thickness. However, as slight changes of the integral cellular refractive index cannot be completely excluded during the cell division process, the obtained values for the cell thickness shown in Figure 6.14 have to be handled carefully [29]. A further increase of the measurement accuracy for the cell thickness may be achieved by methods for simultaneous refractive index determination as, for example, described in Refs. [70,84] and [85]. As the x,y-position data in Figure 6.15 are obtained by the determination of the

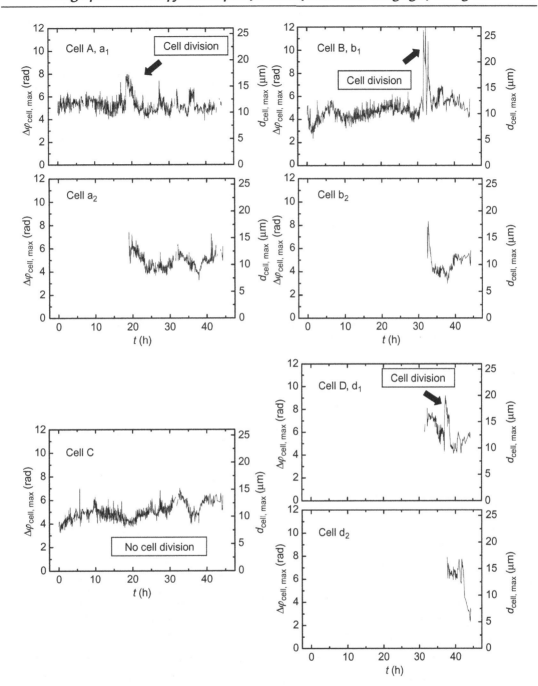

Figure 6.14

Time-dependent maximum phase contrast $\Delta\varphi_{cell,max}$ and time-dependent corresponding maximum cell thickness $d_{cell,max}$ automated obtained from the quantitative contrast images in Figure 6.12. Source: *Modified from Ref.* [82].

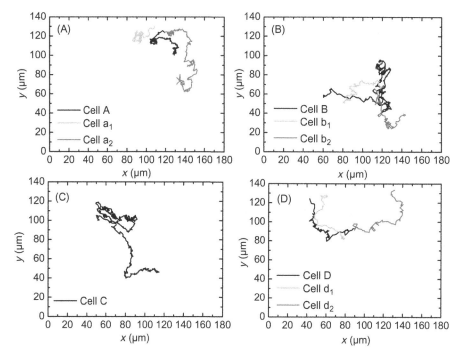

Figure 6.15

2D cell tracking in DHM phase contrast images in Figure 6.11. (A) Trajectories of cell A and daughter cells a_1, a_2 after cell division; (B) trajectories of cell B and daughter cells b_1, b_2 after cell division; (C) trajectory of cell C (no cell division); (D) trajectories of cell D and daughter cells d_1, d_2 after cell division [82].

coordinates of the maximum phase contrast, the precision of the applied algorithm can be estimated to be sensitive to cell thickness fluctuations.

Furthermore, the magnification of the applied experimental setup and disturbances in the phase contrast images due to scattering effects, e.g., by the cell culture medium, have to be taken into account. Thus, the error for the detection of the lateral cell position was estimated by the fluctuations of the curves in Figure 6.15 which are found in the range of 1–2 μm. These values are specific for the applied measurements setup and amount in the range of about 10% of the average lateral cell diameter. This is low in view of the cell shape changes that are observed in the DHM phase contrast images in Figure 6.12 and can be explained as follows: For the cell cycle phases in which the cells adhere on the substrate, the nucleoli predominate the coordinates of the maximum phase contrast. As the nucleoli are located in the nucleus, the resulting x, y values can be expected to be a good approximation of the cell center. The influence of cellular organelles other than the nucleoli is expected to be small due to the applied smooth filter of 5×5 pixels. This is supported by the small fluctuations in the x, y trajectories as in Figure 6.15, as well as by the fact that in the phase contrast images in Figure 6.12, the cellular organelles other than the nucleoli appear only with a marginal contrast. During the cell-rounding process, the maximum phase contrast is well defined by the center of the resulting spherical structure. Thus, also for this period of the cell cycle, a reliable detection of the cell center is possible. For the phases of the cell cycle in which the cells adhere on the substrate with thin morphology, the presented algorithm is expected to be more efficient than the evaluation of the DHM phase contrast images for 2D cell tracking by classical edge detection algorithms. During the periods when the cells adhere, the thickness of the outer cell border areas is $\leq 1-2$ μm. Consequently, the cells appear with a low contrast of the boundaries in the DHM phase contrast images (see Figure 6.12) which would affect the robustness of an edge detection-based determination of the lateral cell position.

In conclusion, the data in Figures 6.12–6.15 demonstrate the applicability of DHM for quantitative monitoring of the cell division processes during long-term time-lapse observations. DHM provides an efficient method for automated cell thickness monitoring, while simultaneously the 2D cell migration trajectories are obtained. Furthermore, subcellular regions with a higher molecular density than that of the surrounding cell compartments, such as nucleus and nucleoli, are visible in the phase contrast images. Possible applications of the method include the minimally invasive quantification of the cell vitality in the research fields of toxin-mediated cell damage and tumor cell migration analysis in cancer research.

6.7 Cell Tracking in 3D Environments

The analysis of cell migration processes is an important aspect for the understanding of morphogenesis and cancer metastasis. Thus, investigations of the applicability of digital holographic autofocus tracking for 3D cell migration monitoring were performed [83]. Therefore, human fibrosarcoma tumor cells (HT-1080) within nondenatured collagen fibers as a 3D tissue model were observed in a Petri dish with a DHM setup as shown in Figure 6.1. The preparation of the sample was carried out by following the approach of Friedl and Brocker [86]. Different from Ref. [86], a slightly diluted collagen concentration of 1.28 mg/ml was used to decrease light-scattering effects. In analogy to the experiments described in Section 6.6, the temperature was stabilized to 37 °C using an incubation chamber. A series of 61 holograms of three HT-1080 cells within the field of view that were identified as separated under white light illumination was recorded with fixed mechanical focus (40× magnification microscope lens) for a period of 180 min ($\Delta t = 3$ min). For each cell, the lateral position is determined from the automatically refocused phase contrast images as described in Section 6.6, while the axial sample displacement is obtained from the autofocus distance (see Section 6.3.4). Figure 6.16 shows the obtained results for $t = 0, 60, 120$, and 180 min. Figure 6.16A shows bright-field images

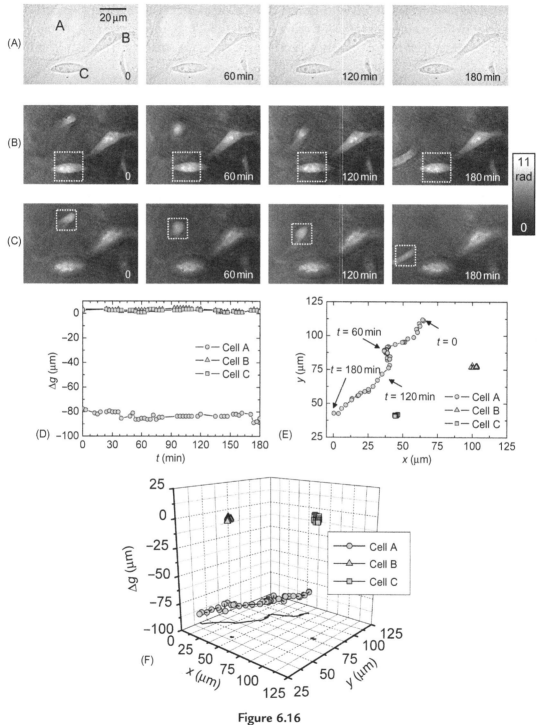

Figure 6.16

(Continued)

of the sample under white light illumination. The corresponding digital holographic phase contrast images are shown in Figure 6.16B and C.

The tracked cells are denoted as A, B, and C. The dotted white squares in the phase contrast images mark the applied ROIs that were used for holographic autofocusing and 2D tracking. In the white light images, cells B and C are only slightly defocused during the experiment. Cell A is located in a different layer and thus appears only marginally visible due to a large defocus. In Figure 6.16A, the ROI for digital holographic autofocusing tracking was set to cell C, which appears sharply focused in the reconstructed quantitative phase contrasts images and even subcellular structures such as the nucleoli are visible. As the axial position of cell B is close to the layer in which cell C is located, this cell is also displayed with comparable image sharpness. In contrast, cell A is far out of focus. In Figure 6.16C, the ROI was set to cell A. After digital holographic autofocusing, cell A is sharply resolved in each phase distribution and cellular motions as well as deformations and thickness changes during the process of migration become clearly visible. Figure 6.16D and E shows the resulting temporal dependencies of the relative z positions Δg as well as the corresponding (lateral) x, y coordinates for all three cells A, B, and C. The 3D trajectories resulting from the combination of the data in Figure 6.16D and E are plotted in Figure 6.16F. In Figure 6.16D–F, data with phase-unwrapping errors were not considered. For cells B and C, 74% (45 of 61 phase images) were faultlessly determined. For cell A, 77% (47 of 61 images) was obtained. The reason for different percentages of phase distributions without unwrapping errors is the mutual influence by unfocused structures outside the ROI. Cells B and C are stationary in position and shape. In contrast, cell A is located in a layer approximately 80 μm below cells B and C and migrates within the collagen. During the observed migration process through the network of collagen fibers, the cell undergoes permanent shape changes in a multistep cascade of coordinated adhesion and contraction which can clearly be observed in Figure 6.16C.

In summary, the results in Figure 6.16 illustrate the ability of DHM for 3D particle and cell tracking. However, it has to be mentioned that due to the underlying principles of the presented DHM configurations integral information is obtained. Thus, specimens in different planes at the same lateral position with axial distances near the depth of field of the applied imaging system cause diffraction patterns. This can lead to misinterpretations in

Time-dependent digital holographic 3D migration monitoring of human fibrosarcoma HT-1080 tumor cells in collagen. (A) Bright-field images under white light illumination; (B) quantitative digital holographic phase contrast images with autofocus tracking ROI set to cell C, as marked with a dashed box; (C) quantitative digital holographic phase contrast images with autofocus tracking ROI set to cell A, as marked with a dashed box; (D) temporal dependency of the axial positions of cells A, B, and C; (E) time-resolved lateral x,y tracking of cells A, B, and C; and (F) resulting 3D trajectories. Source: *Modified from Ref.[83]*.

the image sharpness quantification process that is used for the determination of the axial sample displacements. Furthermore, these disturbances affect the correct determination of the lateral object position from the quantitative phase contrast images. This limits the density of objects under investigations to an amount in which the specimens are imaged laterally separated. Nevertheless, label-free 3D tracking with DHM prospects new application areas of quantitative phase imaging for particle and cell tracking in fluidics and 3D cell migration monitoring in cancer research [54].

6.8 Conclusions

The presented concepts and the experimental results demonstrate that DHM permits reliable multifocus quantitative phase contrast imaging of adherent and suspended cells. DHM in off-axis and self-interference configuration enables a vibration insensitive (hologram capture time in or below the millisecond range) and noncontact full-field measurement of dynamic cellular processes. The results further demonstrate that the presented DHM concepts can be utilized for investigations on dynamic cell morphology changes and for the quantitative analysis of cellular reactions on toxin treatment. The obtained information opens up new quantitative ways for label-free dynamic cell monitoring, and may help to access new parameters, for example, for a minimally invasive support in apoptosis and necrosis recognition or for the label-free quantitative characterization of subcellular structures like nucleoli or vacuoles. Future prospects include the use of DHM for multimodal imaging [87], 3D cell tracking in a 3D environment, further analysis of the quantitative phase images for light scattering (see Ref. [88]), and new evaluation parameters [51]. A higher specificity of the technology and an improvement in the lateral and axial resolutions may be expected by the integration of nonlinear effects like second harmonic generation [89,90]. In addition, recently developed novel algorithms for the deconvolution of the holographically retrieved wave fields [91,92] and the corresponding quantitative phase images [93] prospect a further improvement in the spatial resolution. In conclusion, the presented methods have the potential to form versatile microscopy tools for label-free quantitative analysis in the field of live cell imaging.

Acknowledgments

Financial support by the German Federal Ministry for Education and Research (BMBF) as part of the "Biophotonics" research program and the European Network of Excellence "Photonics for Life (P4L)" is gratefully acknowledged.

References

[1] E.M. Goldys (Ed.), Fuorescence Applications in Biotechnology and the Life Siences, Wiley, New Jersey, 2009.
[2] V. Ntziachristos, Fluorescence molecular imaging, Annu. Rev. Biomed. Eng. 8 (2006) 1–33.

[3] L. Schermelleh, R. Heintzmann, H. Leonhardt, A guide to super-resolution fluorescence microscopy, J. Cell Biol. 190 (2010) 165.
[4] T. Kreis, Holographic Interferometry: Principles and Methods, Akademie-Verlag, Berlin, 1996.
[5] V.P. Shchepinov, V.S. Pisarev, Strain and Stress Analysis by Holographic and Speckle Interferometry, Wiley-Verlag, New York, NY, 1996.
[6] M.-A. Beeck, W. Hentschel, Laser metrology—a diagnostic tool in automotive development processes, Opt. Lasers Eng. 34 (2000) 101.
[7] E. Cuche, P. Marquet, C. Depeursinge, Simultaneous amplitude-contrast and quantitative phase-contrast microscopy by numerical reconstruction of Fresnel off-axis holograms, Appl. Opt. 38 (1999) 6694.
[8] F. Charrière, J. Kühn, T. Colomb, F. Montford, E. Cuche, Y. Emery, et al., Characterization of microlenses by digital holographic microscopy, Appl. Opt. 45 (2006) 829.
[9] D. Carl, B. Kemper, G. Wernicke, G. von Bally, Parameter-optimized digital holographic microscope for high-resolution living-cell analysis, Appl. Opt. 43 (2004) 6536.
[10] P. Marquet, B. Rappaz, P. Marquet, B. Rappaz, P.J. Magistretti, E. Cuche, et al., Digital holographic microscopy: a noninvasive contrast imaging technique allowing quantitative visualization of living cells with subwavelength axial accuracy, Opt. Lett. 30 (2005) 468.
[11] C.J. Mann, L. Yu, C.-M. Lo, M.K. Kim, High-resolution quantitative phase-contrast microscopy by digital holography, Opt. Express 13 (2005) 8693.
[12] B. Kemper, G. von Bally, Digital holographic microscopy for live cell applications and technical inspection, Appl. Opt. 47 (2008) A52.
[13] L.G. Alexopoulos, G.R. Erickson, F. Guilak, A method for quantifying cell size from differential interference contrast images: validation and application to osmotically stressed chondrocytes, J. Microsc. 205 (2001) 125.
[14] A. Barty, K.A. Nugent, D. Paganin, A. Roberts, Quantitative optical phase microscopy, Opt. Lett. 23 (1998) 817.
[15] J. Farinas, A.S. Verkman, Cell volume and plasma membrane osmotic water permeability in epithelial cell layers measured by interferometry, Biophys. J. 71 (1996) 3511.
[16] T. Ikeda, G. Popescu, R.R. Dasari, M.S. Feld, Hilbert phase microscopy for investigating fast dynamics in transparent systems, Opt. Lett. 30 (2005) 1165.
[17] G. Popescu, L.P. Deflores, J.C. Vaughan, K. Badizadegan, H. Iwai, R.R. Dasari, et al., Fourier phase microscopy for investigation of biological structures and dynamics, Opt. Lett. 29 (2004) 2509.
[18] V.P. Tychinskii, Coherent phase microscopy of intracellular processes, Phys. Usp. 44 (2001) 617.
[19] A.D. Aguirre, P. Hsiung, T.H. Ko, I. Hartl, J.G. Fujimoto, High-resolution optical coherence microscopy for high-speed, in vivo cellular imaging, Opt. Lett. 28 (2003) 2064.
[20] E.A. Swanson, J.A. Izatt, M.R. Hee, D. Huang, C.P. Lin, J.S. Schuman, et al., In vivo retinal imaging by optical coherence tomography, Opt. Lett. 18 (1993) 1864.
[21] Y. Zhao, Z. Chen, Z. Ding, H. Ren, J.S. Nelson, Real-time phase-resolved functional optical coherence tomography by use of optical Hilbert transformation, Opt. Lett. 27 (2002) 98.
[22] C.G. Rylander, D.P. Davé, T. Akkin, T.E. Milner, K.R. Diller, A.J. Welch, Quantitative phase-contrast imaging of cells with phase-sensitive optical coherence microscopy, Opt. Lett. 29 (2004) 1509.
[23] A.K. Ellerbee, T.L. Reazzo, J.A. Izatt, Investigating nanoscale cellular dynamics with cross-sectional spectral domain phase microscopy, Opt. Express 15 (2007) 8115.
[24] C. Joo, K.H. Kim, J.F. de Boer, Spectral-domain optical coherence phase and multiphoton microscopy, Opt. Lett. 32 (2007) 623.
[25] M.K. Kim, Springer Series in Optical Sciences Digital Holographic Microscopy, Principles, Techniques and Applications, Springer, New York, NY, 2011.
[26] M.K. Kim, Principles and techniques of digital holographic microscopy, SPIE Rev. 1 (2010) 018005.
[27] B. Kemper, D. Carl, A. Höink, G. von Bally, I. Bredebusch, J. Schnekenbuger, Modular digital holographic microsocopy system form marker-free quantitative phase contrast imaging of living cells, Proc. SPIE 6191 (2006) 61910T.

[28] D.B. Murphy, Fundamentals of Light Microscopy and Electronic Imaging, Wiley, New York, NY, 2001.
[29] B. Kemper, D. Carl, J. Schnekenburger, I. Bredebusch, M. Schafer, W. Domschke, et al., Investigation of living pancreas tumor cells by digital holographic microscopy, J. Biomed. Opt. 11 (2006) 034005.
[30] H. Ding, G. Popescu, Instantaneous spatial light interference microscopy, Opt. Express 18 (2010) 1569.
[31] J. Jang, C.Y. Bae, J.-K. Park, J.C. Ye, Self-reference quantitative phase microscopy for microfluidic devices, Opt. Lett. 35 (2010) 514.
[32] N.T. Shaked, Y. Zhu, N. Badie, N. Bursac, A. Wax, Reflective interferometric chamber for quantitative phase imaging of biological sample dynamics, J. Biomed. Opt. 15 (2010) 030503.
[33] G. Popescu, T. Ikeda, R.R. Dasari, M.S. Feld, Diffraction phase microscopy for quantifying cell structure and dynamics, Opt. Lett. 31 (2006) 775.
[34] P. Bon, G. Maucort, B. Wattellier, S. Monneret, Quadriwave lateral shearing interferometry for quantitative phase microscopy of living cells, Opt. Express 17 (2009) 13080.
[35] B. Kemper, A. Vollmer, C.E. Rommel, J. Schnekenburger, G. von Bally, Simplified approach for quantitative digital holographic quantitative phase contrast imaging of living cells, J. Biomed. Opt. 16 (2011) 026014.
[36] B. Kemper, F. Schlichthaber, A. Vollmer, S. Ketelhut, S. Przibilla, G. von Bally, Self interference digital holographic microscopy approach for inspection of technical and biological phase specimens, Proc. SPIE 8082 (2011) 808207.
[37] T.-C. Poon (Ed.), Digital Holography and Three-Dimensional Display, Springer, New York, 2006.
[38] U. Schnars, W.P.O. Jüptner, Digital recording and numerical reconstruction of holograms, Meas. Sci. Technol. 13 (2002) 85.
[39] L. Yaroslavsky, Digital Holography and Digital Image Processing: Principles, Methods, Algorithms, Kluwer Academic Publishers, Boston, 2004.
[40] T. Kreis, Handbook of Holographic Interferometry: Optical and Digital Methods, Wiley-VCH, Weinheim, 2005.
[41] M.K. Kim, L. Yu, C.J. Mann, Interference techniques in digital holography, J. Opt. A: Pure Appl. Opt. 8 (2006) 518.
[42] I. Yamaguchi, J.-I. Kato, S. Ohta, J. Mizuno, Image formation in phase-shifting digital holography and applications to microscopy, Appl. Opt. 40 (2001) 6177.
[43] B. Kemper, S. Stürwald, C. Remmersmann, P. Langehanenberg, Characterisation of light emitting diodes (LEDs) for application in digital holographic microscopy for inspection of micro and nanostructured surfaces, Opt. Lasers Eng. 46 (2008) 499.
[44] C. Remmersmann, S. Stürwald, B. Kemper, P. Langehanenberg, G. von Bally, Phase noise optimization in temporal phase-shifting digital holography with partial coherence light sources and its application in quantitative cell imaging, Appl. Opt. 48 (2009) 1463.
[45] K. Creath, Temporal phase measurement methods, in: D. Robinson, S. Reid (Eds.), Interferogram Analysis, Institute of Physics Publishing, Bristol, 1993, p. 94.
[46] S. Kosmeier, P. Langehanenberg, G. von Bally, B. Kemper, Reduction of parasitic interferences in digital holographic microscopy by numerically decreased coherence length, Appl. Phys. B 106 (2010) 107.
[47] M. Liebling, T. Blu, M. Unser, Complex-wave retrieval from a single off-axis hologram, J. Opt. Soc. Am. A 21 (2004) 367.
[48] J.W. Goodman, Introduction to Fourier Optics, McGraw-Hill, Singapore, 1996.
[49] T. Colomb, F. Montfort, C. Depeursinge, Small Reconstruction Distance in Convolution Formalism, Digital Holography and Three-Dimensional Imaging, OSA Technical Digest, paper DMA4, 2008.
[50] S. De Nicola, A. Finizio, G. Pierattini, P. Ferraro, D. Alfieri, Angular spectrum method with correction of anamorphism for numerical reconstruction of digital holograms on tilted planes, Opt. Express 13 (2005) 9935.
[51] N.T. Shaked, L.L. Satterwhite, N. Bursac, A. Wax, Whole-cell-analysis of live cardiomyocytes using wide-field interferometric phase microscopy, Biomed. Opt. Express 1 (2010) 706.

[52] B. Kemper, S. Kosmeier, P. Langehanenberg, G. von Bally, I. Bredebusch, W. Domschke, et al., Integral refractive index determination of living suspension cells by multifocus digital holographic phase contrast microscopy, J. Biomed. Opt. 12 (2007) 054009.

[53] J. Klokkers, P. Langehanenberg, B. Kemper, S. Kosmeier, G. von Bally, C. Riethmüller, et al., Atrial natriuretic peptide and nitric oxide signaling antagonizes 2 vasopressin-mediated water permeability in inner medullary 3 collecting duct cells, Am. J. Physiol. Renal Physiol. 297 (2009) PF693-703.

[54] P. Langehanenberg, G. von Bally, B. Kemper, Autofocussing in digital holographic microscopy, 3D Res. 2 (2011) 01004.

[55] P. Marquet, B. Rappaz, F. Charrière, Y. Emery, C. Depeursinge, P.J. Magistretti, Analysis of cellular structure and dynamics with digital holographic microscopy, Proc. SPIE 6633 (2007) 66330F.

[56] G. Nomarski, Differential microinterferometer with polarized waves, J. Phys. Radium. 16 (1955) 9.

[57] L. Miccio, A. Finizio, R. Puglisi, D. Balduzzi, A. Galli, P. Ferraro, Dynamic DIC by digital holography microscopy for enhancing phase-contrast visualization, Biomed. Opt. Express 2 (2011) 331.

[58] B. Kemper, J. Kandulla, D. Dirksen, G. von Bally, Optimization of spatial phase shifting in endoscopic electronic-speckle-pattern-interferometry, Opt. Commun. 217 (2003) 151.

[59] F. Dubois, C. Schockaert, N. Callens, C. Yourassowsky, Focus plane detection criteria in digital holography microscopy by amplitude analysis, Opt. Express 14 (2006) 5895.

[60] P. Langehanenberg, B. Kemper, D. Dirksen, G. von Bally, Autofocusing in digital holographic phase contrast microscopy on pure phase objects for live cell imaging, Appl. Opt. 47 (2008) D176.

[61] Y. Sun, S. Duthaler, B.J. Nelson, Autofocusing in computer microscopy: selecting the optimal focus algorithm, Microsc. Res. Tech. 65 (2004) 139.

[62] F.C. Groen, I.T. Young, G. Ligthart, A comparison of different focus functions for use in autofocus algorithms, Cytometry A 6 (1985) 81.

[63] L. Firestone, K. Cook, K. Culp, N. Talsania, K. Preston, Comparison of autofocus methods for automated microscopy, Cytometry 12 (1991) 195.

[64] M. Bravo-Zanoguera, B.von Massenbach, A.L. Kellner, J.H. Price, High-performance autofocus circuit for biological microscopy, Rev. Sci. Instrum. 69 (1998) 3966.

[65] P. Langehanenberg, B. Kemper, G. von Bally, Autofocus algorithms for digital-holographic microscopy, Proc. SPIE 6633 (2007) 66330E.

[66] A. Ashkin, Optical trapping and manipulation of neutral particles using lasers, Proc. Natl. Acad. Sci. U.S.A. 94 (1997) 4853.

[67] J. Guck, R. Ananthakrishnan, T.J. Moon, C.C. Cunningham, J. Kas, Optical deformability of soft biological dielectrics, Phys. Rev. Lett. 84 (2000) 5451.

[68] R. Barer, Interference microscopy and mass determination, Nature 169 (1952) 366.

[69] B. Rappaz, B. Rappaz, P. Marquet, E. Cuche, Y. Emery, C. Depeursinge, et al., Measurement of the integral refractive index and dynamic cell morphometry of living cells with digital holographic microscopy, Opt. Express 13 (2005) 9361.

[70] W. Choi, C. Fang-Yen, K. Badizadegan, S. Oh, N. Lue, R.R. Dasari, et al., Tomographic phase microscopy, Nat. Methods 4 (2007) 717.

[71] M. Debailleul, V. Georges, B. Simon, R. Morin, O. Haeberlé, High-resolution three-dimensional tomographic directive microscopy of transparent inorganic and biological samples, Opt. Lett. 34 (2009) 79.

[72] S. Kosmeier, B. Kemper, P. Langehanenberg, I. Bredebusch, J. Schnekenburger, A. Bauwens, et al., Determination of the integral refractive index of cells in suspension by digital holographic phase contrast microscopy, in: Proc. SPIE, 2008, p. 699110.

[73] M. Kemmler, M. Fratz, D. Giel, N. Saum, A. Brandenburg, C. Homann, Noninvasive time-dependent cytometry monitoring by digital holography, J. Biomed. Opt. 12 (2007) 064002.

[74] Å. Björk, Numerical Methods for Least Squares Problems, SIAM, Philadelphia, 1996.

[75] A. Bauwens, M. Bielaszewska, B. Kemper, P. Langehanenberg, G. von Bally, R. Reichelt, et al., Differential cytotoxic actions of Shiga toxin 1 and Shiga toxin 2 on microvascular and macrovascular endothelial cells, Thromb. Haemost. 105 (2011) 515.

[76] B. Kemper, L. Schmidt, S. Przibilla, C. Rommel, S. Ketelhut, A. Vollmer, et al., Influence of sample preparation and identification of subcellular structures in quantitative holographic phase contrast microscopy, Proc. SPIE 7715 (2010) 771504.

[77] M. Bielaszewska, A. Bauwens, L. Greune, B. Kemper, U. Dobrindt, J.M. Geelen, et al., Vacuolisation of human microvascular endothelial cells by enterohaemorrhagic *Escherichia coli*, Thromb. Haemost. 102 (2009) 1080.

[78] B. Kemper, P. Langehanenberg, I. Bredebusch, J. Schnekenburger, G. von Bally, Techniques and applications of digital holographic microscopy for life cell imaging, Proc. SPIE 6633 (2007) 66330D.

[79] B. Rappaz, E. Cano, T. Colomb, J. Kühn, C. Depeursinge, V. Simanis, et al., Noninvasive characterization of the fission yeast cell cycle by monitoring dry mass with digital holographic microscopy, J. Biomed. Opt. 14 (2009) 034049.

[80] G. Popescu, Y.K. Park, N. Lue, C. Best-Popescu, L. Deflores, R.S. Dasari, et al., Optical imaging of cell mass and growth dynamics, Am. J. Physiol. Cell Physiol. 295 (2008) C538.

[81] M. Mir, Z. Wang, Z. Shen, M. Bednarz, R. Bashir, I. Golding, et al., Optical measurement of cycle-dependent growth, Proc. Natl. Acad. Sci. U.S.A. 108 (2011) 13124.

[82] B. Kemper, A. Bauwens, A. Vollmer, S. Ketelhut, P. Langehanenberg, J. Müthing, et al., Label-free quantitative cell division monitoring of endothelial cells by digital holographic microscopy, J. Biomed. Opt. 15 (2010) 036009.

[83] P. Langehanenberg, L. Ivanova, B. Bernhardt, S. Ketelhut, A. Vollmer, D. Dirksen, et al., Automated 3D-tracking of living cells by digital holographic microscopy, J. Biomed. Opt. 14 (2009) 014018.

[84] B. Rappaz, F. Charrière, C. Depeursinge, P.J. Magistretti, P. Marquet, Simultaneous cell morphometry and refractive index measurement with dual-wavelength digital holographic microscopy and dye-enhanced dispersion of perfusion medium, Opt. Lett. 33 (2008) 744.

[85] B. Kemper, S. Przibilla, A. Vollmer, S. Ketelhut, G. von Bally, Quantitative phase imaging-based refractive index determination of living cells using incorporated microspheres as reference, Proc. SPIE 8086 (2011) 808607.

[86] P. Friedl, E.B. Brocker, Reconstructing leukocyte migration in 3D extracellular matrix by time-lapse videomicroscopy and computer-assisted tracking, Methods Mol. Biol. 239 (2004) 77.

[87] M. Esseling, B. Kemper, M. Antkowiak, D.J. Stevenson, L. Chaudet, M.A.A. Neil, et al., Multimodal biophotonic workstation for live cell analysis, J. Biophotonics 5 (2012) 9.

[88] H. Ding, X. Liang, Z. Wang, S.A. Boppart, K. Tangella, G. Popescu, Measuring the scattering parameters of tissues from quantitative phase imaging of thin slices, Opt. Lett. 36 (2011) 2281.

[89] E. Shaffer, C. Moratal, P. Magistretti, P. Marquet, C. Depeursinge, Label-free second-harmonic phase imaging of biological specimen by digital holographic microscopy, Opt. Lett. 35 (2010) 4102.

[90] E. Shaffer, P. Marquet, C. Depeursinge, Real time, nanometric 3D-tracking of nanoparticles made possible by second harmonic generation digital holographic microscopy, Opt. Express 18 (2010) 17392.

[91] Y. Cotte, F.M. Toy, C. Arfire, S.S. Kou, D. Boss, I. Bergoënd, et al., Realistic 3D coherent transfer function inverse filtering of complex fields, Biomed. Opt. Express 2 (2011) 2216.

[92] Y. Cotte, M.F. Toy, N. Pavillon, C. Depeursinge, Microscopy image resolution improvement by deconvolution of complex fields, Opt. Express 18 (2010) 19462.

[93] S.D. Babacan, Z. Wang, M. Do, G. Popescu, Cell imaging beyond the diffraction limit using sparse deconvolution spatial light interference microscopy, Biomed. Opt. Express 2 (2011) 1815.

CHAPTER 7

Phase Unwrapping Problems in Digital Holographic Microscopy of Biological Cells

Alexander Khmaladze
Department of Chemistry, University of Michigan, Ann Arbor, MI

Editor: Natan T. Shaked

7.1 Introduction

Quantitative phase mapping in digital holographic microscopy carries the three-dimensional (3D) information about the object. However, phase mapping is ambiguous, as absolute phase is wrapped in the intervals of 2π. This results in a discontinuous imaging, as the phase signals from the points with heights that are exactly integer number of wavelengths apart are the same. The phase unwrapping can be done using a software algorithm that looks for jumps in the phase image and shifts the image segments up or down depending on the surrounding topology. However, when it comes to imaging real objects (and especially biological cells), the typical phase image is noisy and may have areas of low intensity, where the interference fringes are not visible. Many phase unwrapping software algorithms using different methods have been developed in the past [1]. Ultimately, however, the software algorithms rely on surrounding pixels to unwrap the 2π discontinuities. As a result, if a certain area within the phase image proves to be problematic for the unwrapping program, not just the unwrapping from this area is affected, but the error often propagates to other regions as well.

In this chapter, first a method to computationally unwrap the phase is discussed. This method, instead of heavily relying on the immediate surrounding pixels in a single phase image, uses a set of multiple phase images obtained using the angular spectrum method [2,3], with the reconstruction performed at various axial distances. While this method is generally applicable, it was designed for imaging cell-like objects on a flat background, in which case it is especially fast and effective.

Also, the multiple wavelength phase imaging technique, which is based on the comparison of phase maps, acquired using different wavelengths, is discussed. This dual-wavelength technique generates an extended range "beat" wavelength phase image. The beat wavelength phase image is then used to subsequently adjust one of the original single wavelength phase images in order to reduce the measurement uncertainty.

Finally, by comparing the two phase images directly, the dual-wavelength linear regression method accurately unwraps phase images via pixel-by-pixel comparison. The latter unwrapping method is computationally fast, can process complex topologies, and also significantly relaxes the limitations on the total optical height, typically associated with a "traditional" dual-wavelength phase unwrapping.

7.2 Experimental Techniques

7.2.1 Dual-Wavelength Digital Holography Apparatus

Figure 7.1 shows the typical experimental apparatus for simultaneous dual-wavelength digital holographic microscopy [2,3]. It is based on two overlapping Michelson interferometers, which enables the adjustment of location of the first-order components produced by each wavelength in the Fourier space (see later). Here, He–Ne ($\lambda_1 = 633$ nm) and diode-pumped solid-state ($\lambda_2 = 532$ nm) lasers are used as coherent light sources. Both beams are attenuated by neutral density filters and then pass through the microscope

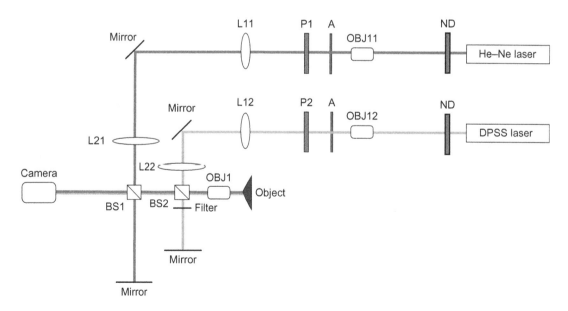

Figure 7.1
Dual-wavelength digital holography apparatus. The beams are collimated between L11 and L21 and between L12 and L22 and again are collimated after 20× OBJ1 microscope objective.

objectives (OBJ11/OBJ12), which together with the apertures A and collimating lenses L11/L12, produce plane waves. Their intensity is further adjusted by the polarizing filters P1 and P2. Beam splitters BS1 and BS2 divide the incoming light into the reference and the object beams. Two separate reference arms are used to fine-tune the location of the first-order diffraction peaks and separate them from each other in the Fourier domain. Lenses L21 and L22 and 20× microscope objective OBJ1 again collimate the beams in the object arm. The wave fronts in both reference arms remain spherical and the resulting curvature mismatch is removed numerically. An interference filter is placed into the reference arm of the diode-pumped solid-state ($\lambda = 532$ nm) laser to allow only this wavelength to pass and block the inverse reflection of the other laser. The interference pattern between the reflected reference and the object waves is recorded by the CCD (Charge-coupled device) camera. A relative angle can be introduced between the object and the reference beams for each wavelength by slightly tilting the reference arm's mirrors. By introducing different tilts in two orthogonal directions for two reference beams, we can separate each spectral component in Fourier space (Figure 7.2B), which allows us to capture both the wavelengths simultaneously.

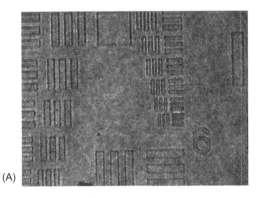

(A)

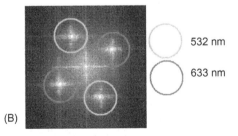

(B)

Figure 7.2
Two wavelength hologram of a USAF resolution target. (A) Digital hologram and (B) its Fourier spectrum with the red and the green wavelengths first-order components shown. (For interpretation of the references to color in this figure legend, the reader is referred to the web version of this book.)

7.2.2 Angular Spectrum Method

Once a hologram is acquired, it is reconstructed by numerically propagating the optical field along the direction perpendicular to the hologram plane (z-direction) in accordance with the laws of diffraction. A Fresnel–Kirchhoff integral can be expressed as [4]:

$$A_0(k_x, k_y; 0) = \iint E_0(x, y; 0) \exp[-i(k_x x + k_y y)] dx dy \quad (7.1)$$

where k_x and k_y are spatial frequencies corresponding to x and y, respectively. Here, $E_0(x,y; z=0)$ is the intensity distribution recorded by the CCD camera. This equation is the expression for Fourier transform and $A_0(k_x, k_y; 0)$ is the angular spectrum of the optical field $E_0(x,y,z=0)$ at the hologram plane $z=0$. The object's angular spectrum consists of a zero order (DC) and a pair of first-order terms: the angular spectrum of the object field and its phase inverted version. Figure 7.2A shows the hologram of a USAF (United States Air Force) resolution target recorded by our dual-wavelength experimental setup. The two crossing interference fringe patterns, formed by each of the two wavelengths, can be clearly seen. Figure 7.2B presents the Fourier spectrum with the two pairs of first-order components, corresponding to the two wavelengths.

Field $E_0(x,y;z=0)$ can be regarded as a projection of many plane waves propagating in different directions in space and with the complex amplitude of each component equal to $A_0(k_x, k_y; 0)$. The angular spectrum can then be propagated in space along the z-axis using the complex transfer function $\exp[ik_z z]$:

$$A(k_x, k_y; z) = A_0(k_x, k_y; 0) \exp[ik_z z] \quad (7.2)$$

where $k_z = \sqrt{k^2 - k_x^2 - k_y^2}$ and $k = 2\pi/\lambda$. Note that there is no requirement for z to be larger than a certain minimum value, as in the case of Fresnel transform or Huygens convolution. The complex field at an arbitrary z can be obtained by performing the inverse Fourier transform:

$$E(x, y; z) = \iint A(k_x, k_y; z) \exp[i(k_x x + k_y y)] dk_x dk_y \quad (7.3)$$

As both integrals in Eqs. (7.1) and (7.3) are computed via the FFT (Fast Fourier transform) algorithm, the angular spectrum method is well suited for the real-time imaging.

Once the complex field is calculated, its phase can be converted to height. If the light wave reflects from an object, its surface is described by a height map $h(x,y)$, which is determined from the phase map $\varphi(x, y)$ of the holographic reconstruction at a given wavelength by:

$$h(x, y) = \frac{\lambda_r}{2\pi} \varphi(x, y) \quad (7.4)$$

On the other hand, if the object is a mostly transparent cell on the reflective substrate (the light propagates through it, reflects from the substrate and propagates back), the physical thickness is

$$h(x, y) = \frac{\lambda_r}{2\pi} \frac{\varphi(x, y)}{(n - n_0)} \tag{7.5}$$

where $(n - n_0)$ is the refractive index difference between the cell and the air. Here, $\lambda_r = \lambda/2$ is half of the wavelength of light, because the light travels through the sample twice.

As mentioned before, the phase $\Delta\varphi$ in these equations can only vary from 0 to 2π, which corresponds to optical thickness variation of 0 to λ. If an object is thicker, then this results in 2π discontinuities in the phase image that need to be unwrapped.

7.2.3 Curvature Correction

The angular spectrum method described earlier is based on the assumption that the reference and object waves are both ideal plane waves. However, in the real setup, each wave has its wave front curvature, resulting in a curvature mismatch [3].

Consider the complex field captured by a CCD array (Figure 7.3). R is the wave's radius of curvature centered at point C, which can be determined experimentally for a given setup, $\bar{r}$ is the vector from the center of the CCD matrix (point O) to an arbitrary point A, and $\bar{r}_0$ is the vector from the center of the CCD matrix to the projection of the center of curvature on the CCD matrix P. Then $|\bar{r}| = \sqrt{x^2 + y^2}$, where x and y are the coordinates of A and $|\bar{r}_0| = \sqrt{x_0^2 + y_0^2}$, where x_0 and y_0 are the coordinates of P.

The phase mismatch can be compensated numerically, by multiplying the original "flat" field $E_0(x, y; z = 0)$ by the phase factor $\exp[i\varphi]$, where $\varphi = kd$ is the phase difference between A and O and d is the optical path difference:

$$d = CA - CO = \sqrt{CP^2 + PA^2} - \sqrt{CP^2 + PO^2} \tag{7.6}$$

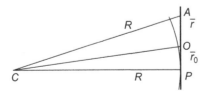

Figure 7.3
Curvature compensation.

From geometry, we obtain:

$$d = \sqrt{R^2 + |\bar{r} - \bar{r}_0|^2} - \sqrt{R^2 + |\bar{r}_0|^2} = \sqrt{R^2 + (x-x_0)^2 + (y-y_0)^2} - \sqrt{R^2 + x_0^2 + y_0^2} \quad (7.7)$$

The difference can be positive or negative, depending on the angle of the curvature we are compensating. Finally, the curvature corrected field:

$$E(x, y; 0) = E_0(x, y; 0) \exp\left[ik\left(\pm\left[\sqrt{R^2 + (x-x_0)^2 + (y-y_0)^2} - \sqrt{R^2 + x_0^2 + y_0^2}\right]\right)\right] \quad (7.8)$$

Equation (7.8) agrees with the approximation from Ref. [5], in the case $R \gg r$ and $\bar{r}_0 \to 0$:

$$k\left(\sqrt{R^2 + r^2} - R\right) = kR\left[\sqrt{1 + \frac{r^2}{R^2}} - 1\right] = \frac{2\pi R}{\lambda}\left[1 + \frac{r^2}{2R^2} - 1\right] = \frac{2\pi}{\lambda}\frac{x^2 + y^2}{2R} \quad (7.9)$$

This is a known expression for Newton's rings, which means that if the object is a plane mirror, the resulting interference pattern would be a set of concentric rings of radius of $\sqrt{mR\lambda}$, where $m = 0, 1, 2, \ldots$.

Figure 7.4 shows the image of the USAF resolution target covered with a layer of aluminum to make it entirely reflective. The pattern on the resolution target is elevated approximately 100 nm above the flat background. Figure 7.4A shows the reconstructed image before the curvature correction and Figure 7.4B is the same image after the curvature correction was applied. In general, if the parameters are chosen correctly, even a substantial curvature mismatch can be compensated.

7.2.4 Additional Phase Background Removal

In practice, numerically compensating the wave front curvature works well for smaller image frames. For larger images, even a small amount of uncompensated background or tilt can be a problem, especially if, for example, many cells are imaged simultaneously with the goal of measuring their total volume. In that case, even a small variation of the background due to not fully compensated curvature will drastically affect the volume measurements [6].

Figure 7.5A shows several simulated cells on the flat substrate. The simulated "imaging" here is done with two wavelengths and the final unwrapped images were obtained using the linear regression method (see later). Even a small amount of uncompensated curvature in a single wavelength image (Figure 7.5B) is present in the final unwrapped image (see Figure 7.5C). To properly calculate the total volume of all cells, this curvature must be completely removed. The ideal method for background subtraction will remove the uncompensated curvature with the minimal user intervention, while retaining the cells'

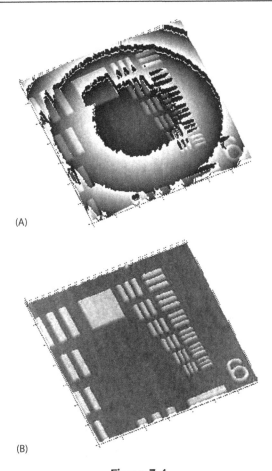

Figure 7.4

USAF resolution target (A) without curvature correction and (B) with curvature correction applied. The images are 174 μm × 174 μm (450 × 450 pixels).

shapes. In order to separate the background from the cell, an algorithm, based on modified polynomial fitting [7], is implemented. The method goes as follows:

1. Fit a polynomial surface to the unwrapped phase image.
2. Compare the polynomial surface with the original image pixel by pixel and retain the lower of two values. Generate a new profile.
3. Substitute the unwrapped phase image with the new profile obtained in step 2.
4. Repeat steps 1, 2, and 3 several times, until the profile converges.

Finally, the processed background is subtracted from the original phase image, yielding a phase profile with a near-null background. This method, however, while substantially reducing the background, tends to "overcompensate" any curvature present in the original unwrapped phase image. To illustrate this point, consider a curved profile (Figure 7.6). The

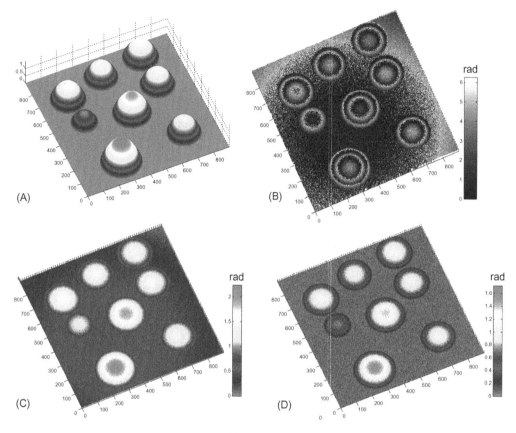

Figure 7.5

Background removal for simulated cells. (A) Simulated cells on the flat substrate, (B) single wavelength (635 nm) simulated phase images with added phase, (C) dual-wavelength unwrapped phase image, and (D) dual-wavelength unwrapped phase image with background removed.

first iteration (dotted line), while removing the original peaks, also "dives" under the original curved profile near the ends of the frame. As a result (dashed line), the curvature is overcompensated. However, it can be fixed by modifying the algorithm's step 2 in the following way:

2a. Compare the polynomial surface with the original image.
2b. Find the maximum difference between the fit and the original profiles.
2c. Add a fraction of that maximum difference to the polynomial surface.

This modification, although making the convergence much slower, does result in a flat background (see Figure 7.6—solid line). The fraction, used in step 2c can be as high as 95% and is limited only by computational rounding error as well as the computational speed (the higher is the number, the longer it takes for the method to converge). The result

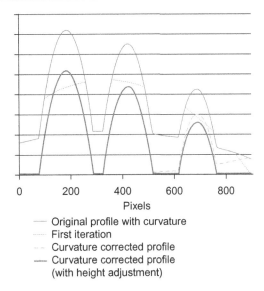

Figure 7.6
Background subtraction by modified polyfit method.

of this modified algorithm applied to the simulated cells in Figure 7.5A is presented in Figure 7.5D.

7.3 Varying Reconstruction Distance Method

In a case when the second wavelength is not available, the software unwrapping can be accomplished by varying the reconstruction distance algorithm [8]. Since the resolution of the digital hologram in *XY*-plane is diffraction limited to the wavelength of light, the resolution in the *z*-direction is on the order of several nanometers. Thus, each pixel of a phase map represents the average height of the object within this pixel. As a result, if one changes the reconstruction distance by a small amount in the angular spectrum algorithm and compares the two reconstructed fields, they appear almost identical, except the 2π discontinuities are shifted. Taking only the continuous parts of each image, the unwrapped phase profile can be recreated. This "digital scanning" is possible because of many advantages of the angular spectrum method, such as its accuracy and ease at which the reconstruction at various distances can be computed.

Consider imaging a slanted surface, which is several wavelengths high (Figure 7.7A). The corresponding phase map will then exhibit multiple phase jumps (Figure 7.7B). We can select a part of the phase map (B) with values between 0 and Δz (Figure 7.7C), which are far from any discontinuities. We then pick a certain area (marked in (C)) as a "starting point" and discard all the other areas that are not adjacent to it, as they are likely from the

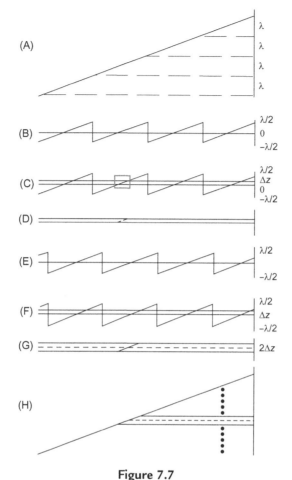

Figure 7.7
Phase unwrapping by varying the reconstruction distance (see text for details).

different phase periods (Figure 7.7D). The distance at which the hologram is digitally reconstructed can be changed by the small amount Δz. By doing so, we effectively move the object by Δz. Then, the phase difference between the object and the reference waves will also change by a constant and the discontinuities in the phase image will appear at the different places (Figure 7.7E). Then, we choose the part of (E) that is between 0 and Δz (Figure 7.7F), shift it up by Δz, add the result to map (D), and discard the areas that are not adjacent (Figure 7.7G). These steps can be repeated (moving up and then down to unwrap one new area during each repetition), while keeping track of where the new pixels are added, until the entire phase image is reconstructed "layer by layer" (Figure 7.7H). The algorithm stops when it finds no new pixels to add to the already unwrapped part of the image. The size of an individual step cannot be too large to include areas with a discontinuity. The smaller step size produces the correct solution, but it is more

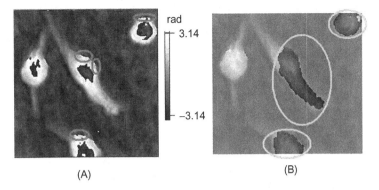

Figure 7.8

RAW cells—quality map guided flood fill unwrapping: (A) initial wrapped phase image (the areas of discontinuous phase gradient are highlighted) and (B) unwrapped phase image. The images are 51.2 µm × 51.2 µm (256 × 256 pixels).

computationally demanding, as each step requires performing one (inverse) FFT. In practice, an "adaptive step" routine can be introduced, which will adjust the step size depending on the phase image topology.

The slanted surface in this example can easily be reconstructed by even the simplest software unwrapping algorithm. However, in the case when an image has a certain amount of phase noise, even the most sophisticated algorithms still often fail in places where the 2π discontinuities are not closed loops [1]. Figure 7.8A shows the wrapped phase image of several living RAW cells. As expected, the areas where the phase jumps are not continuous, closed loops were problematic (see Figure 7.8B). Moreover, the method clearly not only failed in the areas where the phase jump was discontinuous (marked in Figure 7.8A), but the error propagated to the other parts of the image as well.

In comparison, Figure 7.9 demonstrates the results of the varying reconstruction distance method, as it is applied to unwrap the image in Figure 7.8A. Figure 7.9A shows the first step, where a point on the background was selected and then all the pixels with the height of less than $\Delta z = 30$ nm were kept. Figure 7.9B shows the removal of the areas of the image that are not adjacent to the background area and the gray areas correspond to yet to be unwrapped parts of the image. Figure 7.9C shows the intermediate step, where it is already clear that the algorithm has successfully unwrapped areas that were problematic in Figure 7.8A. Consequently, Figure 7.9D and E displays the final result using the varying reconstruction distance method free from unwrapping errors seen in Figure 7.8.

If a phase image, obtained from a particular reconstruction distance, is noisy in a certain area, the images, obtained from other reconstruction distances, may not be noisy in the same area. Thus, by varying the reconstruction distance and examining various z-planes, we can effectively construct the unwrapped image through comparison of multiple phase images.

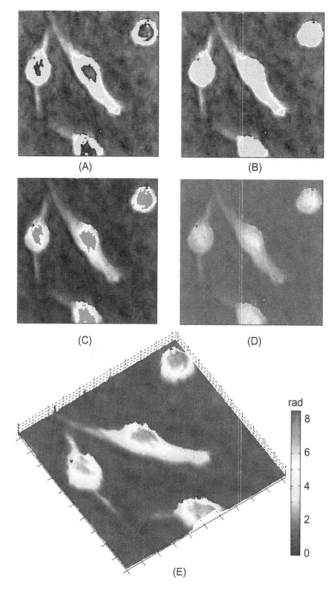

Figure 7.9

RAW cells—varying reconstruction distance unwrapping: (A)–(D) unwrapping stages and (E) 3D rendering of the final unwrapped phase image. The images are 51.2 μm × 51.2 μm (256 × 256 pixels).

While this approach is superior to conventional phase unwrapping methods, it has its own limitations. If the object is many wavelengths thick and/or the phase signal is very noisy, that is, the phase images for different reconstruction planes are inconsistent, the method begins to fail. Figure 7.10 shows holographic image and the reconstruction of onion cells.

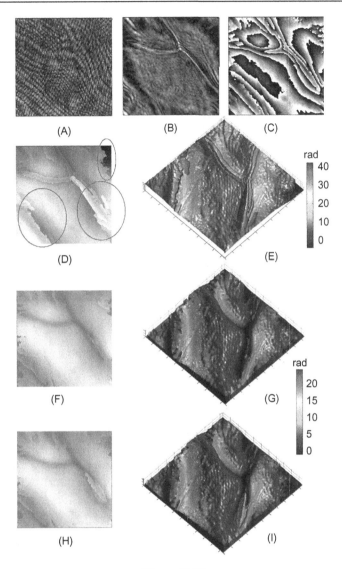

Figure 7.10
Onion cells—unwrapping: (A) hologram, (B) amplitude, (C) wrapped phase, (D) quality map guided flood fill unwrapped phase with artifacts (shown), (E) 3D rendering of (D), (F) phase unwrapping using varying reconstruction distance, (G) 3D rendering of (F), (H) phase unwrapping using the addition of modulo of 2π (some artifacts are still seen), (I) 3D rendering of (H). The images are 100 μm × 100 μm (416 × 416 pixels).

The phase image displays multiple discontinuities and the intensity image shows that some of the areas are either out of focus or too dark to provide a reliable phase signal. As a result, the quality map guided algorithm produces multiple artifacts, which again propagate into the areas of the image that are not noisy (Figure 7.10D and E). In comparison, the varying reconstruction distance unwrapping in Figure 7.10F and G, although it still fails to unwrap the very noisy areas, accurately unwraps the rest of the image.

This approach can be useful even in a general case when the phase images from multiple reconstruction distances are not available. Instead of varying the reconstruction distance, one can simply add a constant to the single phase image followed by the modulo of 2π and repeat all the same steps as if they were different reconstruction distance phase images. This method will lose the advantage of multiple phase plane reconstructions, and therefore the noisy areas in one image will remain noisy in the others, but it can still produce a phase unwrapped result. When it was applied to the hologram in Figure 7.10A, it still resulted in a better (compared to path following method) reconstruction, although not as good as the varying reconstruction distance unwrapping (see Figure 7.10H and I).

Finally, sudden vertical changes in height of an object which correspond to the phase change greater than 2π will not be correctly unwrapped using this method alone. In such a case, dual-wavelength phase imaging must be used to further increase the axial range at which an unambiguous phase imaging can be performed.

7.4 Dual-Wavelength Phase Imaging

7.4.1 Synthetic Wavelength

In general, a multiple wavelength phase imaging technique that removes the 2π discontinuities is based on the comparison of phase maps acquired at different wavelengths, as in this case the discontinuities, while being present in all phase maps, will occur at different places for each wavelength [9–12]. For dual-wavelength phase imaging, the object's height as a function of coordinate can be expressed as:

$$h(x,y) = \frac{\lambda_1}{2\pi}\varphi_1(x,y) + \lambda_1 m_1(x,y) = \frac{\lambda_2}{2\pi}\varphi_2(x,y) + \lambda_2 m_2(x,y) \qquad (7.10)$$

where m_1 and m_2 are the unknown non-negative integer numbers of wavelengths. Let us assume that λ_2 is bigger than λ_1, in which case m_1 cannot be smaller than m_2. Rearranging Eq. (7.10), we obtain:

$$h(x,y) = \frac{\lambda_2 \lambda_1}{\lambda_2 - \lambda_1} \cdot \left[\frac{\varphi_1(x,y) - \varphi_2(x,y)}{2\pi} + m_1(x,y) - m_2(x,y)\right] = \frac{\Lambda_{12}}{2\pi}(\varphi_1(x,y) - \varphi_2(x,y)) + \Lambda_{12} M_{12}$$

$$(7.11)$$

Here, the term $\Lambda_{12} = \lambda_2 \lambda_1 / \lambda_2 - \lambda_1$ is known as the synthetic ("beat") wavelength. Since the height must be a positive number, the term $M_{12} = m_1(x, y) - m_2(x, y)$ must be non-negative. Moreover, if $\varphi_1(x, y) - \varphi_2(x, y) < 0$, then $M_{12} \geq 1$. Note that if we assume that Λ_{12} is bigger than the entire height of the object, then M_{12} cannot exceed 1, so then M_{12} is either

0 or 1 depending on whether $\varphi_1(x, y) - \varphi_2(x, y)$ is positive or negative. Therefore, subtracting the two phase maps and adding 2π whenever the difference is negative effectively yields a new phase map, which corresponds to the beat wavelength Λ_{12}.

Figure 7.11 shows the phase images of the USAF resolution target imaged at a slight angle. The images produced with a single wavelength exhibit multiple phase steps (see Figure 7.11A and B). By comparing the phase images from each wavelength, the 2π phase ambiguities can be resolved. For $\lambda_1 = 633$ and $\lambda_2 = 532$ nm, $\Lambda_{12} = 3334$ nm, which is high enough to remove the discontinuities (see Figure 7.11C).

The negative aspect of this method is that the phase noise is also amplified by the same factor as the range. However, one can then use this dual-wavelength "coarse" map as a guide, together with one of the original phase maps (φ_1 or φ_2), to produce the low noise "fine" phase map [10]. If the noise in the coarse phase map is too excessive, some of the single wavelength segments might still end up being vertically shifted from its correct position by a single wavelength, creating phase image artifacts. These errors can then be corrected in software by simply looking for such jumps and shifting them up or down [3]. In comparison to the coarse map, the noise in the resulting fine map (see Figure 7.11D) is much lower, while the axial range is still the same.

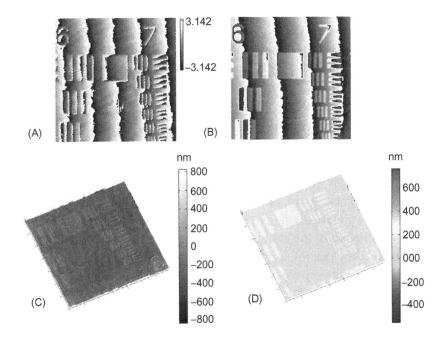

Figure 7.11
Phase maps for (A) $\lambda_1 = 532$ nm and (B) $\lambda_2 = 633$ nm; (C) 3D rendering of synthetic dual-phase map with beat wavelength $\Lambda_{12} = 3334$ nm and (D) reduced noise fine map (the images are 174 μm × 174 μm, 450 × 450 pixels).

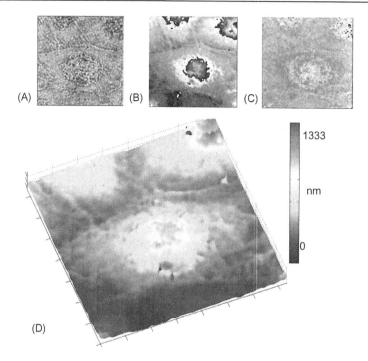

Figure 7.12
Confluent SKOV-3 ovarian cancer cells: (A) amplitude image, (B) reconstructed phase for $\lambda = 532$ nm, (C) dual-wavelength coarse phase image, and (D) 3D rendering of fine map. All images are 92 μm × 92 μm (240 × 240 pixels). The area at the bottom of the images is the exposed part of the gold substrate.

The dual-wavelength phase imaging method has been applied to 3D imaging of ovarian cancer cells (SKOV-3) [3]. Figure 7.12A shows the intensity image of a confluent group of cells, which is similar to what one can see using an ordinary microscope. Figure 7.12B displays a single wavelength wrapped phase image, Figure 7.12C shows the coarse dual-wavelength unwrapped phase image, and Figure 7.12D displays 3D rendering of the final low noise "fine" map. To measure the physical thickness of cells by using Eq. (7.5), we need to make an assumption of the cell's refractive index, $n = 1.375$. Note that while it may not be possible to precisely determine the refractive index of the cell at each individual point, this number is always very close to the refractive index of water.

Figure 7.13 shows the image of a single (SKOV-3) cell, where the cell's nucleus and pseudopodia are clearly seen. By using the phase to thickness conversion (Eq. (7.5)), we can determine the 3D features of the cell. Figure 7.13D displays the line height profile, which indicates, for example, that the overall cell height is about 1.47 μm, with the thickness of the cell's pseudopodia (lamelipodia) at around 270 nm.

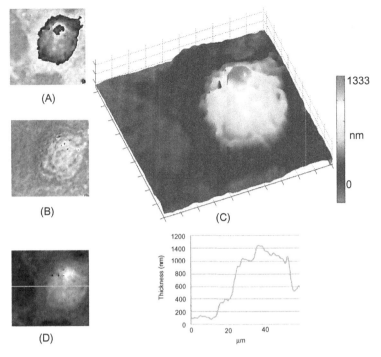

Figure 7.13
A single SKOV-3 cell: (A) reconstructed phase for $\lambda = 633$ nm, (B) dual-wavelength coarse phase image, (C) 3D rendering of fine map, and (D) line thickness profile. All images are 63.5 μm × 59 μm (165 × 153 pixels).

7.4.2 Linear Regression Method

Finally, we present a method that determines m_1 or m_2 directly [9]. Rewriting Eq. (7.10), we obtain:

$$\lambda_1 m_1(x,y) - \lambda_2 m_2(x,y) = \frac{1}{2\pi}(\lambda_2 \varphi_2(x,y) - \lambda_1 \varphi_1(x,y)) \tag{7.12}$$

Although this equation contains two unknowns (m_1 and m_2), they are both non-negative integers. Theoretically, if the wavelengths do not have a common multiplier and the measurement uncertainty is neglected, there is only one set of values of m_1 and m_2 which satisfies this equation exactly. In the presence of phase noise, we can determine the integer values m_1 and m_2 such that the difference between the two sides of the Eq. (7.12) is minimal. Expressing m_1 in terms of m_2, we obtain:

$$m_1(x,y) = \frac{\lambda_2}{\lambda_1} m_2(x,y) + \frac{1}{2\pi}\left(\frac{\lambda_2}{\lambda_1}\varphi_2(x,y) - \varphi_1(x,y)\right) \tag{7.13}$$

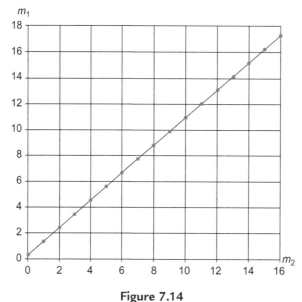

Figure 7.14
Linear regression phase unwrapping: $m_1(m_2)$.

By simply trying different values of m_2, we can pick the value at which this equation yields the result which is closest to a non-negative integer number, thus determining the most likely number of integer wavelengths.

The linear function $m_1(m_2)$ is plotted in Figure 7.14. The ratio of wavelengths can usually be determined very accurately, so the uncertainty of the slope of this line is low. Moreover, it is uniform throughout the image, as both the phases are contained only in the y-intercept term.

Figure 7.15 shows the square of the difference between m_1 and the nearest integer. In this case, $\lambda_1 = 635$ and $\lambda_2 = 675$ nm, so $\Lambda_{12} = 10.7$ μm, which increases the range approximately 16 times. Indeed, the function in Figure 7.2 has a local minimum every 16 points. Within the range of 0–16 (the height of the object is less than the synthetic wavelength), the value of m_2 is easy to select, directly obtaining the lower noise phase unwrapped map. Moreover, as some of these local minima are smaller than others, we can further extend the imaging range by simply selecting the smallest of the local minima, extending the unambiguous phase measurements well beyond the synthetic wavelength.

In practice, if the phase measurements are noisy, this method may fail and pick the wrong local minimum, so the total height (maximum value of m_2) has to be limited to some reasonable number. For example, if the approximate size of the imaged object is known, either through some other technique or simply by looking at a wrapped single wavelength phase map and estimating the approximate number of phase jump, then there is no need to

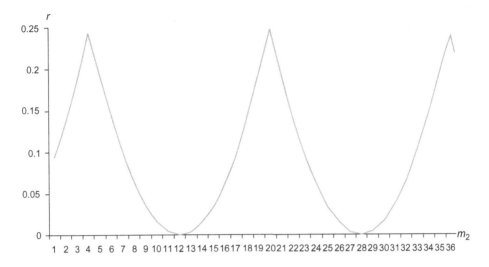

Figure 7.15
Linear regression phase unwrapping: $r(m_2)$—the square of difference between m_1 and the nearest integer. In this example, $r(m_2)$ has its minimum at $m_2 = 11$.

set the limit on m_2 higher. Even if the measurements are very noisy, but the object is only a few wavelengths high, the maximum number of m_2 can be set low (much lower than 16 in the example given earlier), providing a tradeoff between the amount of noise the method can tolerate and the overall height range.

Consider a slanted surface with the total height of 7.5 μm. Simulated phase images of this surface using two wavelengths (532 and 633 nm) of that structure are shown in Figure 7.16A and B. Predictably, each single wavelength phase map shows multiple discontinuities. By using Eq. (7.13), the algorithm correctly guesses the value of m_2 (Figure 7.16C), and the unwrapped profile is then easily reconstructed by Eq. (7.4), yielding the final phase map free of discontinuities (Figure 7.16D). Note that in this case, the overall height of the objects is more than double the synthetic wavelength $\Lambda_{12} = 3.3$ μm.

But the real strength of this method lies in the fact that for each point (x, y), the algorithm does not rely on the surrounding pixels and so even more topologically complicated images can be unwrapped just as well. Consider the case where the object is a cell-like semi-sphere, which has a relatively steep sloped surface on the side. Here, both phase images (Figure 7.17A and B) were also corrupted by random (up to 2 rad) phase noise. Additionally, the object is placed to the edge of the image frame, so for each phase jump, a single wavelength software algorithm has only a few pixels to work with. As evident from Figure 7.17C, it fails to unwrap the phase correctly, while the linear regression method produces a correct profile, without amplifying the noise. It is worth noting that the phase noise level here is much higher than what previous dual-wavelength algorithm can tolerate [10].

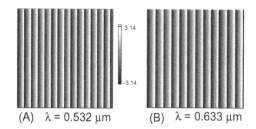
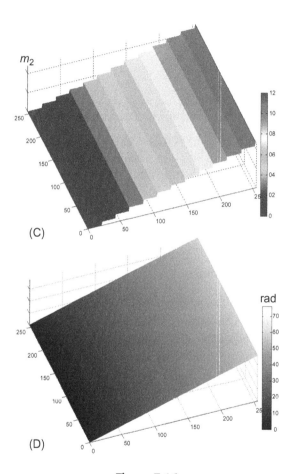

Figure 7.16
Slanted surface: (A) phase image at 532 nm, (B) phase image at 633 nm, (C) determining m_2, (D) final unwrapped phase image. All images are 256 × 256 pixels [6].

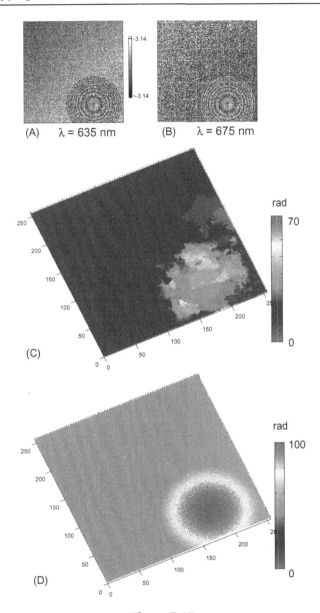

Figure 7.17
Simulated cell: (A) phase image at 635 nm, (B) phase image at 675 nm, (C) unsuccessful software unwrapping, (D) dual-wavelength unwrapped phase image. All images are 256 × 256 pixels.

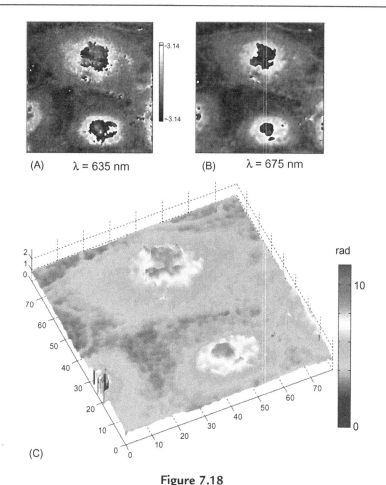

Figure 7.18
3D holographic imaging of cells: (A) phase image at 635 nm, (B) phase image at 675 nm, (C) 3D pseudo-color rendering of dual-wavelength unwrapped phase image. All images are 78 μm × 78 μm.

Figures 7.18 and 7.19 show the phase images of actual cells, acquired by our digital holographic microscope. Examining single wavelength images in Figure 7.18A and B, one could determine that the maximum value of m_2 could be set to 2, since the height map clearly does not extend beyond two wavelengths. By applying the linear regression method, the final unwrapped image free of discontinuities (Figure 7.18C) is obtained.

Finally, while the object in Figure 7.19 is twice the height of the synthetic wavelength Λ_{12}, it is still unwrapped correctly. Therefore, there is generally no need to use three or more wavelengths [12], as adding more wavelengths complicates the optical system and can actually lead to higher noise due to misalignment.

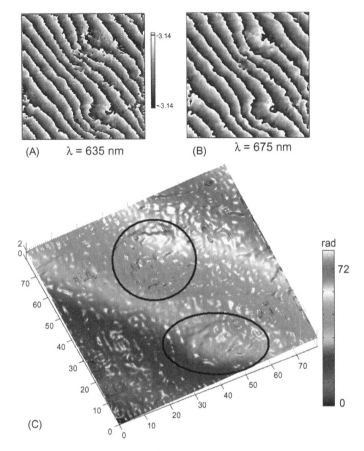

Figure 7.19
Two KB cells (marked) with a substrate inserted at an angle: (A) phase image at 635 nm, (B) phase image at 675 nm, (C) dual-wavelength unwrapped phase image. All images are 78 μm × 78 μm.

7.5 Conclusions

Phase unwrapping of biological cell images presents several challenges. While the cells are typically positioned on a flat substrate, which can be used as a reference for the unwrapping algorithm, the various cells organelles often scatter most of the light, so no coherent signal can be obtained from some parts of an image frame. As a result, these areas often become the source of unwrapping errors, which propagate to the other parts of the image as well. By using the varying reconstruction distance algorithm, we can effectively bypass the areas of phase noise and unwrap the cellular phase images "layer by layer." Also, the cell membrane is sometimes oriented almost perpendicularly to the substrate, resulting in sudden vertical changes in the height of an object, which cannot be correctly unwrapped using the single wavelength phase imaging. In such a case, two wavelengths together allow us to increase the maximum height of the features that can be unambiguously imaged.

The wave front curvature, which may be presented in the holographic image, needs to be removed as well. The methods of curvature correction and background removal, presented here, are simple and effective enough to easily implement the experiment without the microscope objectives in the reference arms of the holographic interferometer. This greatly simplifies the optical setup and makes it much easier to do the optical adjustments of the apparatus. Simultaneous dual-wavelength setup utilized together with the angular spectrum algorithm provides a way to acquire single frame images in real time, which can be used to study cell migration, dynamic cellular volume change, etc. By extending unambiguous optical phase imaging methods to objects of different heights, linear regression dual-wavelength unwrapping method makes phase imaging much more practical, which allow 3D measurements of a wide variety of biological systems and microstructures.

References

[1] D.C. Ghiglia, M.D. Pritt, Two-Dimensional Phase Unwrapping: Theory, Algorithms, and Software, John Wiley & Sons, New York, NY, USA, 1998.
[2] A. Khmaladze, A. Restrepo-Martínez, M.K. Kim, R. Castañeda, A. Blandón, Simultaneous dual-wavelength reflection digital holography applied to the study of the porous coal samples, Appl. Opt. 47 (2008) 3203.
[3] A. Khmaladze, M.K. Kim, C.-M. Lo, Phase imaging of cells by simultaneous dual-wavelength reflection digital holography, Opt. Express 16 (2008) 10900.
[4] M. Born, E. Wolfe, Principles of Optics, Pergamon, Oxford/London/Edinburgh/New York/Paris/Frankfurt, 1964.
[5] P. Ferraro, S. De Nicola, A. Finizio, G. Coppola, S. Grilli, C. Magro, et al., Compensation of the inherent wave front curvature in digital holographic coherent microscopy for quantitative phase-contrast imaging, Appl. Opt. 42 (2003) 1938.
[6] A. Khmaladze, R.L. Matz, J. Jasensky, E. Seeley, M.M. Banaszak Holl, Z. Chen, Dual-wavelength digital holographic imaging with phase background subtraction, Opt. Eng. 51 (2012) 055801.
[7] C.A. Lieber, A. Mahadevan-Jansen, Automated method for subtraction of fluorescence from biological Raman spectra, Appl. Spectrosc. 57 (2003) 1363.
[8] A. Khmaladze, T. Epstein, Z. Chen, Phase unwrapping by varying the reconstruction distance in digital holographic microscopy, Opt. Lett. 35 (2010) 1040.
[9] A. Khmaladze, R.L. Matz, C. Zhang, T. Wang, M.M. Ting, M.M. Banaszak Holl, et al., Dual-wavelength linear regression phase unwrapping in three-dimensional microscopic images of cancer cells, Opt. Lett. 36 (2011) 912.
[10] J. Gass, A. Dakoff, M.K. Kim, Phase imaging without 2π-ambiguity by multiple-wavelength digital holography, Opt. Lett. 28 (2003) 1141.
[11] H. Hendargo, M. Zhao, N. Shepherd, J. Izatt, Synthetic wavelength based phase unwrapping in spectral domain optical coherence tomography, Opt. Express 17 (2009) 5039.
[12] U. Schnars, W. Jüptner, Digital Holography: Digital Hologram Recording, Numerical Reconstruction, and Related Techniques, Springer, Berlin/Heidelberg/New York, 2004.

CHAPTER 8

On-Chip Holographic Microscopy and its Application for Automated Semen Analysis

Ting-Wei Su[1,2], Aydogan Ozcan[1,2,3,4]

[1]Electrical Engineering Department, University of California, Los Angeles, CA [2]ADR, Los Angeles, CA [3]Bioengineering Department, University of California, Los Angeles, CA [4]California NanoSystems Institute, University of California, Los Angeles, CA

Editor: Natan T. Shaked

8.1 Introduction

Lens-based light microscopy provides an indispensable toolset for medicine and biology to investigate biomedical samples at the microscale. As new microscopy techniques are being developed to give this centuries-old imaging technology better resolution, speed, sensitivity, contrast, throughput, and specificity [1–6], these imaging systems are also becoming more complicated, bulky, and expensive, which partially limit their use outside of well-equipped facilities. Furthermore, the relatively small field of view (FOV) of these lens-based imaging platforms also provides a challenge for various applications such as imaging cytometry, rare cell analysis, or model organism phenotyping [7–9], where a wide FOV would be highly desired to generate reliable statistics and improve throughput of analysis.

To mitigate these existing challenges, a compact, cost-effective, and yet wide-field lensless imaging platform that utilizes partially coherent light sources has recently been introduced for conducting microanalysis on a chip [10]. At the core of this technology is a lensless holographic on-chip microscope (Figure 8.1) that measures ~ 4.2 cm $\times$ 4.2 cm $\times$ 5.8 cm and weighs ~ 46 g while still providing an imaging FOV of ~ 24 mm^2 with an effective numerical aperture (NA) of $\sim 0.1-0.2$ [11]. Compared to the FOV of a typical 10× objective lens, this imaging FOV is more than 20-fold larger and therefore provides a significant throughput improvement allowing automated monitoring of thousands of cells all in parallel. This lensless on-chip microscope is based on digital in-line holography, and it

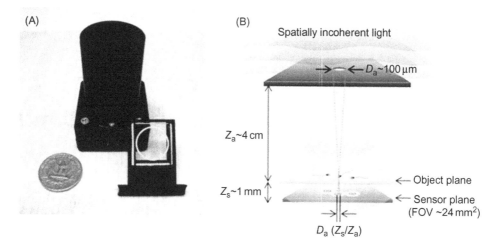

Figure 8.1
(A) A lensless holographic on-chip microscope with dimensions of 4.2 cm × 4.2 cm × 5.8 cm and a weight of 46 g is shown with its sample-loading tray and a US quarter. A USB connection from the side powers both the CMOS image sensor and the embedded light source (an LED filtered by a pinhole). Both amplitude and phase images of the sperms can be generated with this compact on-chip lensfree microscope over an FOV of 24 mm^2 with an effective NA of ~0.2. (B) The relative positions of the light source, the semen sample, and the sensor chip are depicted with a schematic diagram of the lensless holographic microscope shown in (A). This schematic drawing is not to scale.

utilizes an incoherent or partially coherent light source (such as a light-emitting diode—LED) that is filtered by a large aperture of, for example, ~0.1 mm to illuminate the sample of interest as illustrated in Figure 8.1. Over a short propagation distance of ~4 cm, this illumination light picks up partial spatial coherence, after which it scatters from each cell to coherently interfere with the background light, forming lensless holograms of the cells over a large FOV (~24 mm^2). These lensless cell holograms can then be rapidly processed (e.g., <1 s) using a Graphics Processing Unit to reconstruct their microscopic images (both amplitude and phase [11]) as illustrated in Figure 8.2.

Among various applications of this lensfree holographic microscope [11–15], automated semen analysis is one that can especially benefit from this microscope's field portability and high throughput. As an important routine in fertility clinics, semen analysis is extensively practiced for evaluating male fertility [16] and preparing artificial insemination [17]. Semen samples are put into a counting chamber and then the sperm are manually counted through an optical microscope to determine the sperm concentration. Such visual assessment, although quite labor intensive, is still the gold standard that is not only recommended by the World Health Organization (WHO) but also widely used in most semen processing laboratories [18]. Several optical approaches, including turbidimetry

[19–26], laser Doppler velocimetry [27–32], and photon correlation spectroscopy [33–35], have also been proposed to automatically analyze semen and avoid the labor intensive nature of this manual method. However, these approaches can only indirectly estimate the overall sperm concentration or motility and therefore are not widely adopted.

Computer-assisted semen analysis (CASA) systems are currently considered as one of the most promising technologies to replace the traditional manual semen analysis methods with pattern analysis algorithms that automatically process the images recorded with a conventional optical microscope [36–42]. CASA systems are important because of their ability to provide quantitative information about sperm motility (e.g., the speed distribution of individual sperm), which can be used to predict fertilization rate [43,44] and to evaluate the effect of various drugs on sperm quality [45,46]. Although the state-of-the-art CASA systems are very efficient and versatile, their widespread use in fertility clinics is partially hindered by their relatively large dimensions, high costs, and maintenance needs. For the same reason, CASA platforms have limited applications to field use in veterinary medicine such as stud farming and animal breeding [43,47–78].

Through a color change due to chemical staining or labeling of sperm-specific proteins, commercially available male fertility test kits for personal home use, such as FertilMARQ [50] or SpermCheck [51], also aim to indirectly quantify sperm concentration. However, sperm motility or the concentration of motile sperm cannot be quantified by these tests. Recently, an alternative semen analysis platform has also been reported by utilizing a compact microfluidic device that can measure electrical impedance changes due to sperm movement [52]. Unfortunately, only the total number of the sperm in the sample can be provided by this lab-on-a-chip platform, that is, motile and immotile sperm cannot be differentiated from each other.

To conduct automated semen analysis using our lensless holographic microscope (Figure 8.1A), digital summation of sequentially captured lensless frames enables us to rapidly count only the immobile sperm based on the reconstructed phase images. On the other hand, digital subtraction of these consecutive holographic frames permits quantification of the speed and the trajectories of individual motile sperm within an FOV of ~ 24 mm^2.

In this chapter, we first present a detailed analysis on the fundamental principles of on-chip incoherent in-line holography which is at the heart of the presented semen analysis platform. Next, implementation of such a lensless holographic imaging device toward automated semen analysis is explained together with the digital algorithms that we employed for automatic quantification of sperm density and motility. Lastly, we present results of this high-throughput on-chip semen analysis platform and its potential applications. Such a compact and cost-effective automated semen analysis platform running on a wide-field lensless on-chip microscope would be a very valuable tool for andrology laboratories, fertility clinics, male fertility tests, as well as for field use in veterinary medicine.

8.2 Partially Coherent In-Line Holography on a Chip

Holography indirectly records the optical phase information through amplitude oscillations generated by the interference of coherent optical waves. For making use of this phase information toward microscopy, most existing lensless in-line holography systems require a high level of spatial and temporal coherence and therefore utilize a laser source that is filtered through a small aperture, e.g., 1–2 μm wide [6,53–61]. The optical set-up, however, can be made simpler and more robust if a completely incoherent light source filtered through a large aperture (e.g., >100–200λ in diameter) could be used [11,62].

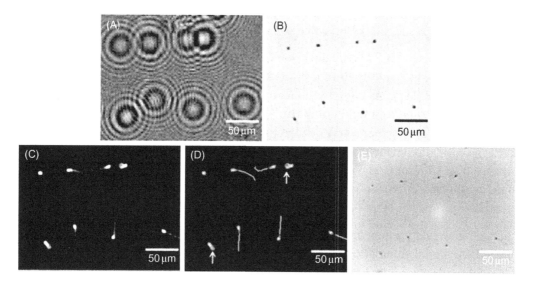

Figure 8.2
(A) A digitally cropped lensless hologram that is captured with the device in Figure 8.1A shows the hologram signatures of eight immobilized sperms. (B) The locations of the heads of the sperms are revealed by an amplitude image reconstructed from the raw hologram shown in (A) for the same FOV. (C) Both the heads and the tails of the sperms are illustrated by the phase image reconstructed from the raw hologram shown in (A) for the same FOV. (D) Automatic characterization results are generated based on the reconstructed phase image in (D). Red circles enclose the elliptical areas corresponding to sperm heads while green lines label the tails. Defective sperms with missing or unusually curved tails (marked with the white arrows in (D)) are not reported toward positive sperm counts. (E) To verify the results in (D), a bright-field microscope image is acquired with a 40× objective lens (NA = 0.65) for the same FOV as in (A). For interpretation of the references to color in this figure legend, the reader is referred to the web version of this book.

One of the key steps for recording high-quality in-line holograms with a spatially incoherent source emanating from a large aperture is to bring the cell plane close to the detector array by ensuring $z_s \ll z_a$, where z_a defines the distance between the incoherently illuminated aperture plane and the object/cell plane, and z_s defines the distance between the object/cell plane and the sensor array (see Figure 8.1). While the total aperture-to-detector distance $(z_s + z_a)$ and the overall device length remain almost unchanged, conventional lensless in-line holography approaches typically choose to utilize $z_a \ll z_s$. Therefore, in addition to an incoherent source used with a large aperture, our choice of $z_s \ll z_a$ is also quite different from the mainstream lensless in-line holographic imaging approaches.

To better quantify the impact of these differences on the detected holograms and their reconstructions, we assume two point scatterers (separated by $2a$) located at the object/cell plane $(z = z_a)$ with a field transmission of $t(x, y) = 1 + c_1\delta(x - a, y) + c_2\delta(x + a, y)$, where the intensities of c_1 and c_2 denote the strength of the scattering process for these two point sources and $\delta(x, y)$ is a Dirac-delta function in space. Subcellular elements that make up a cell can be represented by such point scatterers. Let us further assume that there is a large aperture (of arbitrary shape) that is positioned at $z = 0$ with a transmission function of $p(x, y)$ and that the digital recording device (e.g., a CMOS (complementary-symmetry metal–oxide–semiconductor) or CCD (charge-coupled device) sensor array) is positioned at $z = z_a + z_s$, where typically z_a is 3–10 cm and z_s is 0.5–2 mm.

Assuming that a spatially incoherent light source uniformly illuminates the aperture $p(x, y)$, the cross-spectral density at the aperture plane can be written as:

$$W(x_1, y_1, x_2, y_2, \nu) = S(\nu)p(x_1, y_1)\delta(x_1 - x_2)\delta(y_1 - y_2)$$

where (x_1, y_1) and (x_2, y_2) denotes two arbitrary points on the aperture plane and $S(\nu)$ represents the power spectrum of the incoherent source having its center frequency at ν_0 (corresponding to the center wavelength of λ_0). After propagating over a distance of z_a in free space, one can write the cross-spectral density at the object plane (just before interacting with the cells) as [63]:

$$W(\Delta x, \Delta y, \eta, \nu) = \frac{S(\nu)}{(\lambda z_a)^2} \exp\left(-j\frac{2\pi\nu q}{c z_a}\right) \int\int p(x, y) \exp\left(j\frac{2\pi}{\lambda z_a}(x\Delta x + y\Delta y)\right) dx\, dy$$

where $\Delta x = x'_1 - x'_2$, $\Delta y = y'_1 - y'_2$, $\eta = (x'_1 + x'_2/2)\Delta x + (y'_1 + y'_2/2)\Delta y$, $\lambda = (c/\nu)$, c is the speed of light, and (x_1, y_1) (x_2, y_2) denotes two arbitrary points on the object plane. After interacting with the specimen, that is, $t(x, y)$, the cross-spectral density right after the object plane can be written as:

$$W(\Delta x, \Delta y, \eta, \nu) \cdot t^*(x'_1, y'_1) \cdot t(x'_2, y'_2)$$

Before reaching the detector plan, this cross-spectral density function will also propagate for another distance of z_s. Therefore, the cross-spectral density at the detector plane can be written as:

$$W_D(x_{D1}, y_{D1}, x_{D2}, y_{D2}, \nu)$$
$$= \iiiint W(\Delta x, \Delta y, \eta, \nu) t^*(x_1', y_1') t(x_2', y_2') h_C^*(x_1', x_{D1}, y_1', y_{D1}, \nu) h_C(x_2', x_{D2}, y_2', y_{D2}, \nu) dx_1' \, dy_1' \, dx_2' \, dy_2'$$

where (x_{D1}, y_{D1}) and (x_{D2}, y_{D2}) refer to two arbitrary points on the detector plane (within the hologram region of each cell) and $h_C(x', x_D, y', y_D, \nu) = \frac{1}{j\lambda z_s} \exp(j\frac{2\pi z_s}{\lambda}) \exp\left(j\frac{\pi}{\lambda z_s}\left[(x' - x_D)^2 + (y' - y_D)^2\right]\right)$.

At the sensor plane (x_D, y_D), one can then write the optical intensity $I(x_D, y_D)$ as:

$$I(x_D, y_D) = \int W_D(x_D, y_D, x_D, y_D, \nu) d\nu$$

Assuming $t(x, y) = 1 + c_1 \delta(x - a, y) + c_2 \delta(x + a, y)$, the optical intensity $I(x_D, y_D)$ can be further expanded into four terms, each with a different physical meaning, i.e.:

$$I(x_D, y_D) = I_s(x_D, y_D) + I_C(x_D, y_D) + H_1(x_D, y_D) + H_2(x_D, y_D)$$

where:

$$I_s(x_D, y_D) = D_0 + \frac{|c_1|^2 S_0}{(\lambda_0 z_a z_s)^2} \tilde{P}(0, 0) + \frac{|c_2|^2 S_0}{(\lambda_0 z_a z_s)^2} \tilde{P}(0, 0) \tag{8.1}$$

$$I_C(x_D, y_D) = \frac{c_2 c_1^* S_0}{(\lambda_0 z_a z_s)^2} \tilde{P}\left(\frac{2a}{\lambda_0 z_a}, 0\right) \exp\left(j\frac{4\pi a x_D}{\lambda_0 z_s}\right) + \text{c.c.} \tag{8.2}$$

$$H_1(x_D, y_D) = \frac{S_0}{(\lambda_0 z_a)^2} \left[c_1 \cdot \{p(-x_D \cdot M + a \cdot M \cdot F, -y_D \cdot M)^* h_C(x_D, y_D)\} + \text{c.c.}\right] \tag{8.3}$$

$$H_2(x_D, y_D) = \frac{S_0}{(\lambda_0 z_a)^2} \left[c_2 \cdot \{p(-x_D \cdot M - a \cdot M \cdot F, -y_D \cdot M)^* h_C(x_D, y_D)\} + \text{c.c.}\right] \tag{8.4}$$

where "c.c." and "*" denote the complex conjugate and convolution operations, respectively, $M = (z_a/z_s)$, $F = (z_a + z_s/z_a)$, and P is the 2D spatial Fourier transform of the arbitrary aperture function $p(x, y)$. It should be emphasized that (x_D, y_D) in these equations is restricted to the extent of the cell/object hologram, rather than extended to the entire FOV of the detector array. Furthermore, $h_C(x_D, y_D) = (1/j\lambda_0 \cdot F \cdot z_s) \exp(j(\pi/\lambda_0 \cdot F \cdot z_s)(x_D^2 + y_D^2))$ and it represents the 2D coherent impulse response of free space over a distance of $\Delta z = F \cdot z_s$. For the incoherent illumination source, we have assumed that the spectral bandwidth is much smaller than its center frequency ν_0, i.e., $S(\nu) \cong S_0 \delta(\nu - \nu_0)$. This is an appropriate assumption since the light sources (LEDs) that we have typically used in our experiments have their central wavelengths at 500–650 nm with a spectral FWHM (full width at half maximum) of 10–15 nm.

In these derivations, paraxial approximation has also been assumed to simplify the results since z_a and z_s for our hologram recording geometry are typically much longer than the extent of each cell hologram. However, such an approximation was not used in the numerical reconstruction process of the cell images, which will be further discussed later in this chapter.

D_0 in Eq. (8.1) can be further expanded into:

$$D_0 = \iiiint \int \frac{W(\Delta x, \Delta y, \eta, \nu)}{(\lambda z_s)^2} \exp\left(-j\frac{\pi}{\lambda z_s}\left[(x_1'-x_D)^2 + (y_1'-y_D)^2\right]\right)$$
$$\exp\left(j\frac{\pi}{\lambda z_s}\left[+(x_2'-x_D)^2 + (y_2'-y_D)^2\right]\right) dx_1'\, dy_1'\, dx_2'\, dy_2'\, d\nu$$

which physically represents the background illumination reaching the sensor plane and it carries no spatial information regarding the cells' structure or distribution. For typical illumination configurations, D_0 constitutes a uniform or slowly varying background, and hence can be digitally subtracted out without an issue.

Equations (8.1)–(8.4) are rather important to explain the key parameters in our partially coherent lensless on-chip holography scheme utilizing an incoherent light source emanating through a large aperture. Equation (8.1) includes the background illumination (D_0 term) and the self-interference of the scattered waves (the terms that are proportional to $|c_1|^2$ and $|c_2|^2$), both of which represent the classical diffraction that occurs between the object and the sensor planes under the paraxial approximation. Note also that these self-interference terms in Eq. (8.1) are scaled with $P(0,0)$ as the physical extent of the spatial coherence at the cell plane is not a determining factor for self-interference.

Equation (8.2), on the other hand, represents the interference between these two scatterers located at the object plane. Just like self-interference, the cross-interference term, $I_C(x_D, y_D)$, is also not useful for holographic reconstruction of object images. Since this cross-interference term is proportional to the amplitude of $\tilde{P}((2a/\lambda_0 z_a), 0)$, two scatterers that are far from each other can still interfere effectively if a small aperture size is used (hence wide $\tilde{P}$). The $\tilde{P}((2a/\lambda_0 z_a), 0)$ term predicts that, if $(2a < \lambda_0 z_a/D_a)$ (where D_a is the aperture width), the cross-interference $I_C(x_D, y_D)$ from these two scattered fields will generate strong but undesired cross-interference patterns at the sensor plane. This conclusion is also supported by the fact that the coherence diameter at the object plane is in the order of $\sim(\lambda_0 z_a/D_a)$, as estimated by van Cittert-Zernike theorem. Therefore, as another advantage of using a large aperture that is illuminated by an incoherent light source, this cross-interference noise term, $I_C(x_C, y_D)$, will only contain the contributions of a limited number of cells within the imaging

FOV since the extent of $\tilde{P}((2a/\lambda_0 z_a), 0)$ will be suppressed by a large aperture. Therefore, incoherent illumination through a large aperture will provide better image quality for characterizing dense cell suspensions such as undiluted semen samples.

The final two terms (Eqs. (8.3) and (8.4)) represent the holographic diffraction patterns in the recorded intensity and are the foci of all digital holographic imaging systems, including the on-chip implementation discussed in this chapter. Ideally, these terms should dominate the information content of the recorded intensity, which is typically true for weakly scattering objects. More specifically, $H_1(x_D, y_D)$ is the holographic diffraction of the first scatterer $c_1\delta(x-a, y)$; and $H_2(x_D, y_D)$ is the holographic diffraction of the second scatterer $c_2\delta(x+a, y)$. Since $h_C^*(x_D, y_D)$ creates twin-image artifacts at the reconstruction plane when propagated in the reverse direction, the complex conjugate (c.c.) terms in Eqs. (8.3) and (8.4) represent the source of the twin images. Numerical elimination of these twin-image artifacts will be discussed in Section 8.3.

As indicated by the terms inside the curly brackets in Eqs. (8.3) and (8.4), a scaled and shifted version of the aperture function $P(x, y)$ coherently diffracts around the position of each scatterer. In other words, each point scatterer projects a scaled version of the aperture function (i.e., $p(-x_D \cdot M, -y_D \cdot M)$) to a location shifted by F folds from origin, and the distance between the object and the sensor planes is now also scaled by F folds (i.e., $\Delta z = F \cdot z_s$). It is also important to emphasize that the aperture size is effectively narrowed down by an M fold at the object plane. For $M \gg 1$ (typically 50–200), a spatially incoherent light source through a large aperture can still provide coherent illumination to each cell individually for generating each cell's holographic signature on the sensor plane. This is true as long as the cell's diffraction pattern is smaller than the coherence diameter at the sensor plane. In our geometry, coherence diameter is typically $500\lambda_0 - 1000\lambda_0$ and it is much larger than the practical width of cell holograms on the sensor plane, which is easy to satisfy especially for small z_s values. Consequently, for a completely incoherent source emanating through an aperture width of D_a and illuminating a sensor area of A, the effective width of each point scatterer diffracting toward the sensor plane would be D_a/M and the effective imaging FOV would be A/F^2. Considering typical values for z_a (e.g., $\sim 3-10$ cm) and z_s (0.5–2 mm), the scaling factor (M) becomes >100 and $F \approx 1$, as a result of which even a 50 μm wide pinhole would be scaled down to <500 nm and the entire active area of the sensor array now can be used as the imaging FOV (i.e., $FOV \approx A$).

Although the entire derivation above is based on the formalism of wave theory, the final result is also matched to a scaling factor of ($M = z_a/z_s$) (see Figure 8.1B) predicted by simple geometrical optics. In the case of $M \gg 1$ with partially coherent illumination, each cell hologram only occupies a tiny fraction of the entire FOV and has little cross-talk with other cell holograms. As a result, unlike conventional lensless in-line holography, there is no longer an overall Fourier transform relationship between the sensor and the object planes. Such a spatial Fourier transform relationship exists only between each individual hologram and its corresponding cell.

According to Eqs. (8.3) and (8.4), a narrow enough $p(-x_D M, -y_D M)$ can ensure that the spatial features of the cells are not being washed out by partially coherent illumination through a large aperture, as a result of which the modulation of the holographic term at the detector plane to be approximated as $\sin((\pi/\lambda_0 F z_s)(x_D^2 + y_D^2))$. Such a modulation term suggests that a large fringe magnification (F) allows the use of a larger pixel size of the sensor array for recording the holographic fringes of the objects and effectively increases the NA of hologram recording, which is then only limited by the sensor width. There are, however, also disadvantages for having a large F: (i) a coherent light source emanating through a small aperture is needed to provide high spatial coherence at the sensor plane, which results in a more complicated set-up in terms of alignment of optical components, increasing the overall cost and complexity of the imaging platform; and (ii) the effective imaging FOV is also reduced by a factor of F^2.

8.3 Reconstruction of Microscopic Images and the Phase Recovery Process

In partially coherent lensless on-chip holography with $M \ll 1$ since the incident wave on each cell has very small curvature in its wave front, the recorded cell holograms can be reconstructed assuming plane-wave illumination. To propagate the wave fronts, an angular spectrum approach is used for numerically solving the Rayleigh–Sommerfeld integral. Within a linear and isotropic medium, this calculation can be done by multiplying the Fourier transform of the incident field with a transfer function defining this propagation, i.e.:

$$H_z(f_x, f_y) = \begin{cases} \exp\left(j2\pi z \frac{n}{\lambda}\right)\sqrt{1 - (\lambda f_x/n)^2 - (\lambda f_y/n)^2}, & \sqrt{f_x^2 + f_y^2} < \frac{n}{\lambda} \\ 0, & \text{otherwise} \end{cases}$$

where f_x and f_y are the spatial frequencies along x and y, respectively, and n is the refractive index of the medium.

Among various reconstruction methods used in digital holography literature [64–72], a simple yet effective phase-retrieval approach is chosen to reconstruct the microscopic images of cells (for both the amplitude and the phase profiles) and eliminate the twin-image artifacts introduced by our in-line hologram recording geometry [11,72]. With a finite support constraint defined around each object, this technique can iteratively recover the phase of the diffracted field from a single intensity image recorded by the sensor array. As a result, the complete complex field (both the amplitude and the phase components) of the cell holograms can be back-propagated to the sample plane for reconstruction of the object's image without twin-image contamination. This numerical method can be summarized as follows:

a. In the first iteration, the intensity of the recorded hologram is propagated back to the object plane by a distance of $-z_s$, assuming zero as the initial phase value for the complex field. Object support, S, is defined for each object either by thresholding the field amplitude at the object plane or by finding its local maxima/minima.

b. For the complex field at the object plane, $U^i_{-z_s}(x,y)$, the values outside the object support are substituted with a background value $D_{-z_s}(x,y)$, and the field inside the object support remains unchanged, i.e.:

$$U^i_{-z_s}(x,y) = \begin{cases} m \cdot D_{-z_s}(x,y), x, y \notin S \\ U^{i-1}_{-z_s}(x,y), x, y \in S \end{cases}$$

where $D_{-z_s}(x,y)$ is generated by back-propagating the square root of the background image (obtained in the absence of the objects) and $m = \text{mean}(U^{i-1}_{-z_s}(x,y))/\text{mean}(D_{-z_s}(x,y))$ is a normalization factor.

c. The modified field at the object plane is then propagated forward to the sensor plane, on which the updated phase of the sensor plane field remains unchanged but the amplitude is replaced with the square root of the original recorded hologram intensity. In other words, the measured amplitude of the sensor plane field will be kept constant throughout these iterations to ensure convergence of the phase. Accordingly, the complex diffraction field at the sensor plane after the ith iteration, $U_0^{(i)}(x,y)$, can be written as:

$$U_0^{(i)}(x,y) = |U_0^{(0)}(x,y)|\exp(\varnothing_0^{(i)}(x,y))$$

where the superscripts refer to the iteration step, and $\varnothing_0^{(i)}(x,y)$ denotes the phase of the field at sensor plane after the ith iteration. After this modification, the sensor plane complex field is again propagated backward to the object plane to get the updated object field.

By iterating between steps b and c, the phase values of the sensor field typically converge within 15–20 iterations. Once converged, the complex field is propagated backward to the object plane as the final output of the microscopic cell images containing both amplitude and phase profiles.

8.4 On-Chip Imaging of Sperm Using Partially Coherent Lensfree In-Line Holography

In this section, we are going to explain the implementation of our holographic on-chip imaging technique that is at the heart of the presented semen analysis platform and clarify the digital algorithms that this platform uses for automatic quantification of sperm density and motility.

As the key component of this compact semen analysis platform, a self-contained on-chip microscope (see Figure 8.1A) is designed to record the holographic images of semen samples over an FOV of $\sim$24 mm^2 with an effective NA of $\sim$0.1−0.2 without mechanical scanning [10,11]. Inside this compact microscope assembly (weighs $\sim$46 g and measures $\sim$4.2 cm $\times$ 4.2 cm $\times$ 5.8 cm, see Figure 8.1A), a simple light-emitting diode is butt-coupled to a 0.1 mm pinhole for casting partially coherent illumination over the semen sample, which is $\sim$4 cm away from the aperture plane (see Figure 8.1B). Inside the same assembly, a monochrome CMOS image sensor with an active area of 24 mm^2 records the lensless holograms of the sperm and then transfers these raw images through a USB 2.0 connection to a laptop computer, which could be replaced by a smartphone or personal digital assistant (PDA) for better mobility. The semen samples are dispensed into disposable counting chambers and then are loaded into this microscope with a sliding tray, which holds the chamber at $\sim$1 mm above the active area of the CMOS image sensor (see Figure 8.1A).

For performing automated sperm analysis with this lensless microscope, 20 consecutive lensless holographic frames are recorded for each semen sample at a frame rate of $\sim$2 frames per second. The integration time of each frame is set to <35 ms for minimizing the motion blur of motile sperm. In order to separately characterize the immotile and the motile sperm in the semen sample, two different processing approaches, that is, digital summation and subtraction of these lensless holographic frames, were applied, respectively.

For identification and counting of immotile sperm, all the individual holographic frames are first normalized and summed up digitally. This summation operation not only increases the signal-to-noise ratio (SNR) of the immotile sperm' holographic patterns but also smears out the patterns from the motile sperm. This critical step creates significant contrast improvement on the faint signatures of the sperm' tails and enables automated identification of the immotile sperm through their tail signatures (see Figure 8.2C and D).

Next, this summation hologram is processed by the iterative holographic reconstruction algorithm described in Section 8.3 to generate the microscopic amplitude and phase images of the immotile sperm (see Figure 8.2B and C). For automatic counting of immotile sperm, distinct and bright elliptical heads in the reconstructed phase images are initially isolated from the background as candidate patterns. The isolation process involves a thresholding operation, where pixels above a certain intensity value were grouped together, and a morphological screening process, where several properties of the connected regions such as pixel area, orientation, and circularity are analyzed [12]. After this initial head isolation step, the tail of each sperm also needs to be identified with its length, location, and orientation. Since the presence of a healthy tail is necessary for a viable sperm count according to the WHO laboratory manual [18], the tail in the reconstructed phase image

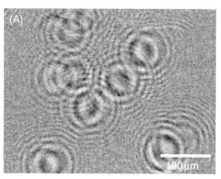 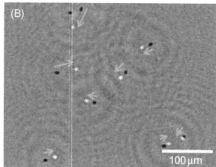

Figure 8.3
(A) A digitally subtracted lensless hologram shows the motion of eight sperms moving between two successive frames (500 ms apart). (B) A microscopic image is digitally reconstructed from the lensless differential hologram shown in (A) to illustrate the positions of eight sperms in two successive frames (black spots indicate the sperms' starting positions and white spots indicate their end positions). The displacement vectors of these sperms are labeled with green arrows. For interpretation of the references to color in this figure legend, the reader is referred to the web version of this book.

must have an adequate length and an orientation aligned with its head to qualify for a positive count. To ensure this, a determinant-of-Hessian filter [73] is applied to enhance the ridge-like features of sperm tails before matching the enhanced ridges to the location of the detected heads. If the length of a tail ridge is outside a typical range (20–60 μm) or its orientation is not aligned with the head, this candidate object will be discarded as negative.

Quantification of motile sperm in semen is a relatively easier task for our technique because motile sperm are the only moving objects within semen samples. Consecutive lensless holographic frames are digitally subtracted from each other to create differential holograms representing the displacement of the sperm (Figure 8.3A). These lensfree differential images are then reconstructed with the same reconstruction algorithm as discussed earlier. As shown in Figure 8.3B, the displacement of each sperm generates one negative (dark) and one positive (bright) spot in these reconstructed holographic images, which respectively represents the start and the end positions of the sperm's motion. As a result, we can simultaneously estimate the speed and the dynamic trajectories of all the motile sperm within our large FOV (~24 mm^2) by quantifying the relative distances between these bright and dark spots. To do this, the locations of these spots were first detected by simple thresholding and then refined by their centers of gravity [74]. After finding the nearest neighbors with opposite polarity for the bright/dark spots, the displacement of individual sperm between frames is recorded for each bright/dark spot pair (see Figure 8.3B). One can next link up these displacements of motile sperm in all consecutive frames and then plot out the dynamic trajectories of all the sperm within the imaging FOV (Figure 8.4A). The average speed of each motile sperm and the speed distribution of the whole population can

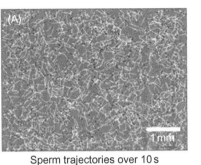
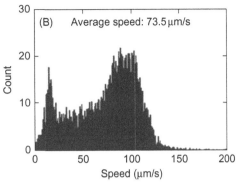

Sperm trajectories over 10 s
(FOV ~24 mm²)

Figure 8.4
(A) Dynamic trajectories of 1831 motile sperms within an FOV of ~24 mm² are automatically tracked over a time span of ~10 s. The yellow lines illustrate the trajectories of the tracked sperms, while the blue spots mark their end positions. (B) The sperm displacements from all the consecutive frames in (A) are summed up and then divided by the total image acquisition duration to provide the speed histogram. For interpretation of the references to color in this figure legend, the reader is referred to the web version of this book.

then be calculated through the magnitudes of these sperm displacements accumulated over the total duration of the frame acquisition, as shown in Figure 8.4B.

8.5 High-Throughput On-Chip Semen Analysis Results

The imaging capabilities of partially coherent lensless on-chip holography toward automated semen analysis can first be investigated through quantification of immobilized sperm. In our experiments, both amplitude and phase images of sperm samples were digitally reconstructed from their lensless holograms over an imaging FOV of ~24 mm², as detailed in Sections 8.3 and 8.4 [10]. Figure 8.2 shows a digitally focused region of this FOV with raw lensless holograms of a few sperm and their corresponding reconstruction results.

An interesting observation in these images is that, while the heads of the sperm can be clearly observed in both the amplitude and the phase images, the sperm tails were only visible in the reconstructed phase images (see Figure 8.2B and C). This is due to the fact that a sperm's tail is a submicron structure which has a relatively weak scattering property. Therefore, such a small structure cannot generate enough scattered field in the reconstructed amplitude image with the limited NA and SNR of our wide-field lensless microscope shown in Figure 8.1. However, a sufficient contrast in the reconstructed phase image can still be created by the refractive index difference between the tail and the surrounding medium, which permitted automated detection of the sperm tails shown in Figure 8.2C and D with the procedures described in Section 8.4. Once identified, the sperm tails in

Figure 8.2D were automatically matched to the orientations of their corresponding heads for quantifying healthy sperm. Through this orientation screening process, round cells or unhealthy sperm were rejected because they lacked matching tails or had unusual curvature on their tails (see Figure 8.2D). These automatic detection results were also validated with a regular microscope image taken on the same FOV, as shown in Figure 8.2E.

As an example of motile sperm quantification with our partially coherent lensless holography platform, Figures 8.3 and 8.4 show differential imaging and automated tracking results from a typical semen sample. Digital subtraction of consecutive lensfree frames took away all the stationary holograms and kept only the differential holograms of the motile sperm, as described in Section 8.4 and illustrated in Figure 8.3A. These differential holograms preserved the information about each motile sperm's displacement between these two consecutive frames, including its magnitude and direction. The holographic reconstruction on these differential holograms revealed two spots for each motile sperm: one dark spot indicating its start position and another bright one for its end position (see Figure 8.3B). Based on these reconstructed differential images, the dynamic trajectories of the motile sperm were quantified over the entire imaging FOV (~ 24 mm^2) as illustrated in Figure 8.4A. To determine the speed distribution of these sperm shown in Figure 8.4A, the displacements of individual motile sperm were linked across all the 20 lensless frames acquired within ~ 10 s and their time-averaged speed distribution was plotted out as a speed histogram as shown in Figure 8.4B. This average speed histogram provided by our lensfree imaging platform is essentially equivalent to the distribution of sperm straight-line velocity (also known as VSL) reported by commercial CASA systems. Since VSL has been reported to highly correlate with the success rate of in vitro fertilization [43], the average sperm speed provided by this platform should also be an effective indicator for male fertility.

Finally, the automated counting accuracy of this on-chip imaging platform should also be validated by comparing its results against manual counting results obtained with an optical microscope, which is still considered as one of the gold standards for semen analysis. Figure 8.5 compares the automatic counting results achieved with our lensless holographic microscope against the manual counting results provided with a conventional bright-field microscope. These two sets of results are based on the same 12 semen samples containing both immotile and motile sperm at various concentration levels. The results of this comparison verified that our automated on-chip semen analysis platform can accurately quantify a sperm density up to ~ 12 million sperm per ml. Such a large dynamic range permits this on-chip imaging platform to reliably analyze human semen samples with a dilution factor of 1–10-fold. Additionally, Figure 8.5 also points to this platform's superior ability on automatic analysis of semen samples with very low sperm concentrations. For example, the lowest concentration in Figure 8.5B, that is, 0.09 million sperm per ml, was defined by 42 sperm tracked by this platform over its 24 mm^2 FOV. With a conventional microscope equipped with a $20\times$ objective lens, one would need to look at almost 100

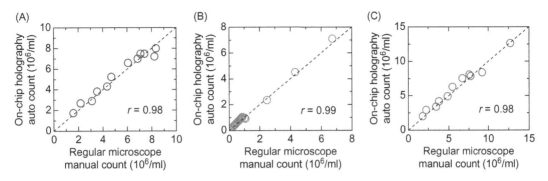

Figure 8.5

Counting accuracy of the presented automated semen analysis platform is verified at various sperm concentrations up to 12.5×10^6/ml for (A) immotile sperms; (B) motile sperms; and (C) both the motile and the immotile sperms. The manually counted sperm concentrations using a conventional bright-field microscope are plotted along the x-axes, while the sperm concentrations automatically counted by the presented lensless semen analysis platform for the same semen samples are plotted along the y-axes. The total counts in (C) are the summed concentrations of the immotile sperms in (A) and the motile sperms in (B). Correlation coefficients (r) of these characterization results shown in (A), (B), and (C) further confirm the accuracy of our semen analysis platform that is based on a compact and lightweight holographic lensless microscope.

different FOVs to maintain the same statistical characterization accuracy for such a low density semen sample. Therefore, applications requiring quantitative results with extremely low sperm densities, such as post-vasectomy checkups [75,76], would greatly benefit from the large FOV of this lensless on-chip imaging platform.

Potential application areas of this lensfree on-chip holography-based automatic semen analysis platform include high-throughput motility analysis in andrology laboratories, automated sperm counting in fertility clinics, semen quality evaluation in sperm banks, personal male fertility tests at home, post-vasectomy tests either at home or in clinics, stud performance assessment at animal breeding centers, and field monitoring of stud health in farms. Additionally, by replacing the compact CMOS image sensor with a larger area explained in p0070. chip [13], this platform can also be transformed into a drug screening system with significantly higher throughput, where sperm' response to various drugs or chemicals of different concentrations can be analyzed simultaneously. Furthermore, this platform can also be extended into the third dimension with additional light sources illuminating from multiple angles [74], hence the vertical position and the 3D trajectory of each sperm could be resolved with submicron accuracy to better understand sperm dynamics in less constrained space. On the top of these, this optical technology with its automated analysis capability is also able to characterize cells within bodily fluids such as whole blood samples [13,77] and detect bacteria/parasites in drinking water [12,14]. Therefore, the compact, lightweight, and cost-effective design of this imaging platform and its connectivity to PDAs or smartphones may allow the users to perform various microanalysis tasks in field settings with minimal resources.

8.6 Conclusions

In this chapter, we reviewed the fundamental principles of partially coherent lensless on-chip holographic microscopy and its application toward automated semen analysis. With a weight of ~46 g and dimensions of ~4.2 cm × 4.2 cm × 5.8 cm, such a wide-field lensless on-chip microscope fills up an important gap between simple qualitative male fertility test kits and sophisticated quantitative CASA systems. Being fully automated with digital image processing algorithms, this platform can perform comprehensive semen analysis tasks including measuring the absolute concentrations of both motile and immotile sperm, as well as resolving the trajectories and the speed distributions of motile sperm within the sample. Since factors in addition to sperm concentration are now required to accurately predict male fertility [78], such a compact and versatile semen analysis tool based on wide-field lensless on-chip microscopy could be especially important for andrology laboratories, fertility clinics, personal male fertility tests, as well as for field use in veterinary medicine such as in stud farming and animal breeding applications.

Acknowledgment

A. Ozcan gratefully acknowledges the support of NSF (CAREER Award on BioPhotonics), the Office of Naval Research under the Young Investigator Award 2009 and the NIH Director's New Innovator Award—Award Number DP2OD006427 from the Office of The Director, National Institutes of Health. The authors also acknowledge the support of the Okawa Foundation, Vodafone Americas Foundation, and NSF BISH program (under Awards #0754880 and 0930501).

References

[1] Beyond the diffraction limit, Nat. Photonics, 3 (2009) 361.
[2] W.R. Zipfel, R.M. Williams, W.W. Webb, Nonlinear magic: multiphoton microscopy in the biosciences, Nat. Biotech. 21 (2003) 1369–1377.
[3] C.L. Evans, X. Xu, S. Kesari, X.S. Xie, S.T.C. Wong, G.S. Young, Chemically-selective imaging of brain structures with CARS microscopy, Opt. Express 15 (2007) 12076–12087.
[4] R.P.J. Barretto, B. Messerschmidt, M.J. Schnitzer, In vivo fluorescence imaging with high-resolution microlenses, Nat. Methods 6 (2009) 511–512.
[5] K. Goda, K.K. Tsia, B. Jalali, Serial time-encoded amplified imaging for real-time observation of fast dynamic phenomena, Nature 458 (2009) 1145–1149.
[6] W. Choi, C. Fang-Yen, K. Badizadegan, S. Oh, N. Lue, R.R. Dasari, et al., Tomographic phase microscopy, Nat. Methods 4 (2007) 717–719.
[7] A. Chung, S. Karlan, E. Lindsley, S. Wachsmann-Hogiu, D.L. Farkas, In vivo cytometry: a spectrum of possibilities, Cytometry A 69A (2006) 142–146.
[8] R.T. Krivacic, A. Ladanyi, D.N. Curry, H.B. Hsieh, P. Kuhn, D.E. Bergsrud, et al., A rare-cell detector for cancer, Proc. Natl. Acad. Sci. U.S.A. 101 (2004) 10501–10504.

[9] K. Chung, M.M. Crane, H. Lu, Automated on-chip rapid microscopy, phenotyping and sorting of *C. elegans*, Nat. Methods 5 (2008) 637–643.

[10] T.-W. Su, A. Erlinger, D. Tseng, A. Ozcan, Compact and light-weight automated semen analysis platform using lensfree on-chip microscopy, Anal. Chem. 82 (2010) 8307–8312.

[11] O. Mudanyali, D. Tseng, C. Oh, S.O. Isikman, I. Sencan, W. Bishara, et al., Compact, light-weight and cost-effective microscope based on lensless incoherent holography for telemedicine applications, Lab. Chip. 10 (2010) 1417–1428.

[12] S. Seo, T.-W. Su, D.K. Tseng, A. Erlinger, A. Ozcan, Lensfree holographic imaging for on-chip cytometry and diagnostics, Lab. Chip. 9 (2009) 777–787.

[13] S. Seo, S.O. Isikman, I. Sencan, O. Mudanyali, T.-W. Su, W. Bishara, et al., High-throughput lens-free blood analysis on a chip, Anal. Chem. 82 (2010) 4621–4627.

[14] O. Mudanyali, C. Oztoprak, D. Tseng, A. Erlinger, A. Ozcan, Detection of waterborne parasites using field-portable and cost-effective lensfree microscopy, Lab. Chip. 10 (2010) 2419–2423.

[15] S.O. Isikman, I. Sencan, O. Mudanyali, W. Bishara, C. Oztoprak, A. Ozcan, Color and monochrome lensless on-chip imaging of *Caenorhabditis elegans* over a wide field-of-view, Lab. Chip. 10 (2010) 1109–1112.

[16] L. Björndahl, D. Mortimer, C.L.R. Barratt, J.A. Castilla, R. Menkveld, U. Kvist, et al., A Practical Guide to Basic Laboratory Andrology, 1st Ed., Cambridge University Press, New York, 2010.

[17] R.H. Foote, The history of artificial insemination: selected notes and notables, J. Anim. Sci. 80 (2002) 1–10.

[18] World Health Organization, WHO Laboratory Manual for the Examination of Human Semen and Sperm–Cervical Mucus Interaction, fourth ed., Cambridge University Press, Cambridge, UK, 1999.

[19] J.E. Sokoloski, L. Blasco, B.T. Storey, D.P. Wolf, Turbidimetric analysis of human sperm motility, Fertil. Steril. 28 (1977) 1337–1341.

[20] R.W. Atherton, F.L. Jackson, G. Bond, E.W. Radany, R.M. Kitchin, K. Polakoski, A correlation between a spectrophotometric quantitation of rabbit spermatozoan motility and velocity, Arch. Androl. 3 (1979) 301–308.

[21] R.W. Atherton, E.W. Radany, K.L. Polakoski, Spectrophotometric quantitation of mammalian spermatozoon motility I. Human, Biol. Reprod. 18 (1978) 624–628.

[22] R.M. Levin, S.H. Greenberg, A.J. Wein, Clinical use of the turbidimetric analysis of sperm motility: comparison with visual techniques, Fertil. Steril. 35 (1981) 332–336.

[23] R.M. Levin, J.A. Hypolite, A.J. Wein, Clinical use of the turbidimetric analysis of sperm motility: an update, Andrologia 16 (1984) 434–438.

[24] B.E. Morton, R. Sagadraca, Quantitation of sperm population migration: capillary scanning assay, Arch. Androl. 7 (1981) 219–227.

[25] R.M. Levin, J. Shofer, S.H. Greenberg, A quantitative method for determining the effects of drugs on spermatozoal motility, Fertil. Steril. 33 (1980) 631–635.

[26] W. Halangk, R. Bohnensack, Quantification of sperm motility by a turbidimetric assay. Correlation to cellular respiration, Biomed. Biochim. Acta 45 (1986) 331–341.

[27] P. Jouannet, B. Volochine, P. Deguent, C. Serres, G. David, Light scattering determination of various characteristic parameters of spermatozoa motility in a series of human sperm, Andrologia 9 (1977) 36–49.

[28] G.P. Naylor, J.S. Martin, E.N. Chantler, Apparatus for the study of motile sperm using microprocessor analysis of scattered laser light, Med. Biol. Eng. Comput. 20 (1982) 207–214.

[29] T. Craig, F.R. Hallett, B. Nickel, Motility analysis of circularly swimming bull spermatozoa by quasi-elastic light scattering and cinematography, Biophys. J. 38 (1982) 63–70.

[30] M.W. Woolford, J.D. Harvey, Light-scattering studies of bull spermatozoa. II. Interaction and concentration effects, Biophys. J. 40 (1982) 7–16.

[31] R. Rigler, P. Thyberg, Rotational and translational swimming of human spermatozoa: a dynamic laser light scattering study, Cytometry 5 (1984) 327–332.

[32] J.C. Earnshaw, G. Munroe, W. Thompson, A.I. Traub, Automated laser light scattering system for assessment of sperm motility, Med. Biol. Eng. Comput. 23 (1985) 263–268.

[33] J. Frost, H. Cummins, Motility assay of human sperm by photon correlation spectroscopy, Science 212 (1981) 1520–1522.

[34] M.C. Wilson, J.D. Harvey, Twin-beam laser velocimeter for the investigation of spermatozoon motility, Biophys. J. 41 (1983) 13–21.

[35] M.C. Wilson, J.D. Harvey, P. Shannon, Aerobic and anaerobic swimming speeds of spermatozoa investigated by twin beam laser velocimetry, Biophys. J. 51 (1987) 509–512.

[36] S.T. Mortimer, CASA—practical aspects, J. Androl. 21 (2000) 515–524.

[37] S. Mortimer, A critical review of the physiological importance and analysis of sperm movement in mammals, Hum. Reprod. Update 3 (1997) 403–439.

[38] A. Agarwal, R.K. Sharma, Automation is the key to standardized semen analysis using the automated SQA-V sperm quality analyzer, Fertil. Steril. 87 (2007) 156–162.

[39] M.J. Tomlinson, K. Pooley, T. Simpson, T. Newton, J. Hopkisson, K. Jayaprakasan, et al., Validation of a novel computer-assisted sperm analysis (CASA) system using multitarget-tracking algorithms, Fertil. Steril. 93 (2010) 1911–1920.

[40] L. Ramió, M.M. Rivera, A. Ramírez, I.I. Concha, A. Peña, T. Rigau, et al., Dynamics of motile-sperm subpopulation structure in boar ejaculates subjected to "in vitro" capacitation and further "in vitro" acrosome reaction, Theriogenology 69 (2008) 501–512.

[41] L. Maree, S.S. du Plessis, R. Menkveld, G. van der Horst, Morphometric dimensions of the human sperm head depend on the staining method used, Hum. Reprod. 25 (2010) 1369–1382.

[42] K. Coetzee, A. de Villiers, T.F. Kruger, C.J. Lombard, Clinical value of using an automated sperm morphology analyzer (IVOS), Fertil. Steril. 71 (1999) 222–225.

[43] D.Y. Liu, G.N. Clarke, H.W. Baker, Relationship between sperm motility assessed with the Hamilton-thorn motility analyzer and fertilization rates in vitro, J. Androl. 12 (1991) 231–239.

[44] C.L. Barratt, M.J. Tomlinson, I.D. Cooke, Prognostic significance of computerized motility analysis for in vivo fertility, Fertil. Steril. 60 (1993) 520–525.

[45] Y.L. Kuo, W.L. Tzeng, H.K. Chiang, R.F. Ni, T.C. Lee, S.T. Young, New system for long-term monitoring of sperm motility: EDTA effect on semen, Arch. Androl. 41 (1998) 127–133.

[46] D.R.J. Glenn, C.M. McVicar, N. McClure, S.E.M. Lewis, Sildenafil citrate improves sperm motility but causes a premature acrosome reaction in vitro, Fertil. Steril. 87 (2007) 1064–1070.

[47] D.M. McCurnin, J.M. Bassert, Clinical Textbook for Veterinary Technicians, fifth ed., Philadelphia, US, 2002.

[48] C.F. Shipley, Breeding soundness examination of the boar, Swine Health Prod. 7 (1999) 117–120.

[49] R.M. Turner, Current techniques for evaluation of stallion fertility, Clin. Tech. Equine Pract. 4 (2005) 257–268.

[50] The BabyStart Male Infertility Test (FertilMARQ), (accessed 22.08.11).

[51] SpermCheck® Fertility, (accessed 22.08.11).

[52] L.I. Segerink, A.J. Sprenkels, P.M. ter Braak, I. Vermes, A. van den Berg, On-chip determination of spermatozoa concentration using electrical impedance measurements, Lab. Chip. 10 (2010) 1018–1024.

[53] N.T. Shaked, L.L. Satterwhite, N. Bursac, A. Wax, Whole-cell-analysis of live cardiomyocytes using wide-field interferometric phase microscopy, Biomed. Opt. Express 1 (2010) 706–719.

[54] L. Granero, V. Micó, Z. Zalevsky, J. García, Superresolution imaging method using phase-shifting digital lensless Fourier holography, Opt. Express 17 (2009) 15008–15022.

[55] P. Marquet, B. Rappaz, P.J. Magistretti, E. Cuche, Y. Emery, T. Colomb, et al., Digital holographic microscopy: a noninvasive contrast imaging technique allowing quantitative visualization of living cells with subwavelength axial accuracy, Opt. Lett. 30 (2005) 468–470.

[56] A. Khmaladze, R.L. Matz, C. Zhang, T. Wang, M.M. Banaszak Holl, Z. Chen, Dual-wavelength linear regression phase unwrapping in three-dimensional microscopic images of cancer cells, Opt. Lett. 36 (2011) 912–914.

[57] B. Kemper, G. von Bally, Digital holographic microscopy for live cell applications and technical inspection, Appl. Opt. 47 (2008) A52–A61.

[58] L. Miccio, A. Finizio, R. Puglisi, D. Balduzzi, A. Galli, P. Ferraro, Dynamic DIC by digital holography microscopy for enhancing phase-contrast visualization, Biomed. Opt. Express 2 (2011) 331–344.

[59] N.T. Shaked, Y. Zhu, M.T. Rinehart, A. Wax, Two-step-only phase-shifting interferometry with optimized detector bandwidth for microscopy of live cells, Opt. Express 17 (2009) 15585–15591.

[60] C.A. Sciammarella, F.M. Sciammarella, Measurement of mechanical properties of materials in the micrometer range using electronic holographic moireé, Opt. Eng. 42 (2003) 1215–1222.

[61] B. Rappaz, F. Charrière, C. Depeursinge, P.J. Magistretti, P. Marquet, Simultaneous cell morphometry and refractive index measurement with dual-wavelength digital holographic microscopy and dye-enhanced dispersion of perfusion medium, Opt. Lett. 33 (2008) 744–746.

[62] F. Dubois, C. Yourassowsky, N. Callens, C. Minetti, P. Queeckers, T. Podgorski, et al., Digital holographic microscopy working with a partially spatial coherent source, in: P. Ferraro, A. Wax, Z. Zalevsky (Eds.), Coherent Light Microscopy, Springer, Berlin, Heidelberg, 2011, pp. 31–59.

[63] D.J. Brady, Optical Imaging and Spectroscopy, John Wiley & Sons, Inc., Hoboken, NJ, 2009.

[64] W.S. Haddad, D. Cullen, J.C. Solem, J.W. Longworth, A. McPherson, K. Boyer, et al., Fourier-transform holographic microscope, Appl. Opt. 31 (1992) 4973–4978.

[65] W. Xu, M.H. Jericho, I.A. Meinertzhagen, H.J. Kreuzer, Digital in-line holography for biological applications, Proc. Natl. Acad. Sci. U.S.A. 98 (2001) 11301–11305.

[66] G. Pedrini, H.J. Tiziani, Short-coherence digital microscopy by use of a lensless holographic imaging system, Appl. Opt. 41 (2002) 4489–4496.

[67] L. Repetto, E. Piano, C. Pontiggia, Lensless digital holographic microscope with light-emitting diode illumination, Opt. Lett. 29 (2004) 1132–1134.

[68] J. Garcia-Sucerquia, W. Xu, M.H. Jericho, H.J. Kreuzer, Immersion digital in-line holographic microscopy, Opt. Lett. 31 (2006) 1211–1213.

[69] G. Situ, J.T. Sheridan, Holography: an interpretation from the phase-space point of view, Opt. Lett. 32 (2007) 3492–3494.

[70] G.C. Sherman, Application of the convolution theorem to Rayleigh's integral formulas, J. Opt. Soc. Am. 57 (1967) 546–547.

[71] G. Koren, F. Polack, D. Joyeux, Iterative algorithms for twin-image elimination in in-line holography using finite-support constraints, J. Opt. Soc. Am. A. 10 (1993) 423–433.

[72] J.R. Fienup, Reconstruction of an object from the modulus of its Fourier transform, Opt. Lett. 3 (1978) 27–29.

[73] T. Lindeberg, Edge detection and ridge detection with automatic scale selection, Int. J. Comput. Vis. 30 (1998) 117–156.

[74] T.-W. Su, S.O. Isikman, W. Bishara, D. Tseng, A. Erlinger, A. Ozcan, Multi-angle lensless digital holography for depth resolved imaging on a chip, Opt. Express 18 (2010) 9690–9711.

[75] A. Chawla, Vasectomy follow-up: clinical significance of rare nonmotile sperm in postoperative semen analysis, Urology 64 (2004) 1212–1215.

[76] N.B. Dhar, A. Bhatt, J.S. Jones, Determining the success of vasectomy, BJU Int. 97 (2006) 773–776.

[77] W. Bishara, U. Sikora, O. Mudanyali, T.-W. Su, O. Yaglidere, S. Luckhart, et al., Holographic pixel super-resolution in portable lensless on-chip microscopy using a fiber-optic array, Lab. Chip. 11 (2011) 1276–1279.

[78] S.E.M. Lewis, Is sperm evaluation useful in predicting human fertility? Reproduction 134 (2007) 31–40.

CHAPTER 9

Synthetic Aperture Lensless Digital Holographic Microscopy (SALDHM) for Superresolved Biological Imaging

Vicente Micó[1], Zeev Zalevsky[2], Luis Granero[3], Javier García[1]

[1]Departamento de Óptica, Universitat de Valencia, Burjassot, Spain [2]School of Engineering, Bar-Ilan University, Ramat-Gan, Israel [3]AIDO: Instituto Tecnológico de Óptica, Color e Imagen, Parc Tecnològic, Paterna, Spain

Editor: Natan T. Shaked

9.1 Introduction

Holography was invented by Dennis Gabor in 1948 when proposing a scheme to improve image quality in electron microscopy by avoiding lenses in the experimental setup since, at that time, resolution in electron microscopes were generally limited by spherical aberration of magnetic electron lenses [1–3]. In a few words, Gabor's concept implies the addition, in a given plane, of an imaging wave caused by diffraction at the sample plane and a reference wave incoming from the nondiffracted light passing through the sample. The resulting light distribution contains information about the interference of both waves and it is recorded on a photographic film. The reconstruction is obtained by illuminating the film with a reconstruction wave which is identical to the reference wave that was used in the recording process. The diffracted electron field of the sample is reconstructed without spherical aberrations by a suitable arrangement of lenses. This two-step holographic electron microscopy method is unique but suffered from three major drawbacks: the reconstructed image is affected by coherent noise, the twin image problem also affects the final image quality, and a restricted sample range due to the weak diffraction assumption (samples must be essentially transparent) is needed for preserving the holographic behavior of the method (otherwise, the sample excessively blocks the reference beam and diffraction rules the recording process, preventing the accurate recovery of the sample's complex wave front).

The holographic principle proposed by Gabor was immediately applied to the visible range by Rogers [4]. However, its applicability was limited due to the low coherence of the available light sources in the visible spectrum yielding holograms with a low quality interference contrast. It was with the invention of the laser [5] when Leith and Upatnieks alleviated some of the drawbacks provided by the Gabor's concept and extended the applicability of holography. Leith and Upatnieks [6–8] abandoned the in-line configuration proposed by Gabor and externally reinserted the reference beam at the recording plane in an angle (off-axis geometry). The off-axis holographic method resolved the twin image problem since in this configuration the actual and twin conjugate images were separated in space after the reconstruction. At the same years, Denisyuk [9,10] proposed reflection holography where signal and reference waves strike the photographic plate from opposite sides, which also suppressed the conjugate image. Also in 1965, Stroke [11] reported on a new way to record a different type of hologram named lensless Fourier transform (FT) hologram. Here, the reference beam diverges from the same plane where the object is placed, and the reconstruction requires a focusing lens or a mirror system to provide the Fourier transformation of the recorded hologram. Finally, Gabor and Goss [12] reported on an interference microscope in which the reference beam is externally inserted in on-axis geometry and two quadrature phase-shifted interferograms are recorded in a single photographic plate. This was the first evidence of phase-shifting holography.

From its first evidence [13] till now [14], electronic image recording devices (typically a Charge-Coupled Device (CCD) or a Complementary Metal Oxide Semiconductor (CMOS) camera) have replaced holographic recording media, resulting in a new technology named digital holography (DH). DH avoids chemical processing and other time-consuming procedures ascribed to classical holography, and enables numerous capabilities from the numerical post-processing of the recorded hologram carried out digitally. Just as an example, DH has mirrored classical holographic approaches such as the Gabor's approach [15,16], implementations based on the Leith and Upatnieks' setups [17,18], as well as Fourier holographic architectures [19,20], and phase-shifting algorithms [21,22]. However, although most of the remarkable properties of DH were previously known [23], their practical implementations have been restricted because of the strong requirements for the computer's capabilities concerning digital image acquisition and processing.

One of the remarkable applications of DH is provided in the field of digital in-line holographic microscopy (DIHM) [15,16,24]. In DIHM, an electronic device records the in-line hologram incoming from the superposition of a reference beam (the portion of the propagating spherical wave that travels without being diffracted by the sample) and an imaging beam (the portion of the propagating wave which is diffracted by the sample). By "in-line" we mean that both reference and imaging beams are generated along the same propagation direction and travel together along almost the same optical path. In other words, DIHM supposes a practical implementation of the Gabor's concept in the visible range enhanced with numerical processing for retrieving the full 3D sample information. Several applications have risen up from this kind of microscopy allowing 3D imaging with micrometer resolution [25]. Some examples include underwater observations [26–28], tracking of moving objects and particles [29,30], and study of erosion processes in coastal sediments [31,32].

As in conventional microscopy, resolution in DIHM is a combination of several connected parameters. First, the wavelength of the illumination source used in the setup affects the resulted resolution since resolution (R) is proportional to that value ($R \sim \lambda/\text{NA}$, being λ the illumination wavelength and NA the numerical aperture). Second, the pinhole diameter (playing the role of both the illumination diaphragm and the condenser lens in the illumination module of a microscope) controls the spatial coherence, the useful amount of light and the divergence of the illumination beam. Third, the NA (classically defined by the microscope objective) is determined in DIHM by both the transversal dimensions of the electronic recording device and the sample-to-detection plane distance. To optimize the experimental setup in terms of light efficiency, the pinhole diameter is adjusted to provide a cone of light according with the NA defined by the electronic device. Typical achievable NA values in DIHM are in the range of 0.4 [33]. And fourth, the characteristics of the electronic recording device (number of pixels, pixel size, and dynamic range) affect the sampling of the interferometric fringes and the noise level in the recorded in-line hologram.

For fixed illumination wavelength, there are two key points in DIHM when trying to achieve high NA, and thus a low spatial resolution limit. On one hand, the first one concerns the definition of a high optical magnification in the experimental setup as a requirement to circumvent the limitation imposed by the sampling incoming from the pixel size of the electronic device. Since magnification (M) in DIHM can be defined as the geometrical projection at the detection plane of the sample being illuminated from the pinhole, a highly magnified diffraction pattern will be recorded by controlling the distance between the pinhole and the sample (z_S) which will be much lower than the distance between the sample and the detection plane (z_D) in the form of $M = (z_S + z_D)/z_S$. Typical values for z_S and z_D are below 1 mm and in the cm range, respectively, and a high M value can be easily fulfilled. Nevertheless, the high magnification implies a restriction proportional to M^2 in the reconstructed object field of view provided by the method.

On the other hand, the second point is related to the digital processing of the recorded hologram and becomes the hardest task in DIHM, being nowadays basically resolved thanks to the optimum power of available computers. Once the in-line hologram of the sample is recorded (sample hologram: U_S), the sample is removed from the setup and another in-line hologram is stored at the computer's memory only considering the illumination beam (reference hologram: U_R). Then, a synthetic hologram (U'_S) is derived from the previously recorded holograms in the form of $U'_S = (U_S - U_R)/U_R^{1/2}$. In addition, a coordinate transformation must be applied to the synthetic hologram to avoid geometric distortion when recording holograms at high NA (outside paraxial approximation). This procedure is crucial for obtaining optimal reconstruction incoming from the numerical propagation of the synthetic hologram U'_S to focus the sample.

Another different strategy that can be conducted in DIHM was reported by the research group of Prof. Ozcan at the UCLA and named lensless on-chip holographic microscopy (LOHM) [34]. Briefly, instead of placing the illumination pinhole close to the sample to achieve high magnification, the sample in LOHM is placed on the top of the electronic recording device defining typical values for z_S and z_D as between 5 and 10 cm and around 1 mm, respectively [35]. This configuration defines a magnification factor close to 1, so the resolution limit becomes restricted by the pixel size of the electronic device limiting the minimum period of the interferometric fringes that can be sampled by the method. Although it is theoretically possible to have a value of NA which is close to 1 or even higher, in practice the typical values of NA are around 0.2 since the sample is placed very close to the recording device and the space between them is filled with several materials (plastic covers and glass protective layers) having a refractive index higher than 1. This modest NA value also limits the spatial resolution provided by LOHM. However, the entire area of the electronic sensor is dedicated for imaging purposes, enabling a wide object field of view in comparison with the one provided in DIHM. LOHM has been applied in several applications such as cytometry [36] and blood analysis [37], fluorescent imaging [38,39] and DIC microscopy [40], portable applications [41,42] as well as for detection of waterborne parasite [43] and semen analysis [44]. Chapter 8 explains the LOHM principle and presents the semen analysis results.

However, both DIHM and LOHM lack high values of NA, especially the latter one. In order to improve the NA value in DIHM, Garcia-Sucerquia et al. [45] replaced the air gap between the sample and the CCD sensor by a high refractive index medium. This strategy directly increases the NA value since the refractive index of the surrounding medium explicitly defines the NA. NA improvement from 0.39 to 0.55 was validated. Instead of modifying the experimental setup, Kanka et al. [46–48] have proposed several approaches to provide high-contrast high-resolution images in DIHM based on improvements in the numerical reconstruction procedure. Using this methodology, Kanka et al. [46–48] have demonstrated imaging capabilities in DIHM incoming from NA values between 0.6 and 0.7. Concerning LOHM, Ozcan et al. [49–51] also have demonstrated improvements in the NA value in an approach based on the concept of multi-angle illumination and pixel superresolution, where the NA value has been approximately enhanced from 0.2 to 0.5 [50] and to 0.4 [51].

In the current chapter, we report on a different direction for improving the NA value in lensless holographic microscopy. The proposed method, named synthetic aperture lensless digital holographic microscopy (SALDHM), provides the generation of a synthetic aperture (SA) that expands the spatial cutoff frequency of the imaging system and, thus, its spatial resolution limit, in comparison with the case where no SA enlargement is considered. The SA generation is based on the combination of angular- and time-multiplexing incoming from tilted beam illumination (angular-) sequentially performed (time-) in order to multiplex different object's spatial regions. This strategy has been successfully validated to improve the NA value in DHM [52] and now it is applied to LDHM.

The proposed SALDHM concept is validated using two different approaches. The first one concerns the case of imaging essentially transparent (low concentration dilutions) and static (slow dynamics) samples [53]. In other words, it is applied in the Gabor's regime where holography rules the electronic recording process. However, as it was stated in Ref. [24], diffraction dominates the process preventing an accurate recovery of the object's complex wave front when the amount of light blocked by the object, for high density samples, is significant. For these kinds of samples, a different strategy must be used to permit imaging capabilities in LDHM. The use of spatial light modulators in combination with a phase-shifting algorithmic is a possibility [54,55]. Moreover, time multiplexing SA generation incoming by CCD shift at the recording plane has also been reported [56].

Nevertheless, the classical way to circumvent the Gabor's concept is by reinserting an external reference beam at the recording plane in a way similar to the one reported by Leith and Upatnieks [6–8]. In that case, a beam splitter usually is placed in front of the digital recording device (typically a CCD camera) is usually placed to allow a reference beam for holographic recording. The inclusion of a beam splitter in the setup prevents the sample from being placed close to the CCD camera (at least at a distance equal to the face length of the beam splitter cube). This fact, in addition to the restricted CCD sensor size, limits the experimental NA value that can be achieved. Thus, our second reported approach operates in this situation of working for any type of sample due to the external addition of a reference beam [57].

In both presented approaches [53,57], preliminary tests and calibrations are performed using a synthetic object (1951 USAF resolution test target) in order to quantify the resolution improvement. Then, the approaches are implemented on biological specimens (sperm cells and red blood cells, respectively).

9.2 SALDHM Inside the Gabor's Regime

In this section, we demonstrate the generation of an SA having a high SNA value in DIHM incoming from time and angular multiplexing of the sample's spectrum. The experimental setup is depicted in Figure 9.1 for the two cases, (A) on-axis and (B) off-axis illuminations. Essentially, a divergent spherical wave illuminates the sample and a CCD device records

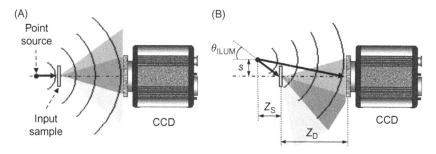

Figure 9.1
Experimental setup for SALDHM in the Gabor's regime: (A) on-axis illumination case and (B) a possible off-axis illumination case, where the main system's parameters are defined on the figure.

the geometrically magnified diffraction pattern (Figure 9.1A). The illumination emanates from a point source that is both closely placed to the sample and assembled onto a translation stage in order to provide high magnification factor and to shift it to the off-axis positions, respectively. The key point is to have an NA of the illumination higher than the NA defined by the imaging system (defined by the position and size of the CCD sensor).

Thus, when on-axis illumination is used (Figure 9.1A), the experimental setup provides the recording of an in-line hologram of the sample having a given NA, magnification, and resolution depending on the system's parameters. To further improve the NA value, we propose the implementation of SA generation strategies. In this approach, the point source is shifted to several off-axis positions in order to provide different tilted beam illuminations over the input sample. The shifting distance (s) of the point source is then adjusted with both the distance z_S and the NA of the imaging system (NA_{CCD}) in order to provide an oblique illumination angle (θ_{ILUM}) that doubles the one provided by NA_{CCD}. Obviously, the point source NA must be high enough to allow this condition. In this case, the object becomes obliquely illuminated and will on-axis diffract; in addition an additional spatial-frequency range of the sample's spectrum will be captured. This new spectral content interferes with the nondiffracted light of the illumination beam that passes through the sample without being distorted since we are under Gabor's assumptions (Figure 9.1B). Such reference beam will arrive at the CCD plane forming an angle with the optical axis. Then, to satisfy Nyquist sampling criterion in the digitally recorded hologram, the s distance must fulfill the paraxial condition given by $s_m \leq \lambda(z_S + z_D)/2p$, where p is the pixel size of the CCD device and s_m is the maximum distance by which the point source can be shifted.

In order to generate an SA having a synthetic cutoff frequency as high as possible, the recovered elementary apertures must be contiguous at the Fourier domain. Since the CCD is typically rectangular, the setup must be designed to provide contiguous elementary apertures in the shorter direction and, thus, overlapping in the larger one. This fact can easily be fulfilled by either placing the CCD at a given distance from the input sample so that the angle defined between the optical axis and the shorter CCD direction (typically the vertical one) will be half the oblique illumination angle (θ_{ILUM}), or by selecting θ_{ILUM} to be twice the angle defined by the NA in the shorter CCD direction while satisfying the Nyquist condition imposed by the s distance. In any case, a contiguous frequency band will be diffracted on-axis when vertical tilted illumination impinges on the input object. Thus, the new vertical cutoff frequency becomes expanded three times or, equivalently, the resolution is improved by a factor of 3 in the vertical Fourier direction. However, when applying the proposed method in other directions, the resolution gain will be slightly lower than 3 because the CCD is normally wider in the horizontal than in the vertical axis.

After angular multiplexing the object's spectrum with different sequential tilted beam illuminations, each recorded hologram is numerically processed to synthesize the expanded SA. The digital manipulation implies, first, a coordinate transformation to remove the nonlinear geometrical distortion incoming from the violation of the Fresnel approximation [15,20]. Second, an input plane numerical back propagation can be achieved by implementing the convolution approach in the Rayleigh–Sommerfeld equation [20,54–56]. Third, a correct reallocation of the spectral content of each bandpass image while generating the SA [57]. Finally, the SR image is obtained by FT of the information contained in the SA.

In order to validate the proposed method experimentally, we present two experiments. In the first one, we propose the use of a Blu-ray source as illumination light ($\lambda = 0.405$ μm) and a positive high-resolution 1951 USAF test target as input object to calibrate the setup. Note that the test target must be positive (clear background) instead of negative (opaque background) to fulfill the Gabor's regime. A CCD camera (Basler A312f, 582 × 782 pixels, 8.3 pixel size, 12 bits/pixel) is used to record the in-line holograms. A commercial grade standard microscope lens (DIN 40×, 0.65NA) is used to provide point source divergent high NA illumination in the setup. The illuminator (Blu-ray source with the microscope lens) is placed onto a motorized linear translation stage to allow off-axis displacement. The point-source-test and point-source-CCD distances are $z_S = 400$ and $z_D = 11$ mm, respectively, so the magnification of the setup is $M = 28.5$. With this M value, a sample's detail of 1 μm occupies 28.5 μm at the CCD plane and becomes properly sampled by the 8.3 μm size CCD pixel.

The point source is shifted 175 μm (s distance) from the optical axis to allow two off-axis recordings for the vertical (V) and horizontal (H) orthogonal directions, respectively. Note that this distance is lower than the one defined by the Nyquist criterion ($s_m = 0.268$ μm) and it is the proper distance to transmit quasi-contiguous spectral content in the shorter (V) CCD direction. As a result, the NA of the tilted illumination is $NA_{ILUM} = 0.4$. Since the USAF test target mainly has diffracted components in the horizontal (H) and vertical (V) directions, we have only considered off-axis illuminations into the H and V planes. Figure 9.2 images the central part of both the four recovered bandpass images when tilted

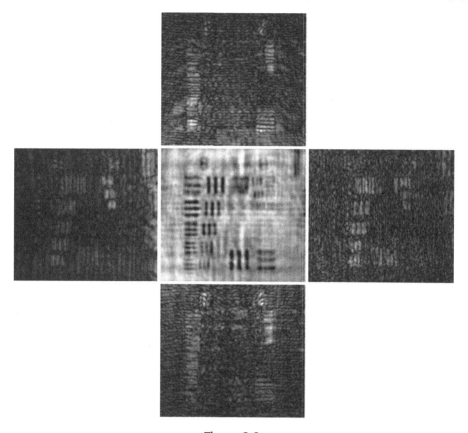

Figure 9.2
Recovered bandpass images of the proposed SALDHM incoming from on-axis (central image) and off-axis (four outer images) illuminations.

beam illumination is performed and after applying the numerical reconstruction procedure (outer images), and the conventional low-resolution image incoming from the use of on-axis illumination (inner image). According to the experimental setup, the NA values for the V and H directions are: $NA_{VER} = 0.21$ and $NA_{HOR} = 0.28$. These NA values define resolution limits given by $R_{VER} = 1.93$ μm and $R_{HOR} = 1.45$ μm. As seen in the central image in Figure 9.2, the last resolved elements in the test target for the V and H directions are Group 9-Element 1 (named as G9-E1 from now on) and G9-E3, respectively, corresponding with test target's periods of 1.95 μm (512 lp/mm) and 1.55 μm (645 lp/mm), respectively. Those experimental values perfectly agree with the theoretical ones.

Then, the four off-axis bandpass images are joined together with the on-axis one to synthesize a superresolved image. In other words, an SA is generated by coherent addition of the five individual rectangular apertures. Figure 9.3A and B depicts the conventional rectangular aperture when using on-axis illumination and the generated SA, respectively.

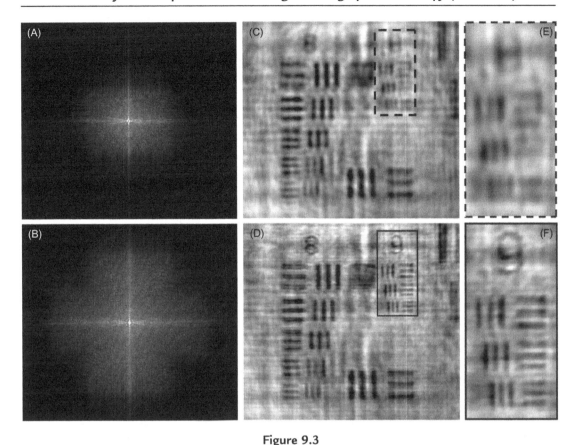

Figure 9.3
Positive 1951 USAF test target results using SALDHM: (A,B) the conventional and the SA, respectively; (C,D) the conventional and the superresolved image, respectively. (E,F) Magnified image of the finest test elements of cases (C) and (D), which clearly shows the resolution improvement.

Cases (C) to (F) compare the low-resolution conventional image ((C) and (E)) with the superresolved image ((D) and (E)) obtained by FT of the information contained in images (A) and (B), respectively. We have included again the low-resolution image for easier comparison with the superresolved one. We can see that all the elements in the Group 9 of the test target are now perfectly resolved in the superresolved image. In addition, the number 9 representative of the Group 9 of the test target can be also resolved in the superresolved image while it looks as an eight more than as a nine in the conventional image. According to the theory, the new superresolution (SR) limits are $SR_{VER} = 0.66$ μm and $SR_{HOR} = 0.59$ μm incoming from $SNA_{VER} = NA_{ILUM} + NA_{VER} = 0.61$ and $SNA_{HOR} = NA_{ILUM} + NA_{HOR} = 0.68$, respectively.

In our second experiment, SALDHM is applied to a biological sample. The sample is an unstained swine sperm that is dried up allowing fixed sperm cells for the experiments. The

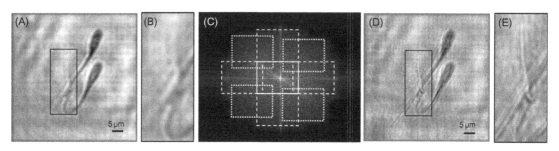

Figure 9.4
Imaging swine sperm cells using SALDHM: (A,B) the conventional low-resolution image and a magnification area of the sperm's tail, respectively; (C) the generated SA incoming from the addition of eight off-axis apertures plus the on-axis one and where the nine elementary apertures are outlined with white lines for clarity. (D,E) The superresolved image and a magnification area of the sperm's tail (the same one as in (B)), respectively.

sperm cells have an ellipsoidal head of 6 μm × 9 μm and a tail's width of around 2 μm on the head side and below 1 μm on its end. Figure 9.4 shows the experimental results. Now, not only four off-axis elementary apertures are considered but eight ones as we can see in the generated SA (Figure 9.4C). Thus, full 2D spatial-frequency space is covered in the generated SA. We observe that the thinner part of the tail which is not visible under conventional DIHM (Figure 9.4A and B) becomes resolved after applying the proposed SALDHM approach (Figure 9.4D and E). In addition, we see in Figure 9.4C that the elementary apertures in the V direction are quasi-contiguous with the central one, thus satisfying the maximum expansion condition for the SA generation.

9.3 SALDHM Outside the Gabor's Regime

When the sample is not weakly diffractive, an external reference beam must be added at the recording plane to allow holographic recording. In this case, our proposed experimental setup is depicted in Figure 9.5A and B for the on-axis and off-axis illumination cases, respectively. It is based on a Mach-Zehnder interferometric configuration in which a He–Ne laser beam is split in two branches. In the first one (imaging branch), the object under test is illuminated in transmission mode and its diffracted Fresnel pattern recorded by a CCD. This diffracted pattern is combined at the CCD plane with a second beam incoming from the reference branch by using a beam splitter cube. The reference beam is an off-axis spherical divergent wave front having the particularity that the distance "d" between the object and the CCD (see Figure 9.5) is equal to the distance between the reference point source (focal plane of the Fourier transforming lens in Figure 9.5) and the CCD. This configuration defines a digital lensless Fourier transforming holographic setup where the entire information about the complex diffracted wave front coming from the input object

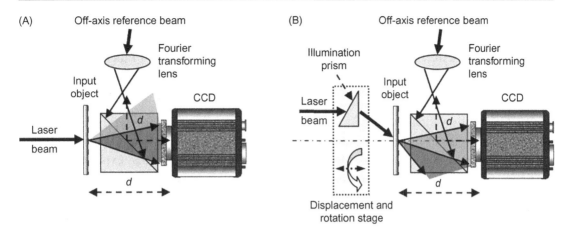

Figure 9.5
Experimental setup for SALDHM out of the Gabor's regime: (A) on-axis illumination case and (B) one possible off-axis illumination case.

can be recovered at the Fourier domain of the recorded hologram since off-axis recording avoids the overlapping between the different holographic diffraction orders.

Using this experimental assembly, the FT of the recorded hologram performs the recovery of the focused bandpass image of the object under test. When on-axis illumination is used (Figure 9.5A), the focused image at the Fourier domain becomes the conventional image provided by the proposed imaging system architecture. That image will be limited in resolution due to the low NA that is defined by the imaging system and which results from both the reduced CCD size and the distance between the input object and the CCD. When angular multiplexing incoming from the tilted beam illumination (Figure 9.5B) is considered, a different bandpass image of the input sample will be accessible at the Fourier domain. The illumination stage is composed from a prism that can be shifted and rotated to provide sequential oblique illumination onto the input sample, thus, the shift allows tilted beam illumination at a given angle (defined by the deviation angle of the prism) and the rotation allows 2D spatial-frequency space coverage around a circle where the deviation angle of the prism is constant. Let us call this circle an illumination ring. Thus, once on-axis illumination is considered, the wedge prism is shifted to off-axis positions and rotated according to such positions in order to illuminate obliquely the input object. This illumination procedure permits the sequential recovery of different spectral bands of the object's spectrum when performing digital FT of every off-axis recorded hologram and a filtering process over one of the diffraction hologram orders. Thus, different rectangular elementary apertures containing a different spatial-frequency range of the object's spectrum are recovered in time sequence. By providing tilted beams around the illumination ring, 2D Fourier space coverage is ensured. This additional information can be properly managed to generate an SA allowing superresolved imaging by a simple digital FT of the information contained in the SA. Figure 9.6 depicts the SA generation incoming from the addition of the elementary apertures. In addition, we have also considered a second illumination ring incoming from a higher value of the oblique illumination angle which can be implemented by considering a wedge prism having a higher deviation angle.

To experimentally validate this second SALDHM approach, we first present initial calibration using a negative USAF resolution test target. Here, the test target must be negative instead of positive in order to break the Gabor's regime. For that experiment, we have used an He—Ne laser (632 nm emitting wavelength) as our illumination source, a CCD (Basler A312f, 582 × 782 pixels, 8.3 μm pixel size, 12 bits/pixel) as our imaging device, a wedge prism with 10° deviation angle as the illumination prism for the first illumination ring ($NA_{ILUM} = 0.17$), a beam splitter cube (20 mm × 20 mm side, BK7) to allow holographic recording, and a doublet lens (80 mm focal length and 60 mm diameter) to provide a spherical divergent off-axis reference beam. Optic mounts, beam expanders, neutral density filter wheels to equalize the beam aspect ratio, and micrometric translation stages complete the experimental setup.

The test target is placed at 36 mm in front of the CCD but the presence of a beam splitter cube brings its image through it closer to the CCD. Under paraxial approximation, the displacement Δs originated by a plano-parallel plate is given by $\Delta s = e(1 - 1/n)$, where e and n are the width and the refractive index of the plate, respectively. Since the refractive

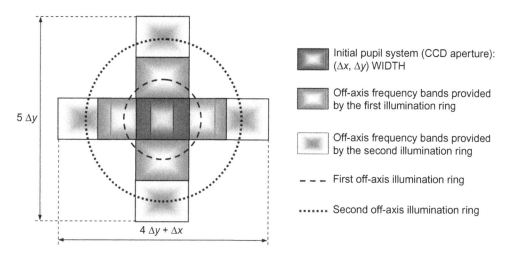

Figure 9.6
Hypothetical SA incoming from the addition of eight off-axis elementary apertures plus the on-axis one. The off-axis apertures are provided by two illumination rings incoming from two different values of the prism's deviation angle. ($\Delta x, \Delta y$) are the width and height, respectively, of the CCD elementary aperture.

index of BK7 at the He—Ne wavelength is 1.515, the displacement introduced by the beam splitter cube is around $\Delta s = 6.6$ mm. Thus, the effective distance between the test target and the CCD is 29.4 mm. Obviously, the reference beam is also forced to diverge from a distance equal to 29.4 mm in front of the CCD in order to get imaging at the Fourier domain.

Using this effective distance and the theoretical specifications of the CCD, the NA and the resolution limits in the H and V directions are $NA_{HOR} = 0.11$ and $NA_{VER} = 0.08$, and $R_{HOR} = 5.74$ μm (174 lp/mm) and $R_{VER} = 7.72$ μm (130 lp/mm), respectively. Figure 9.7 shows the recorded hologram, its FT, and a magnified image of the region of interest when on-axis illumination is used. We can see as the smallest resolved details (marked with white arrows in (C)) have a size of 6.21 μm (G7-E3 with 161 lp/mm) and 7.81 μm (G7-E1 with 128 lp/mm) in the H (vertical bars) and V (horizontal bars) directions, respectively. Although these values are a little bit lower than the theoretical ones, they are in good concordance with the theoretical values since the following element in the resolution test is below the diffraction limit (G7-E4 and G7-E2 have a details size of 181 and 144 lp/mm, respectively). Additionally, the illuminated object's field of view must be limited. Otherwise, the twin image should affect the real image and vice versa. In our case, we have used a 2D square aperture to limit the extension of the illumination beam at the input plane.

Then, we perform the proposed superresolution approach by shifting and rotating the illumination prism to the off-axis positions. Also, the double slit is moved together with the prism to maintain the field of view limitation. To provide tilted beam illumination, the incident laser beam is expanded by using a pinhole in combination with a collimation lens. Thus, incident laser light is obtained over the whole set of off-axis positions where the illumination prism is moved. Similarly to Figure 9.2, Figure 9.8A shows both the four recovered bandpass images when tilted beam illumination is performed (outer images) and

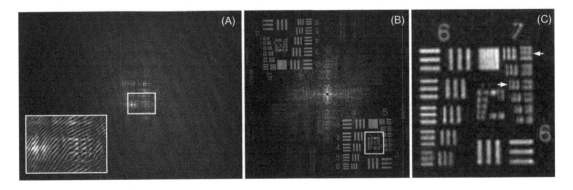

Figure 9.7
Negative 1951 USAF test target results using SALDHM: (A) recorded hologram with the inset showing the fringes; (B) FT of (A). (C) Magnification of the test region of interest (marked with a white rectangle in (B)).

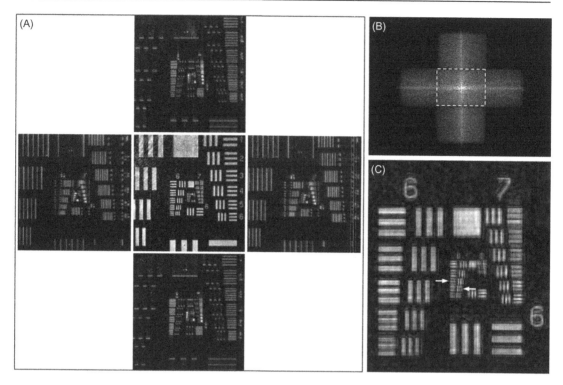

Figure 9.8
(A) Recovered bandpass images of the proposed SALDHM incoming from on-axis (central image) and off-axis (four outer images) illuminations: (B) generated SA and (C) superresolved image. The off-axis illuminations are provided by the first illumination ring.

the conventional low-resolution image incoming from the use of on-axis illumination (inner image). The SA and the superresolved image are depicted in Figure 9.8B and C, respectively. According with theoretical calculations, the SNA and the superresolution limits in H and V directions are given by $SNA_{HOR} = NA_{HOR} + NA_{ILUM} = 0.28$ and $SNA_{VER} = NA_{VER} + NA_{ILUM} = 0.25$, and $SR_{HOR} = 2.26\ \mu m$ (442 lp/mm) and $SR_{VER} = 2.53\ \mu m$ (395 lp/mm), respectively. From Figure 9.8C, we can see that the last resolved elements (marked with a white arrow) are G8-E5 (406 lp/mm) and G8-E4 (362 lp/mm) in H and V, respectively. Once again, these experimental values agree with the theoretical ones since the next elements in the USAF test are below the superresolved diffraction limit: G8-E6 (456 lp/mm) and G8-E5 (406 lp/mm) for H and V directions, respectively. With these values, the resolution gain factors are 2.5 and 3 for both H and V directions, respectively.

In addition, we present a second experiment with the USAF test target to demonstrate improved performance. By considering a second illumination ring at twice the oblique illumination angle of the first ring, it is possible to expand up the synthetic cutoff frequency by

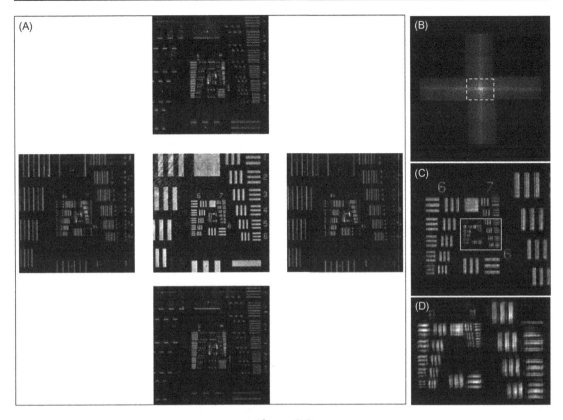

Figure 9.9
(A) Recovered bandpass images of the proposed SALDHM incoming from on-axis (central image) and off-axis (four outer images) illuminations: (B) generated SA, (C) superresolved image, and (D) magnified area of the inner part marked with a solid white line rectangle in case (C). The off-axis illuminations are provided by the second illumination ring.

addition of a new rectangular pupil. This second illumination ring is produced by combining two identical wedge prisms like in a beam steering application. With this procedure, a set of tilted beams having an oblique illumination angle of 20° are produced over the input plane defining our second illumination ring. The new set of bandpass images (outer images) provided by the second illumination ring is included in Figure 9.9A, in addition again with the one obtained using the on-axis illumination (inner image). The SA incoming from the addition of the eight off-axis elementary apertures plus the on-axis one is presented in Figure 9.9B. And finally, the superresolved image is included in Figure 9.9C while a magnification of its central part is depicted in Figure 9.9D. We can see that the three last elements of the test target are resolved in both H and V directions for the superresolved image.

According to the theory, the SNA and the superresolution limits in H and V directions are given by $SNA_{HOR} = 0.45$ and $SNA_{VER} = 0.42$, and $SR_{HOR} = 1.41$ μm (709 lp/mm) and

$SR_{VER} = 1.51$ μm (664 lp/mm), respectively, taken into account that now the NA of the illumination is $NA_{ILUM} = 0.34$. These values are good enough to resolve the last USAF elements defined by G9-E3 (1.55 μm or 645 lp/mm). With these new values, the resolution gain factors in H and V directions are around 4 and 5, respectively, when comparing with the conventional imaging mode.

Our second experimental study for the case outside Gabor's regime uses fixed human red blood cells as our input sample. Red blood cells are stained using a specially prepared mixture of methylene blue and eosin in methanol (Wright stain). Now, SA generation must cover the whole spatial-frequency plane by considering the addition of elementary apertures in all the possible directions (0−45−90−135−180−225−270−315° tilted illuminations) since the sample contains useful information on every Fourier plane direction. Once again, the theoretical values defined by the experimental setup are the same ones as previously reported for the USAF test target case ($NA_{HOR} = 0.11$ and $NA_{VER} = 0.08$, and $R_{HOR} = 5.74$ μm and $R_{VER} = 7.72$ μm).

Figure 9.10A shows the low-resolution image provided by on-axis illumination with the previous values. We can see that the image presents higher resolution in the H rather than the V direction in respect to rectangular size of CCD. Moreover, since the red blood cells are around 6−7 μm in diameter, they are not imaged in the V direction but can be barely seen in the limit for the H one. Figure 9.10B depicts the generated SA by coherent addition of the eight off-axis rectangular pupils where the conventional rectangular aperture is marked with a dashed white rectangle for clarity and Figure 9.10C depicts the superresolved image.

Note that, in some cells, a faint black point appears in the center of the cells. This black point is originated by deviation of the light at the center of the cells due to the typical donut-like shape of the red blood cells. According to the theory, the SNA values and the superresolution limits are again those presented in the USAF validation, that is: $SNA_{HOR} = 0.28$, $SNA_{VER} = 0.25$, $SNA_{OBLIQUE} = 0.285$, and $SR_{HOR} = 2.26$ μm, $SR_{VER} = 2.53$ μm, $SR_{OBLIQUE} = 2.22$ μm, values that enable image formation of the cells.

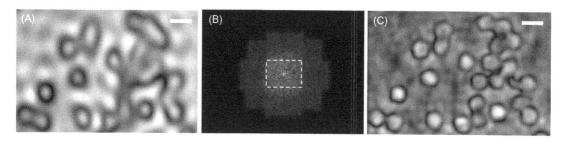

Figure 9.10
Imaging red blood cells using SALDHM. (A) The conventional low-resolution image. (B) The generated SA incoming from the addition of eight off-axis apertures plus the on-axis one. (C) The superresolved image. Scale bars in (A) and (C) are 10 μm.

9.4 Conclusions

In this chapter, we have described and experimentally validated two different approaches capable of improving resolution in biological LDHM by SA generation (SALDHM). Both approaches are based on the combination of angular- and time-multiplexing incoming from tilted beam illumination (angular-) sequentially performed (time-) in order to multiplex different object's spatial regions. Since holography rules the recording of the images, the proposed multiplexing allows the recovery of different elementary apertures containing different spatial-frequency content of the sample's spectrum. Those elementary apertures are coherently added into the expanded SA and a superresolved image is obtained as FT of the information contained in the SA. As a consequence of the temporal multiplexing, the main drawback of both approaches is limitation to samples that must be static or having a slow variation in time (a few seconds).

In the first presented approach, weak diffractive samples are considered. The experimental setup becomes extremely simple since we are inside the Gabor's regime and verifies the definition of an SNA above 0.6 in every direction. The full procedure can be realized in a few seconds by automating the displacement of the source to the off-axis positions and synchronizing the capture of the holograms. We have calibrated and quantified the method using standard 1951 USAF resolution test target achieving a submicron resolution limit. Later on, we have reported on the application of the proposed approach to imaging of a sperm cell to imaging of a sperm cell demonstrating demonstrating its feasibility for the analysis of biological samples. The proposed approach can be applied for submicron resolution lensless imaging purposes when considering essentially transparent (low confluence) samples.

On the contrary, any type of samples can be analyzed using the second presented approach since the reference beam is externally introduced assuring the holographic recording at the CCD plane. The experimental setup can be easily adapted to any system's requirements, which means a versatile, simple, and customized method to improve the resolution in digital lensless Fourier holographic microscopy imaging. We have presented experimental results using a single or two illumination rings, allowing 0.45, 1.41 µm, and 5 as best values for SNA, superresolution limit, and resolution gain factor, respectively.

References

[1] D. Gabor, A new microscopic principle, Nature 161 (1948) 777–778.
[2] D. Gabor, Microscopy by reconstructed wavefronts, Proc. R. Soc. Lond. A197 (1949) 454–487.
[3] D. Gabor, Microscopy by reconstructed wavefronts II, Proc. Phys. Soc. Lond. B64 (1951) 449–469.
[4] G.L. Rogers, Experiments in diffraction microscopy, Proc. R. Soc. Edinb. 63A (1952) 193–221.
[5] A.L. Schawlow, C.H. Townes, Infrared and optical masers, Phys. Rev. 112 (1958) 1940–1949.
[6] E.N. Leith, J. Upatnieks, Reconstructed wavefronts and communication theory, J. Opt. Soc. Am. 52 (1962) 1123–1130.

[7] E.N. Leith, J. Upatnieks, Wavefront reconstruction with continuous tone objects, J. Opt. Soc. Am. 53 (1963) 1377–1381.

[8] E.N. Leith, J. Upatnieks, Wavefront reconstruction with diffused illumination and three-dimensional objects, J. Opt. Soc. Am. 54 (1964) 1295–1301.

[9] Y.N. Denisyuk, On the reproduction of the optical properties of an object by the wave field of its own scattered radiation, Opt. Spect. 15 (1963) 279–284.

[10] Y.N. Denisyuk, On the reproduction of the optical properties of an object by the wave field of its own scattered radiation II, Opt. Spect. 18 (1965) 152–157.

[11] G.W. Stroke, Lensless Fourier-transform method for optical holography, Appl. Phys. Lett. 6 (1965) 201–203.

[12] D. Gabor, W.P. Goss, Interference microscope with total wavefront reconstruction, J. Opt. Soc. Am. 56 (1966) 849–858.

[13] J.W. Goodman, R.W. Lawrence, Digital image formation from electronically detected holograms, Appl. Phys. Lett. 11 (1967) 77–79.

[14] M.K. Kim, Principles and techniques of digital holographic microscopy, SPIE Rev. 1 (2010) 018005-1-50.

[15] W. Xu, M.H. Jericho, I.A. Meinertzhagen, H.J. Kreuzer, Digital in-line holography for biological applications, Proc. Natl. Acad. Sci. U.S.A. 98 (2001) 11301–11305.

[16] L. Repetto, E. Piano, C. Pontiggia, Lensless digital holographic microscope with light-emitting diode illumination, Opt. Lett. 29 (2004) 1132–1134.

[17] U. Schnars, Direct phase determination in hologram interferometry with use of digitally recorded holograms, J. Opt. Soc. Am. A 11 (1994) 2011–2015.

[18] E. Cuche, P. Marquet, C. Depeursinge, Spatial filtering for zero-order and twin-image elimination in digital off-axis holography, Appl. Opt. 39 (2000) 4070–4075.

[19] W.S. Haddad, D. Cullen, J.C. Solem, J.W. Longworth, A. McPherson, K. Boyer, et al., Fourier-transform holographic microscope, Appl. Opt. 31 (1992) 4973–4978.

[20] Y. Takaki, H. Ohzu, Fast numerical reconstruction technique for high-resolution hybrid holographic microscopy, Appl. Opt. 38 (1999) 2204–2211.

[21] A. Decker, Y. Pao, P. Claspy, Electronic heterodyne recording and processing of optical holograms using phase modulated reference waves, Appl. Opt. 17 (1978) 917–921.

[22] I. Yamaguchi, T. Zhang, Phase-shifting digital holography, Opt. Lett. 22 (1997) 1268–1270.

[23] P. Hariharan, Optical Holography: Principles, Techniques, and Applications, second ed., Cambridge University Press, New York, 1996.

[24] W. Xu, M.H. Jericho, I.A. Meinertzhagen, H.J. Kreuzer, Digital in-line holography of microspheres, Appl. Opt. 41 (2002) 5367–5375.

[25] J. Garcia-Sucerquia, W. Xu, S.K. Jericho, P. Klages, M.H. Jericho, H.J. Kreuzer, Digital in-line holographic microscopy, Appl. Opt. 45 (2006) 836–850.

[26] J. Watson, S. Alexander, G. Craig, D.C. Hendry, P.R. Hobson, R.S. Lampitt, et al., Simultaneous in-line and off-axis subsea holographic recording of plankton and other marine particles, Meas. Sci. Technol. 12 (2001) L9–L15.

[27] P.R. Hobson, J. Watson, The principles and practice of holographic recording of plankton, J. Opt. A: Pure Appl. Opt. 4 (2002) S34–S49.

[28] S.K. Jericho, J. Garcia-Sucerquia, W. Xu, M.H. Jericho, H.J. Kreuzer, Submersible digital in-line holographic microscope, Rev. Sci. Instrum. 77 (2006) 043706-1-10.

[29] G. Pedrini, H.J. Tiziani, Short-coherence digital microscopy by use of a lensless holographic imaging system, Appl. Opt. 41 (2002) 4489–4496.

[30] W. Xu, M.H. Jericho, I.A. Meinertzhagen, H.J. Kreuzer, Tracking particles in 4-D with in-line holographic microscopy, Opt. Lett. 28 (2003) 164–166.

[31] H. Sun, R.G. Perkins, J. Watson, M.A. Player, D.M. Paterson, Observations of coastal sediment erosion using in-line holography, J. Opt. A: Pure Appl. Opt. 6 (2004) 703–710.

[32] E. Malkiel, J.N. Abras, J. Katz, Automated scanning and measurements of particle distributions within a holographic reconstructed volume, Meas. Sci. Technol. 15 (2004) 601–612.

[33] J. Garcia-Sucerquia, D.C. Alvarez-Palacio, M.H. Jericho, H.J. Kreuzer, Comment on "Reconstruction algorithm for high-numerical-aperture holograms with diffraction-limited resolution", Opt. Lett. 31 (2006) 2845–2847.

[34] The Ozcan Research Group at the University of Carolina, Los Angeles (UCLA). < http://innovate.ee.ucla.edu/welcome.html>

[35] A. Ozcan, U. Demirci, Ultra wide-field lens-free monitoring of cells on-chip, Lab. Chip. 8 (2008) 98–106.

[36] S. Seo, T. Su, D.K. Tseng, A. Erlinger, A. Ozcan, Lensfree holographic imaging for on-chip cytometry and diagnostics, Lab. Chip. 9 (2009) 777–787.

[37] S. Seo, S.O. Isikman, I. Sencan, O. Mudanyali, T. Su, W. Bishara, et al., High-throughput lensfree blood analysis on a chip, Anal. Chem. 82 (2010) 4621–4627.

[38] A.F. Coskun, T. Su, A. Ozcan, Wide field-of-view lens-free fluorescent imaging on a chip, Lab. Chip. 10 (2010) 824–827.

[39] H. Zhu, O. Yaglidere, T. Su, D. Tseng, A. Ozcan, Cost-effective and compact wide-field fluorescent imaging on a cell-phone, Lab. Chip. 11 (2011) 315–322.

[40] C. Oh, S.O. Isikman, B. Khademhosseini, A. Ozcan, On-chip differential interference contrast microscopy using lensless digital holography, Opt. Express 18 (2010) 4717–4726.

[41] D. Tseng, O. Mudanyali, C. Oztoprak, S.O. Isikman, I. Sencan, O. Yaglidere, et al., Lensfree microscopy on a cell-phone, Lab. Chip 10 (2010) 1787–1792.

[42] S.O. Isikman, W. Bishara, U. Sikora, O. Yaglidere, J. Yeah, A. Ozcan, Field-portable lensfree tomographic microscope, Lab. Chip 11 (2011) 2222–2230.

[43] O. Mudanyali, C. Oztoprak, D. Tseng, A. Erlinger, A. Ozcan, Detection of waterborne parasites using field-portable and cost-effective lensfree microscopy, Lab. Chip 10 (2010) 2419–2423.

[44] T. Su, A. Erlinger, D. Tseng, A. Ozcan, Compact and light-weight automated semen analysis platform using lensfree on-chip microscopy, Anal. Chem. 82 (2010) 8307–8312.

[45] J. Garcia-Sucerquia, W. Xu, M.H. Jericho, H.J. Kreuzer, Immersion digital in-line holographic microscopy, Opt. Lett. 31 (2006) 1211–1213.

[46] M. Kanka, R. Riesenberg, H.J. Kreuzer, Reconstruction of high-resolution holographic microscopic images, Opt. Lett. 34 (2009) 1162–1164.

[47] M. Kanka, A. Wuttig, C. Graulig, R. Riesenberg, Fast exact scalar propagation for an in-line holographic microscopy on the diffraction limit, Opt. Lett. 35 (2010) 217–219.

[48] A. Wuttig, M. Kanka, H.J. Kreuzer, R. Riesenberg, Packed domain Rayleigh–Sommerfeld wavefield propagation for large targets, Opt. Express 18 (2010) 27036–27047.

[49] T. Su, S.O. Isikman, W. Bishara, D. Tseng, A. Erlinger, A. Ozcan, Multi-angle lensless digital holography for depth resolved imaging on a chip, Opt. Express 18 (2010) 9690–9711.

[50] W. Bishara, T.W. Su, A.F. Coskun, A. Ozcan, Lensfree on-chip microscopy over a wide field-of-view using pixel super-resolution, Opt. Express 18 (2010) 11181–11191.

[51] W. Bishara, H. Zhu, A. Ozcan, Holographic opto-fluidic microscopy, Opt. Express 18 (2010) 27499–27510.

[52] V. Micó, Z. Zalevsky, J. García, Optical superresolution: imaging beyond Abbe's diffraction limit, J. Hologr. Speckle 5 (2009) 110–123.

[53] V. Micó, Z. Zalevsky, Superresolved digital in-line holographic microscopy for high resolution lensless biological imaging, J. Biomed. Opt. 15 (2010) 046027-1-5.

[54] V. Micó, J. García, Z. Zalevsky, B. Javidi, Phase-shifting Gabor holography, Opt. Lett. 34 (2009) 1492–1494.

[55] V. Micó, J. García, Common-path phase-shifting lensless holographic microscopy, Opt. Lett. 35 (2010) 3919–3921.

[56] V. Micó, L Granero, Z. Zalevsky, J. García, Superresolved phase-shifting Gabor holography, J. Opt. A: Pure Appl. Opt. 11 (2009) 125408-1-6.

[57] L. Granero, V. Micó, Z. Zalevsky, J. García, Synthetic aperture superresolved microscopy in digital lensless Fourier holography by time and angular multiplexing of the object information, Appl. Opt. 49 (2010) 845–857.

CHAPTER 10

Combining Digital Holographic Microscopy with Microfluidics: A New Tool in Biotechnology

Lisa Miccio[1], Pasquale Memmolo[1], Francesco Merola[1], Melania Paturzo[1], Roberto Puglisi[2], Donatella Balduzzi[2], Andrea Galli[2], Pietro Ferraro[1]

[1]*Istituto Nazionale di Ottica del CNR—UOS di Napoli, Pozzuoli (NA) Italy*
[2]*Istituto Sperimentale Italiano "Lazzaro Spallanzani," Località La Quercia, Cremona, Italy*

Editor: Zeev Zalevsky

10.1 Introduction

In recent years, biomicrofluidic technology has grown significantly [1,2] resulting in high demand for advancements in multifunctional methods for observing processes, for their characterization, and for manipulation of particles in microfluidic environments. Numerous methods have been suggested and implemented for qualitative imaging, phase contrast quantitative analysis [3–13], handling and trapping [14–16], and for the accurate tracking of the paths of nano/microparticles (i.e., nanodrops, carbon nanotubes, biocells, quantum dots, dielectric spheres, and for metallic spheres) in microfluidic lab-on-a-chip devices [17–21].

Optical tweezers have facilitated a rapid growth in particle manipulation approaches [22–24]. Particles can be trapped or moved along certain trajectories or relocated by using appropriate intensity field distribution of the single laser beam. Exploiting Gaussian or Laguerre–Gaussian laser beams modes, the particles can be induced to move in straight or curved trajectories. Thanks to the advent of spatial light modulators, which work as programmable diffraction optical elements, multiparticle trapping has become a reality [24–26]. Optical tweezers allow the study of the properties of the fluid in which the beads are embedded based on the dynamic position of the particle as it is driven along a definite path. Some approaches exist for tracking single or multiple trapped particles [27–30]. Numerous techniques achieve the tracking analysis with high accuracy using a single particle while others do so by means of statistical localization algorithms [34,35]. Moreover, techniques for tracking of micro-objects need to be developed for measuring velocity fields, trajectory patterns, motility of cancer cell, and so forth [3,36].

Analysis of microfluidic environments requires methods of investigation and measurement that are noncontact, full-field, and non-invasive. Consequently, optical and photonics techniques are most suitable. In this context, phase contrast imaging for lab-on-a-chip devices allows quantitative information to be obtained in situ. Digital holography (DH), among others, allows both retrieval of the phase contrast quantitative map. Numerical reconstruction of digital hologram allows, in addition to quantitative optical path length mapping and synthetic differential image contrast imaging [37] that is also as occurs in classical microscopes. DH flexibility permits optimization of phase contrast which improves, for example, visualization of cells in both static and dynamic conditions. In Ref. [38], tracking of individual colloidal particles undergoing three dimensional motions with nanometer resolution can be obtained by an in-line holographic microscope. Real-time tracking of optically trapped micrometer-sized particles with microsecond time resolution has also been achieved [39]. A digital shearing technique to extract three-component velocity in particle image velocimetry also has been demonstrated.

Recently simultaneous trapping and tracking experiments have been performed, but, in order to achieve this double function, two coupled lasers systems have been adopted [38,40,41].

In this chapter, we present a completely new concept: a compact holographic microscope that ensures multifunctionality, by simultaneous accurate 3D tracking, optical manipulation, and quantitative phase contrast analysis in a single configuration. Experimental results are presented and discussed for cells grown in vitro in a microfluidic device. The system is very simple and compact and is based on twin laser beams coming from a single laser source. Through this simple conceptual design we show how two different functionalities can be accomplished by the same optical setup, that is, 3D tracking of a micro-object and quantitative phase contrast imaging.

10.2 Drive and Analyze

In this section, we demonstrate that particles floating in microfluidic environment [1] can be driven by optical forces along preferential directions [42]. Particles are analyzed using a holographic microscope setup to generate quantitative phase contrast maps [43–52]. Consequently, this method allows examination of a large number of particles individually by continuos recording of digital hologram video. This allows examination of each particle at a time in a microfluidic environment. In addition, using the optical trap, particles can be moved to specific locations to facilitate analysis, by analogy with a photographer selecting a specific pose. The intriguing feature is that the same laser source is used for two functions, termed "drive and analyze" [53]. The intriguing feature here is that the same laser source is used for making two functions: drive and analyze [53]. To do this, two laser beams are used from a single laser source. The two beams are slightly noncollinear, differing from each other by a small angle. The interference between the two laser beams generates a sequence of digital holograms while one of the beams creates the driving force. Methods and experimental results using latex spheres and living cells in vitro are reported in later sections.

10.2.1 Trapping Theory

Here, an experiment is shown to prove trapping and driving operations. Latex particles (diameter $D_p = 9.7$ μm) dispersed in a liquid medium are employed. If $D_p > 10\lambda$, a ray approach can be applied, neglecting diffraction effects. Since the net momentum must be conserved in the process, the change in the momentum generates forces that can be divided into two types: the scattering and the gradient force. The gradient force, F_{grad}, is proportional to the gradient of the laser beam's intensity. As the gradient force points toward the high-intensity region, it acts as a force that attracts the particle. On the contrary, the scattering force, F_{scat}, acts like a repulsive force and points in the direction of propagation of the ray. The two forces can be expressed by the following equations [42],

$$F_{scat} = \frac{n_0}{2c} \int_0^{2\pi} \int_0^{\pi/2} I(r,z) \left[1 + R\cos 2\theta - T^2 \frac{\cos 2(\theta - \vartheta) + R\cos 2\theta}{1 + R^2 + 2R\cos 2\vartheta} \right]$$
$$\times \left(\frac{D_p}{2} \right)^2 \sin 2\theta \, d\theta \, d\Theta \quad (10.1)$$

$$F_{grad} = -\frac{n_0}{2c} \int_0^{2\pi} \int_0^{\pi/2} I(r,z) \left[R\sin 2\theta - T^2 \frac{\sin 2(\theta - \vartheta) + R\sin 2\theta}{1 + R^2 + 2R\cos 2\vartheta} \right]$$
$$\times \left(\frac{D_p}{2} \right)^2 \sin 2\theta \cos\Theta \, d\theta \, d\Theta \quad (10.2)$$

where n_0 is the refractive index of the medium, c is the speed of light, and r is the radial offset of the sphere from the Gaussian beam center axis. Moreover, z is the axial distance from the minimum beam waist while R and T are the Fresnel's coefficients of reflectance and transmittance, respectively, and θ is the incident angle of the photon stream with respect to the normal direction of the sphere surface. $\vartheta = \sin^{-1}(n_0/n_1)\sin\theta$ is the refraction angle obtained by Snell's law, n_1 being the refractive index of the particle, and Θ is the polar angle. Beam intensity profile, $I(r, z)$, is given by:

$$I(r,z) = \frac{2P}{\omega^2(z)} \exp\left(-\frac{2r^2}{\omega^2(z)}\right) \tag{10.3}$$

where P is the power of the beam and ω is the beam radius at the axial distance z, given by:

$$\omega(z) = \omega_0 \left[1 + \left(\frac{\lambda z}{\pi \omega_0^2}\right)^2\right]^{1/2} \tag{10.4}$$

where ω_0 is the minimum beam waist diameter.

Thus, the particle is trapped into the light beam by the gradient force F_{grad} and directed along the optical axis due to the scattering force F_{scat}.

10.2.2 Experimental Setup

The optical configuration shown in Figures 10.1 and 10.2 uses the same laser source to drive particles into a chosen position and to record digital holograms simultaneously.

We present results using three types of micro-objects: 1) 9.7 micron latex microspheres, 2) the preadipocyte mouse cell (3I3-F442A) undergoing differentiation, and 3) glutaraldehyde-fixed bovine spermatozoa. Bovine spermatozoa were prepared by the Lazzaro Spallanzani Institute, and fixed in suspension using 0.2% glutaraldehyde in PBS without calcium and magnesium (1:3 v/v).

The latex spheres or cells were injected into a microfluidic chamber. The chamber dimensions were 5 mm × 35 mm × 0.3 mm, and made of two cover glasses of 0.15 mm thickness that were assembled together with a space made from double-sided sticky tape

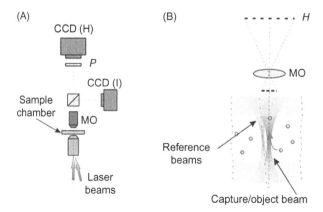

Figure 10.1
(A) Optical configuration adopted; (B) the "in-axis" (or object) beam traps and drives the particle, the "off-axis" (or reference) generates the hologram in the H plane.

(0.3 mm thick). From a solid-state laser source (wavelength of 532 nm and 250 mW), two beams enter the oil immersion Nikon 1003 1.2 NA microscope objective (MO). The first beam is directed along the optical axis while the second beam is slightly off-axis, resulting in a 4 degree angle between the two beams. Another microscope objective (20×) is used as an imaging lens to obtain an image on the CCD plane. We consider the off-axis beam as the reference beam in holographic setup. We used two CCD cameras (Figure 10.1A). The first camera, CCD(I), is used to image the plane proximal to the beam waist (i.e., the focal plane of the first MO) to visualize the optical trapping process. The second camera, CCD(H), records the interferometric pattern (Figure 10.1B). The hologram is recorded and numerically reconstructed by well-known algorithms [54] to obtain the whole complex wave front from which the amplitude and phase map of the particle can be retrieved easily.

To clarify the setup geometry in Figure 10.3, images recorded through the camera CCD(I) and CCD(H) are displayed. Figure 10.3A shows the image from CCD(I) at the image plane near to the MO focal plane. The two beams' waists are clearly separated and visible. Figure 10.3B is an image of the same plane but a latex particle is captured in the object beam. Finally, Figure 10.3C displays a plane well behind the focal plane of the 100× MO and is recorded by the camera CCH(H) at the hologram plane H as shown in Figure 10.1B, and where two interfering beams produce good contrast fringe patterns.

In Figure 10.4, several images illustrate the imaged volume comprised of the focal plane (CCD(I)) and the hologram plane (CCD(H)).

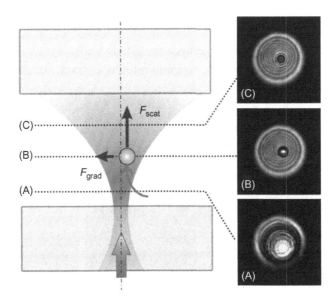

Figure 10.2
Sketch of the forces acting on the particle. (A–C) Three different positions of the particle while guided in the funnel.

A latex particle floating inside the chamber is captured by the highly collimated laser beam. From Figure 10.4A to Figure 10.4D, it can be seen that the particle is moving in the light beam up to the trapping site. From Figure 10.4E to Figure 10.4I, images of the planes between the trapping site and the hologram plane are displayed.

The on-axis beam of the optical tweezers (Figure 10.2) is used to trap or direct micro-objects along preferred directions [42,55].

In Figure 10.2 A-C show images of a particle at three different depths as it is positioned by the scattering force of the light beam.

The object beam has the role of capturing the particle and driving it along the optical funnel to reach the "holographic position" [55]. As a result, the setup parameters are fixed in such a way that the gradient force holds the particle on the beam axis, while the scattering force tends to move it in the direction of the optical beam. Therefore, it can be considered a 2D confinement and all the particles that fall into this "weak trap" are pushed along the desired trajectory. Another way to drive particles in the preferred corridor is to focus the object beam under the sample cell. Indeed, if the beam waist is not fully into the chamber, then the particle cannot be trapped, as illustrated in Figure 10.2. The simple sketch of the sample allows us to explain the optical forces acting on the particle. A Gaussian beam from the solid-state laser emitting at 532 nm is mildly focused by means of the 100× microscope objective. The focal plane is placed inside the glass slide, that is, outside the sample chamber.

10.2.3 Experimental Results

In our "drive and analyze" holographic setup, the on-axis beam acts as object beam while the off-axis one has the role of holographic reference beam. As illustrated in

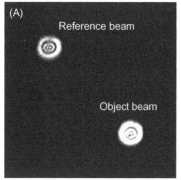 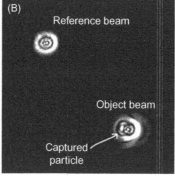 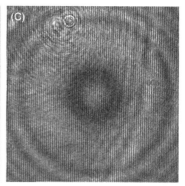

Figure 10.3
(A) Image of the two separated beams. (B) A particle is trapped by the object on-axis beam. (C) Frame acquired in a plane well behind the focal plane of the 100× MO, the hologram fringe pattern is clearly visible.

previous sections and figures, the two non-collinear beams generate the digital hologram that the second (20×) MO images on the CCD arrays. The quantitative phase contrast map can be calculated by the proposed method as we demonstrate in the following paragraphs.

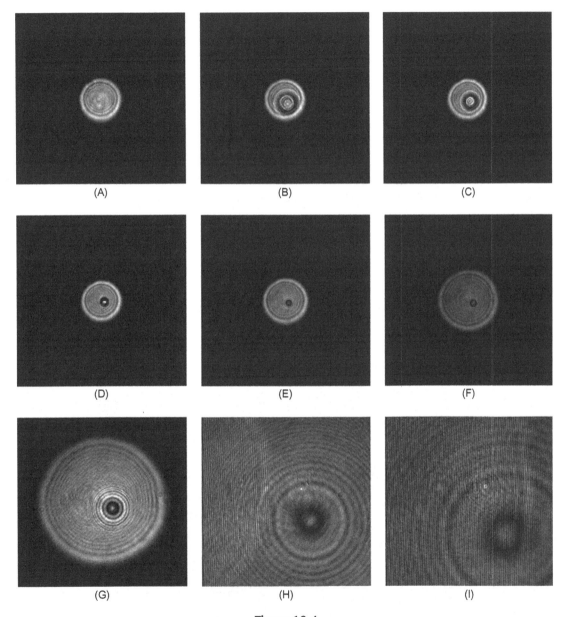

Figure 10.4
Typical behavior for a particle being trapped. After trapping is realized, a hologram is recorded in the suitable plane.

The setup, as clearly depicted in the above figures, can be defined as a quasi-Fourier digital holographic configuration as the point reference beam lies, approximately, on the same plane as the object plane [54]. In the numerical reconstruction process, the spherical reference wave and the chirp function are, respectively, $\overline{R} = \exp\{i\pi/(\lambda r) \cdot [(\xi \rho_\xi)^2 + (\eta \rho_\eta)^2]\}$ and $W = \exp\{-i\pi/(\lambda d) \cdot [(\xi \rho_\xi)^2 + (\eta \rho_\eta)^2]\}$, where d is the reconstruction distance, r is the curvature radius of the beam, and ρ_ξ and ρ_η are the pixel sizes in ξ and η directions (i.e., hologram plane). The fast Fourier transform (FFT) calculated for $d = r + \delta$, with $\delta \ll r$ (i.e., the case of quasi-Fourier configuration), can be approximated to the FFT calculated

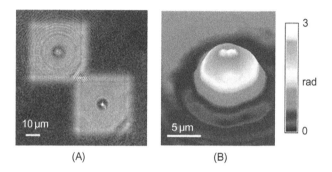

Figure 10.5
(A) Reconstructed amplitude and (B) phase of the complex wavefield diffracted from a latex particle.

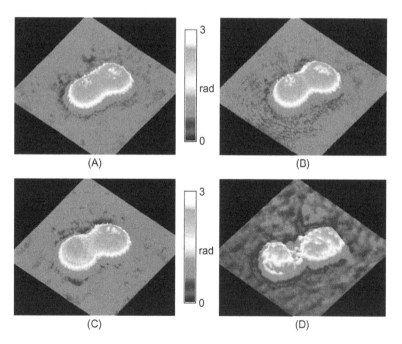

Figure 10.6
Quantitative phase contrast maps of a mouse cell during the different phases.

for $d \cong r$. Moreover, we adopted the subtraction approach of the double exposure to remove aberrations in the phase map [56].

With the aim to test the proposed holographic configuration, we performed different experiments with two kinds of micro-objects, a latex 9.7 μm sized microsphere and in vitro cells. Figure 10.5A shows the amplitude reconstruction of a latex particle, where a good lateral circular cross-section of the sphere can be observed. Instead, Figure 10.5B represents the phase of the complex wavefield coming from the microparticle. In the case of latex beads, the phase map is not a quantitative measurement but it is reported to demonstrate the capability of our setup. Quantitative phase contrast maps are obtained for transparent biological samples as it is shown in Figures 10.6 and 10.7. Figure 10.6 is the phase distribution of a mouse cell; the peaks in the phase maps indicate the presence of lipid particles into the cell, as already shown in a previous study [57].

The cell was driven against the upper glass wall of the chamber and maintained in this position until it attached to the surface. After some hours, the cell entered mitosis as indicated by a continuous change in cell shape that occurs in anaphase and telophase, that was captured by the DH microscope. In Figure 10.6A–D, some frames showing the final stage of the evolution during the duplication process are illustrated. The lateral as well as

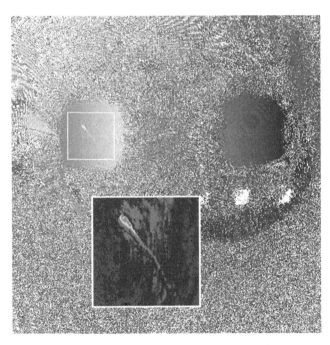

Figure 10.7
Phase map of a bull spermatozoon obtained with the same optical configuration.

the vertical resolution of the proposed holographic configuration is common to all other Fourier DH setups.

As final example, we report, in Figure 10.7, the well-recognizable phase map signature of a sperm cell obtained with the same configuration to show that holographic configuration works in the proper way also for more complex objects.

10.3 Particle Tracking in 3D

3D tracking is performed using the similar experimental setup described in the previous section by evaluating the double out-of-focus projections of the particles due to the twin beams onto the array detector plane. As shown in Figure 10.1, each particle forms two shadows on the CCD array. The separation between the two shadows is a function of the longitudinal position of the particle [58].

10.3.1 Modeling for 3D Tracking

In our model, beams on two distinct planes are sent through a microscope objective in a configuration depicted in Figure 10.8. From a mathematical point of view, we can sketch the imaged volume as composed by two cones, transmitted into the microfluidic sample. The two cones are partially superimposed in the image space. On the CCD digital sensor, the intersection between the beam and the sensor plane of each cone is a circle ellipse with radius $r \ll d_c$, where d_c is the distance between the digital sensor and the vertex of the cone. We consider a reference system of coordinates $Oxyz$, where the plane xy is on the digital sensor with O in the center and we define the vertex of both cones $C_1(x_1,y_1,z_1)$ and $C_2(x_2,y_2,z_2)$, where (x_1,y_1) and (x_2,y_2) are given by the center of two circles in the sensor plane and $z_1 = z_2 = -d_c$. We suppose that the principal axes of cones are parallel and a microscopic object, which can be considered a point with coordinates $P(x_p,y_p,z_p)$, is in the volume defined by the union of the two cones, that is $P \in (C_1,\Omega_1) \cup (C_2,\Omega_2)$, where Ω_1, Ω_2 are the cones' the solid angles. In this scenario, we have three possible situations:

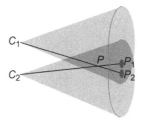

Figure 10.8
Geometrical sketch of two interfering laser beams for 3D tracking.

1. A projection of the point P on the sensor plane in the circle relative to cone 1.
2. A projection of the point P on the sensor plane in the circle relative to cone 2.
3. Two projections of the point P on the sensor plane, that is, each twin laser beam makes a projection of the particle on the CCD array detector.

The last situation is interesting because we demonstrate that it is possible to find the path of a point $P \in (C_1, \Omega_1) \cap (C_2, \Omega_2)$ using the information on the projections' coordinates, by the knowledge of the position of the two vertices C_1, C_2. Suppose that the projections' of P in the two circles are the points $P_1(x_{p1}, y_{p1}, 0)$ and $P_2(x_{p2}, y_{p2}, 0)$, respectively, the point P can be estimated as

$$\hat{P} = \overline{C_1 P_1} \cap \overline{C_2 P_2} \qquad (10.5)$$

where $\overline{AB}$ denote the segment joining the points $A(x_A, y_A, z_A)$ and $B(x_B, y_B, z_B)$:

$$\begin{cases} x' = x_A + l_x t \\ y' = y_A + l_y t \\ z' = z_A + l_z t \end{cases} \qquad (10.6)$$

where $(x', y', z') \in \overline{AB}$, $t \in [0,1]$ and $l_{(\cdot)} = (\cdot)_B - (\cdot)_A$.

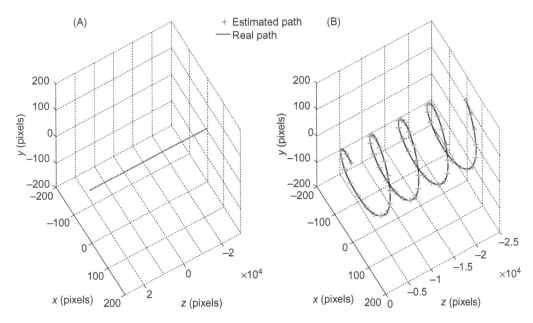

Figure 10.9
Comparison between real and estimated path for (A) linear and (B) coil paths.

The hypothesis $P \in (C_1, \Omega_1) \cap (C_2, \Omega_2)$ assures that only one intersection's point $\hat{P}$ exists.

Numerical simulation is performed to prove how the tracking functionality works. We set $C_1(-112, 76, -d_c)$, $C_2(112, -76, -d_c)$ with $d_c = 44478$, $r = 700$ and the units are in pixels.

We report here simulation of two different paths of particle P. A linear path, in the direction of the z-axis, and helix paths, respectively.

For both simulated paths, we compute the mean error and the standard deviation error on the difference between the real path and the path estimated with the proposed method.

From the simulated images, we evaluate the corresponding points $P1$ and $P2$ in the sensor plane for both paths. By the knowledge of the $P1$ and $P2$ coordinates and through Eq. (10.1), we reconstructed the paths. Figure 10.9 shows the comparison between the

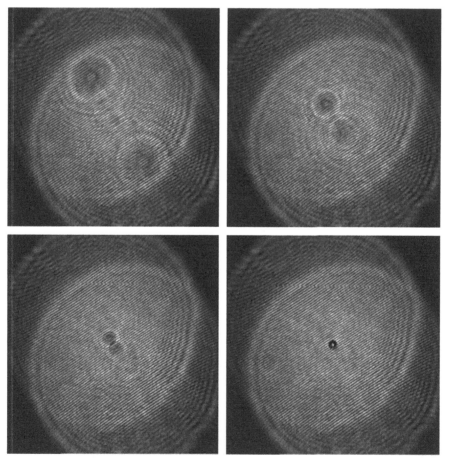

Figure 10.10
Four frames from the recorded sequence of the z-axis motion of a microscopic particle.

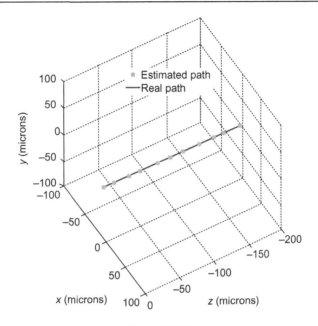

Figure 10.11
Comparison between real and estimated path for real sequence.

simulated and the reconstructed paths for the linear and the coil path, respectively. The results show very good agreement between the simulated values and the retrieved paths.

10.3.2 Experimental Results

We report results from tracking latex microspheres and cells in vitro. In the first experiment, the spherical particle was displaced along the optical axis with known step to create a situation similar to the linear path previously simulated. We made an experimental calibration where a microscope liquid chamber with a fixed spherical latex particle was translated axially along the z-direction with multiple and known steps. In Figure 10.10, the images are recorded at different distances.

By applying the above equations (10.5) and (10.6), particle trajectory is calculated estimating the centroid of the two microsphere images for each hologram recorded. The result of this experiment is shown in Figure 10.11.

In the second experiment, we show how the method can be applied for 3D paths tracking of multiple motile in vitro cells. A real experimental situation in which we had different cells floating into a microfluidic chamber, without a priori information, before they fall to the ground, i.e., before they attach to the bottom of the chamber. We recorded, with the

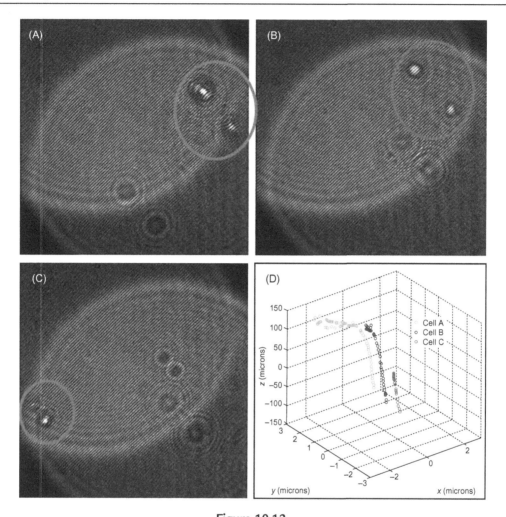

Figure 10.12
(A–C) Three frames corresponding to the passage of three different mouse cells across the imaged volume. (D) The estimated paths for random motion of the three cells.

experimental configuration in Figure 10.1, a sequence of images by the CCD array; some of these frames are shown in Figure 10.12A–C. Also in this case, we estimated cell centroids by an image processing algorithm for each frame of the recorded sequences.

The results of the retrieved 3D paths of the cells were simply estimated by the proposed method. The 3D plot of the paths are shown in Figure 10.12D. From the last result, it is clear that all three cells follow paths along the same streamlines in the microfluidic flux. Cells experience a displacement mainly along the longitudinal axis. This is due to the twin beams exerting on them the driving effect described before. Nevertheless, the optical configuration can be designed to cancel the effect of the light on the paths. In fact, by using a trivial stroboscopic illumination, depending also on the velocity fields of the particles, the influence of the light on the paths can be reduced to become totally negligible, making this method a full noninvasive 3D tracking approach.

10.4 Conclusions

Twin laser beams, passing through the same microscope objective, allow simple and compact digital holographic microscope in off-axis configuration. The effectiveness of this configuration has been proved. It is particularly suitable for phase contrast analysis and 3D imaging of biological samples in a microfluidic environment. We experimentally implemented the setup, and demonstrated the ability to drive micro-objects or track them in the sample volume. An advantage of this arrangement is to measure the phase contrast signature of the above-mentioned particle by digital holographic recording. For tracking experiments, we validated the model either numerically or from the experimental point of view. We demonstrated the 3D tracking and quantitative phase contrast microscopic analysis on multiple cells that were moving in a microfluidic environment. The optical configuration is very simple as it is made by only two beams. The beams could be also produced by various configurations such as diffraction grating at the input pupil of the microscope objective or even by using a spatial light modulator.

References

[1] D. Psaltis, S.R. Quake, C. Yang, Developing optofluidic technology through the fusion of microfluidics and optics, Nature 442 (2006) 381−386.
[2] J.W. Hong, S.R. Quake, Integrated nano-liter systems, Nat. Biotechnol. 21 (2003) 1179−1183.
[3] F. Dubois, C. Yourassowsky, O. Monnom, J.C. Legros, O. Debeir, P. Van Ham, et al., Digital holographic microscopy for the three-dimensional dynamic analysis of in vitro cancer cell migration, J. Biomed. Opt. 11 (5) (2006) 054032.
[4] B. Rappaz, F. Charrière, C. Depeursinge, P.J. Magistretti, P. Marquet, Simultaneous cell morphometry and refractive index measurement with dual-wavelength digital holographic microscopy and dye-enhanced dispersion of perfusion medium, Opt. Lett. 33 (2008) 744−746.
[5] W.M. Ash, M.K. Kim, Digital holography of total internal reflection, Opt. Express 16 (2008) 9811−9820.
[6] B. Kemper, A. Vollmer, C.E. Rommel, J. Schnekenburger, G. von Bally, Simplified approach for quantitative digital holographic phase contrast imaging of living cells, J. Biomed. Opt. 16 (2011) 026014.
[7] J. Garcia-Sucerquia, W. Xu, S.K. Jericho, M.H. Jericho, H.J. Kreuzer, 4-D imaging of fluid flow with digital in-line holographic microscopy, Optik 114 (9) (2007) 419−423.
[8] G. Coppola, G. Di Caprio, M. Gioffré, R. Puglisi, D. Balduzzi, A. Galli, et al., Digital self-referencing quantitative phase microscopy by wavefront folding in holographic image reconstruction, Opt. Lett. 35 (2010) 3390−3392.
[9] C. Song, T.-D. Luong, T. Fook Kong, N.-T. Nguyen, A.K. Asundi, Disposable flow cytometer with high efficiency in particle counting and sizing using an optofluidic lens, Opt. Lett. 36 (2011) 657−659.
[10] M. Antkowiak, M.L. Torres-Mapa, K. Dholakia, F.J. Gunn-Moore, Quantitative phase study of the dynamic cellular response in femtosecond laser photoporation, Biomed. Opt. Express 1 (2010) 414−424.

[11] W. Bishara, T.-W. Su, A.F. Coskun, A. Ozcan, Lensfree on-chip microscopy over a wide field of view using pixel superresolution, Opt. Express 18 (2010) 11181−11191.

[12] N.T. Shaked, L.L. Satterwhite, N. Bursac, A. Wax, Whole-cell-analysis of live cardiomycocytes using wide-field interferometric phase microscopy, Biomed. Opt. Express 1 (2010) 706−719.

[13] Z. Wang, L. Millet, M. Mir, H. Ding, S. Unarunotai, J. Rogers, M.U. Gilette, G. Popescu, Spatial light interference microscopy (SLIM), Opt. Express 19 (2011) 1016−1026.

[14] L. Friedrich, A. Rohrbach, Improved interferometric tracking of trapped particles using two frequency detuned beams, Optics Lett. 35 (11) (2010) 1920−1922.

[15] M.J. Padgett, R. Di Leonardo, Holographic optical tweezers and their relevance to lab on chip devices, Lab. Chip. 11 (2011) 1196−1205.

[16] P. Schiro, C. DuBois, A. Kwok, Large capture-range of a single beam gradient optical trap, Opt. Express 11 (2003) 3485−3489.

[17] A. Rohrbach, H. Kress, E. Stelzer, Three dimensional tracking of small spheres in focussed laser beams: influence of the detection angular aperture, Opt. Lett. 28 (6) (2003) 411−413.

[18] M.D. McMahon, A.J. Berglund, P. Carmichael, J.J. McClelland, J.A. Liddle, 3D particle trajectories observed by orthogonal tracking microscopy, ACS Nano. 3 (3) (2009) 609−614.

[19] E. Toprak, H. Balci, B.H. Blehm, P.R. Selvin, Three dimensional paritcle tracking via bifocal imaging, Nano Lett. 7 (7) (2007) 2043−2045.

[20] H.P. Kao, A.S. Verkman, Tracking of single fluorescent particles in three dimensions: use of cylindrical optics to encode paritcle position, Biophys. J. 67 (1994) 1291−1300.

[21] S.R.P. Pavani, R. Piestun, Three dimensional tracking of fluorescent microparticles using a photon-limited double-helix response system, Opt. Express 16 (2008) 22048−22057.

[22] T. Čižmár, M. Mazilu, K. Dholakia, In situ wavefront correction and its application to micromanipulation, Nat. Photonics 4 (2010) 388−394.

[23] M. Antkowiak, M.L. Torres-Mapa, K. Dholakia, F.J. Gunn-Moore, Quantitative phase study of the dynamic cellular response in femtosecond laser photoporation, Biomed. Opt. Express 1 (2010) 414−424.

[24] M. DaneshPanah, S. Zwick, F. Schaal, M. Warber, B. Javidi, W. Osten, 3D holographic imaging and trapping for non-invasive cell identification and tracking, J. Disp. Technol. 6 (2010) 490−499.

[25] D.G. Grier, A revolution in optical manipulation, Nature 424 (2003) 810−816.

[26] M.J. Padgett, R. Di Leonardo, Holographic optical tweezers and their relevance to lab on chip devices, Lab. Chip. 11 (2011) 1196−1205.

[27] F.C. Cheong, B.J. Krishnatreya, D.G. Grier, Strategies for three-dimensional particle tracking with holographic video microscopy, Opt. Express 18 (2010) 13563−13573.

[28] H. Yang, N. Halliwell, J. Coupland, Application of the digital shearing method to extract three-component velocity in holographic particle image velocimetry, Meas. Sci. Technol. 15 (2004) 694.

[29] Y. Park, G. Popescu, K. Badizadegan, R.R. Dasari, M.S. Feld, Fresnel particle tracing in three dimensions using diffraction phase microscopy, Opt. Lett. 32 (2007) 811−813.

[30] C. Song, T.-D. Luong, T. Fook Kong, N.-T. Nguyen, A.K. Asundi, Disposable flow cytometer with high efficiency in particle counting and sizing using an optofluidic lens, Opt. Lett. 36 (2011) 657−659.

[31] S. Ram, P. Prabhat, E.S. Ward, R.J. Ober, Improved single particle localization accuracy with dual objective multifocal plane microscopy, Opt. Express 17 (8) (2009) 6881−6898.

[32] A.J. Berglund, M.D. McMahon, J.J. McClelland, J.A. Liddle, Fast, bias-free algorithm for tracking single particles with variable size and shape, Opt. Express 16 (18) (2008) 14064−14075.

[33] Y. Park, G. Popescu, K. Badizadegan, R.R. Dasari, M.S. Feld, Fresnel paritcle tracing in three dimensions using diffraction phase microscopy, Opt. Lett. 32 (2007) 811−813.

[34] M.J. Mlodzianoski, M.F. Juette, G.L. Beane, J. Bewersdorf, Experimental characterization of 3D localization techniques for particle-tracking and super-resolution microscopy, Opt. Express 17 (10) (2009) 8264−8277.

[35] M. DaneshPanah, S. Zwick, F. Schaal, M. Warber, B. Javidi, W. Osten, 3D holographic imaging and trapping for non-invasive cell indentification and tracking, J. Disp. Technol. 6 (2010) 490−499.

[36] H. Yang, N. Halliwell, J. Coupland, Application of the digital shearing method to extract three-component velocity in holographic particle image velocimetry, Meas. Sci. Technol. 15 (2004) 694.

[37] L. Miccio, A. Finizio, R. Puglisi, D. Balduzzi, A. Galli, P. Ferraro, Dynamic DIC by digital holography microscopy for enhancing phase-contrast visualization, Biomed. Opt. Express 2 (2011) 331−344.

[38] F.C. Cheong, B.J. Krishnatreya, D.G. Grier, Strategies for three-dimensional particle tracking with holographic video microscopy, Opt. Express 18 (2010) 13563−13573.

[39] O. Otto, F. Czerwinski, J.L. Gornall, G. Stober, L.B. Oddershede, R. Seidel, et al., Real-time particle tracking at 10,000 fps using optical fiber illumination, Opt. Express 18 (2010) 22722−22733.

[40] D.B. Conkey, R.P. Trivedi, S.R.P. Pavani, I.I. Smalyukh, R. Piestun, Three-dimensional parallel particle manipulation and tracking by integrating holographic optical tweezers and engineered point spread functions, Opt. Express 19 (2011) 3835−3842.

[41] B. Kemper, P. Langehanenberg, A. Hoink, G. von Bally, F. Wottowah, S. Schinkinger, et al., Monitoring of laser micromanipulated optically trapped cells by digital holographic microscopy, J. Biophotonics 3 (2010) 425−431.

[42] A.A. Lall, A. Terray, S.J. Hart, On-the-fly cross flow laser guided separation of aerosol particles based on size, refractive index and density−theoretical analysis, Opt. Express 18 (2010) 26775−26790.

[43] N.T. Shaked, L.L. Satterwhite, N. Bursac, A. Wax, Whole-cell-analysis of live cardiomyocytes using wide-field interferometric phase microscopy, Biomed. Opt. Express 1 (2010) 706−719.

[44] F. Dubois, C. Yourassowsky, O. Monnom, J.C. Legros, O. Debeir, P. Van Ham, et al., Digital holographic microscopy for the three-dimensional dynamic analysis of in vitro cancer cell migration, J. Biomed. Opt. 11 (2006) 054032.

[45] B. Rappaz, F. Charrière, C. Depeursinge, P.J. Magistretti, P. Marquet, Simultaneous cell morphometry and refractive index measurement with dual-wavelength digital holographic microscopy and dye-enhanced dispersion of perfusion medium, Opt. Lett. 33 (2008) 744−746.

[46] B. Kemper, A. Vollmer, C.E. Rommel, J. Schnekenburger, G. von Bally, Simplified approach for quantitative digital holographic phase contrast imaging of living cells, J. Biomed. Opt. 16 (2011) 026014.

[47] G. Coppola, G. Di Caprio, M. Gioffré, R. Puglisi, D. Balduzzi, A. Galli, et al., Digital self-referencing quantitative phase microscopy by wavefront folding in holographic image reconstruction, Opt. Lett. 35 (2010) 3390−3392.

[48] L. Miccio, D. Alfieri, S. Grilli, P. Ferraro, A. Finizio, L. De Petrocellis, et al., Direct full compensation of the aberrations in quantitative phase microscopy of thin objects by a single digital hologram, Appl. Phys. Lett. 90 (2007) 041104.

[49] Z. Wang, L. Millet, M. Mir, H. Ding, S. Unarunotai, J. Rogers, et al., Spatial light interference microscopy (SLIM), Opt. Express 19 (2011) 1016−1026.

[50] S.O. Isikman, W. Bishara, H. Zhu, A. Ozcan, Optofluidic tomography on a chip, Appl. Phys. Lett. 98 (2011) 161109.

[51] A. El Mallahi, F. Dubois, Dependency and precision of the refocusing criterion based on amplitude analysis in digital holographic microscopy, Opt. Express 19 (2011) 6684−6698.

[52] W.M. Ash, M.K. Kim, Digital holography of total internal reflection, Opt. Express 16 (2008) 9811−9820.

[53] F. Merola, L. Miccio, M. Paturzo, A. Finizio, S. Grilli, P. Ferraro, Driving and analysis of micro-objects by digital holographic microscope in microfluidics, Opt. Lett. 36 (2011) 3079−3081.

[54] S. Grilli, P. Ferraro, S. De Nicola, A. Finizio, G. Pierattini, R. Meucci, Whole optical wavefields reconstruction by digital holography, Opt. Express 9 (2001) 294−302.

[55] J.A. Rodrigo, A.M. Caravaca-Aguirre, T. Alieva, G. Cristóbal, M.L. Calvo, Microparticle movements in optical funnels and pods, Opt. Express 19 (2011) 5232−5243.

[56] P. Ferraro, S. De Nicola, A. Finizio, G. Coppola, S. Grilli, C. Magro, G. Pierattini, Compensation of the Inherent Wave Front Curvature in Digital Holographic Coherent Microscopy for Quantitative Phase-Contrast Imaging, Appl. Opt. 42 (2003) 1938.

[57] P. Ferraro, D. Alferi, S. De Nicola, L. De Petrocellis, A. Finizio, G. Pierattini, Quantitative phase-contrast microscopy by a lateral shear approach to digital holographic image reconstruction, Opt. Lett. 31 (2006) 1405.

[58] P. Memmolo, A. Finizio, M. Paturzo, L. Miccio, P. Ferraro, Twin-beams digital holography for 3D tracking and quantitative phase-contrast microscopy in microfluidics, Opt. Express 19 (2011) 25833−25842.

CHAPTER 11
Holographic Motility Contrast Imaging of Live Tissues

David D. Nolte, Ran An, Kwan Jeong, John Turek

Dept. of Physics and Dept. of Basic Medical Sciences, Purdue University, West Lafayette IN 47907

Editor: Natan T. Shaked

11.1 Introduction and Review

Phase sensitivity in optical imaging opens a window on the many subtle features that characterize living systems, displaying them with a sharp contrast that goes beyond what only scattered intensities provide. The phase of light transmitted or reflected from biological specimens uses the wavelength as the ruler against which everything is measured, achieving sensitivities of a small fraction of that scale. That is why phase-contrast microscopy, which is a form of imaging interferometry, has such power in biological applications [1].

High-resolution phase-contrast imaging requires the average phase to vary slowly across a lateral distance of a wavelength or at least to have regions of slowly varying phase that may have sharp boundaries. Sharp boundaries and phase heterogeneities diffract light, which produce diffraction patterns in the image that obscure structure. When the spatial scale of the heterogeneities gets too small, and if the phase modulation gets too large, the diffraction patterns overlap and generate speckle—the random intensity patterns that appear as noise in coherent or phase-sensitive imaging systems.

Speckle has played a persistent and controversial role in the history of imaging science. On the one hand, it is a noise source, always present in any system that uses partially coherent light. From the very early days of the laser, it was already viewed as a source of background that limited the sensitivities of interferometry and holography [2]. On the other hand, and also from early on, speckle was recognized as a highly interesting and complex phenomenon that carried information about the optical system and targets in its statistical fluctuations [3,4]. A very early success of speckle statistics, even prior to the invention of the laser, was the

Hanbury-Brown—Twiss stellar interferometer that was used to measure the size of nearby stars [5]. The size of a star is measured through the speckle spatial statistics. For instance, the nearby star Sirius casts a speckle pattern across the face of the earth with an average speckle size of several tens of meters. Experiments like these launched the field of statistical optics.

Speckle interferometry and holography provide a statistical approach to biological specimens and applications. These techniques are closely related to phase-contrast imaging, but they have a very different character. Speckle images tend to have low spatial information content. A theorem in statistical optics states that a speckle field with unity contrast contains no spatial information. Therefore, fully developed speckle provides no structural information from a specimen, and hence provides no image at all. Structural information comes from spatial variations in the speckle statistics. But what speckle images lack in low resolution, they make up for with extremely high sensitivity to phase disturbances. Indeed, substantial information comes from time-dependent phase changes and how these changes affect the time development of speckle fields.

The analysis of dynamic speckle is the subject of dynamic light scattering (DLS). The basic concept of DLS is an ensemble of scattering objects that are in motion. Light is scattered as partial waves, and the motions of the objects produce phase changes in the scattered partial waves. Because the objects are distributed in space, the relative phases are random, and if the motions are uncorrelated, the changes in the phases are also random, leading to a complex scrambled phase of dynamic speckle. However, the way that the objects move is encoded in the frequency content of the speckle intensity fluctuations. For instance, uniform motion of the scattering objects produces a Doppler shift with a well-characterized frequency. Random diffusion has no discrete frequency but does have a well-defined knee frequency that depends on the diffusion coefficient. Therefore, by analyzing the frequency content of dynamic speckle across a wide range of frequencies, many different types of motion can be studied as they are perturbed by outside influences.

In the case of living biological tissue, the scattering objects are mitochondria, nuclei, organelles, and cell membrane. Each of these biological components has different types of motion. In some cases, the motion is driven by molecular motors fed by ATP from active metabolism. In other cases, the motion is driven by external forces, such as when a newly divided cell jostles its neighbors to make room for itself. The different types of scattering objects and their different types of motion determine the fluctuating temporal statistics of the dynamic speckle. Clearly, there is a strong overlap in frequencies and in the sizes of scattering objects, which produce essentially featureless fluctuation power spectra, and it is difficult to distinguish the individual contributions. However, these power spectra can be measured by applying external perturbations, such as altered physiological conditions or application of a pharmaceutical drug, which cause changes in the power spectra and they carry signatures specific to the changes in the internal motion of the live tissue.

This chapter describes the technology and underlying biological physics of holographic motility contrast imaging (MCI) of live tissue. MCI is a statistical optics form of phase-contrast imaging [6]. It grew out of holographic optical coherence imaging (OCI) that uses a holographic coherence gate [1,7] to capture full-frame sections inside tissue of up to 1 mm thick. Multimode illumination of the tissue samples generates highly developed speckle that is sectioned in three dimensions based on optical path length. In live samples, the speckle is highly dynamic, and subcellular motions provide an imaging contrast that differentiates the outer proliferating shell of a tumor from its hypoxic or necrotic core. Changes in the motility contrast and in the fluctuation frequency content caused by applied xenobiotics may provide a new form of high-content drug screening for early drug discovery.

11.2 Optical Coherence Imaging

OCI is a full-frame form of optical coherence tomography (OCT) [8]. It uses holography both as the coherence gate for detection and as a means to perform spatial demodulation and image reconstruction. OCI is characterized by high-contrast speckle because of the wide-field illumination of many spatial modes that self-interfere. The speckle size is also the spatial coherence length, and a temporally coherent reference wave interferes with the speckle field to produce interference (holographic) fringes within each speckle. These fringes enable coherence-gated detection of path-matched light originating primarily from a specified depth in the tissue. Therefore, understanding the performance of OCI requires an understanding of the speckle field properties, the role of scattering, and the coherence gate in the presence of multiple scattering.

The optical setup for the OCI system is shown in Figure 11.1. The probe beam is from a Superlum superluminescent diode system with a wavelength of 840 nm and a bandwidth of 50 nm. The beam size at the target is about 1 mm. The target is a multicellular tumor spheroid that can grow as large as 1 mm in diameter, but we typically work with tumors of diameters around 0.5 mm. The backscattered light is collected through a polarizing beamsplitter and a first Fourier lens L1. A second lens relays the image to the first image plane IP1 which is the object plane of the Fourier transform lens L3. The signal arm passes through a beamsplitter where it is joined by an off-axis reference beam with a crossing angle of about 1.3°. The reference and signal beams combine on the charge-coupled device (CCD) chip with a fringe spacing of approximately 25 μm.

The speckle diameter at the CCD camera depends on the Fourier optics of the lenses and apertures, shown in Figure 11.1B. For a target diameter D_{Samp}, the speckle diameter at the first Fourier plane FP1 is given by:

$$d_c^{FP1} = \frac{4}{\pi} \frac{f_1 \lambda}{D_{samp}} \tag{11.1}$$

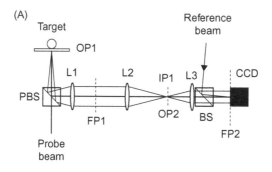
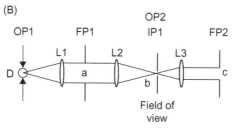

Figure 11.1
OCI system. (A) The layout of the lenses, beamsplitters, and CCD camera in an off-axis holography configuration. (B) The apertures, Fourier and image planes of the OCI system. OP1, first optical plane; FP1, first Fourier plane; IP1 and OP2, first image plane and second object plane; and FP2, second Fourier plane coincident on the face of the CCD camera.

for the wavelength λ. The subtended solid angle for the collecting lens L1 is approximately $4\pi/100$. The aperture diameter at FP1 is set equal to a, and the speckle diameter at the first image plane IP1 is

$$d_c^{IP1} = \frac{4 f_2 \lambda}{\pi \; a} \tag{11.2}$$

The magnification at the first image plane IP1 is

$$M = \frac{f_2}{f_1} \tag{11.3}$$

and the size of the sample image on the image plane before the third lens (FT lens) is equal to MD_{samp}. Therefore, the speckle diameter at the second Fourier plane FP2 (after lens 3) is

$$d_c^{FP2} = \frac{4}{\pi} \frac{f_3 \lambda}{MD_{samp}} = \frac{4}{\pi} \frac{f' \lambda}{D_{samp}} \tag{11.4}$$

where the effective focal length of the three-lens configuration is

$$f' = \frac{f_1 f_3}{f_2} \tag{11.5}$$

In digital holography, the optimum condition for image reconstruction from the Fourier plane uses three CCD pixels per interference fringe period Λ and three fringes per speckle. This sets the condition for the speckle size on the CCD chip as

$$d_c = 3\Lambda = 9d_{pix} \tag{11.6}$$

The required minimum focal length of the three-lens system to achieve this speckle size is

$$f' = \frac{9\pi}{4} \frac{d_{pix} D_{samp}}{\lambda} \tag{11.7}$$

An interline pixel pitch of $d_{pix} = 8$ μm and a sample diameter $D_{samp} = 0.5$ mm for a wavelength of 840 nm requires at least a 3 cm effective focal length. In our current configuration, $f_1 = 15$ cm, $f_2 = 15$ cm, and $f_3 = 5$ cm with $f' = 5$ cm which gives a larger speckle diameter and a safety factor to prevent washout of the fringes or aliasing.

The holographic fringes on the CCD plane provide the coherence gate that depth-sections the three-dimensional speckle field returning from the biological sample. In the limit of dilute samples dominated by single scattering, only the partial waves from scattering objects that are path-matched to the reference wave form holographic fringes in the speckles. However, live tissues are optically dense. The optical thickness of a tumor that is 500 μm in diameter is approximately $\tau = 5$, and in the backscatter direction $\tau = 10$ from the distal portions of the sample. Therefore, the coherence gate is only approximate and each pixel captures photons scattered from a cone-shaped volume with the vertex at the selected depth but with some photons originating from the shallower depths. Nonetheless, much of the multiple scattering is by small forward angles. Our superluminescent diode light source has a bandwidth of 50 nm, with an associated coherence length of 10 μm in water and an OCT axial resolution of 5 μm, but multiple scattering spreads this to about 20 μm.

An example of a series of coherence-gated sections of a multicellular tumor spheroid is shown in Figure 11.2. The spheroid diameter is approximately 850 μm. The sections are acquired with an external delay-stage step of 10 μm/frame, and the frame numbers are sequential. There are 120 frames from the bottom to the top of the tumor for 1200 μm of external mirror shift. When this is divided by the refractive index of the tumor ($n = 1.4$), the tumor diameter is 850 μm. The data show reflectance on a logarithmic color scale (dB). The necrotic core is more optically heterogeneous and shows stronger reflectance. The proliferating shell is more optically homogeneous and has lower reflectance.

These data are volumetric, captured as a stack of 120 frames that span the diameter of the tumor spheroid. Therefore, the data can be visualized in many ways. Examples are shown in Figure 11.3. The full external surface is shown on the left. When the isosurface values are eroded to show only the strongest reflectances, the necrotic core emerges on the right of the figure, casting its shadow on the underlying Petri dish. (For these data, the beam was incident from the top.)

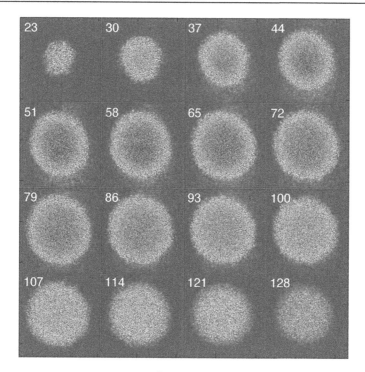

Figure 11.2
Coherence-gated *xy* sections of a multicellular tumor spheroid that is 850 μm in diameter. The data are acquired at 10 μm intervals, with one frame in eight being displayed. The reflectance color scale is logarithmic (dB) showing the bright reflections from the necrotic core and the dimmer reflections from the proliferating shell. (For interpretation of the references to color in this figure legend, the reader is referred to the web version of this book.)

11.3 Motility Contrast Imaging

The OCI data in Figures 11.2 and 11.3 are strongly speckled, which is a consequence of the broad-field illumination of the sample with spatially coherent light. However, at a fixed depth, the speckle is highly dynamic with large intensity fluctuations in time. These fluctuations are a direct consequence of the intracellular motions of live tissue. There are different types of motion for different constituents of the cells and tissues. Membranes undulate and move, driven by active forces induced by the changing cytoskeleton; organelles are transported by molecular motors along microtubules; thermal motions and Brownian diffusion occur for small vesicles; and very large-scale motions take place when a cell divides. The statistical properties of the different motions lead to fluctuations on many different time scales. By capturing the degree of fluctuations within the dynamic speckle, motility maps of tissue can be constructed. In this sense, internal motion of a live tissue provides the imaging contrast.

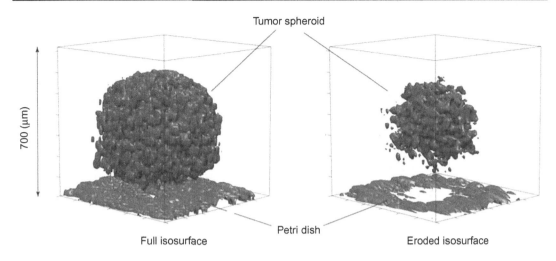

Figure 11.3
Three-dimensional visualizations of OCI data from a multicellular tumor spheroid resting on a Petri dish.

The drifting intensities at a fixed depth inside a tumor spheroid are shown in Figure 11.4A. The horizontal axis is a row of reconstructed pixels, and the vertical axis is time. For a selected pixel, the intensity varies randomly (but with an average correlation time). The data in Figure 11.4B are for a punctuated flythrough for which 40 frames are acquired, then the coherence gate is moved deeper and 40 frames are acquired again, and the process is repeated.

For each fixed-depth stack of images, the statistical properties of the fluctuating intensities for each pixel are evaluated. One of the simplest statistical measures of temporal fluctuations is the normalized standard deviation (NSD), also known as temporal speckle contrast. For a given pixel, the standard deviation in the intensity values is divided by the mean intensity. Values of the NSD approaching unity represent highly variable pixels. An example of the NSD at several depths inside a tumor spheroid is shown in Figure 11.5. The data are color coded with red denoting highly variable pixels and blue denoting smaller deviations. Around the midsection, the proliferating shell is clearly observed as strongly variable pixels surrounding the blue core which has much less motion. The core of a tumor is composed of dead or dying cells. However, even in regions where there are intact cells in the core, the tissue is hypoxic because oxygen is consumed by the surrounding proliferating shell. The hypoxic tissue has low metabolic activity which is reflected by lower intracellular motions. The stack of motility maps for increasing depth into the tumor spheroid can be combined into a three-dimensional visualization of the motility in the tumor tissue. The volumetric motility contrast of a tumor that is 800 μm in diameter is shown in Figure 11.6 [6].

The motility contrast is a quantitative measure of intracellular motion and can be used to monitor changes in internal motions induced by changing environmental conditions or

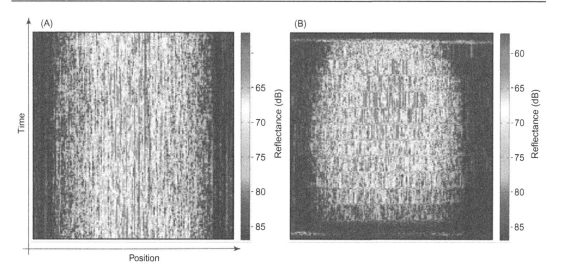

Figure 11.4
Space and time representation of a single line of speckle values. The horizontal axis spans the pixels of the image, while the vertical axis is time. The pixel intensities vary randomly with time. The data in (A) are from a fixed depth. The data in (B) are for a punctuated flythrough in which 40 frames are acquired at a fixed depth, then the coherence gate is moved deeper and 40 frames are acquired again, and the process is repeated.

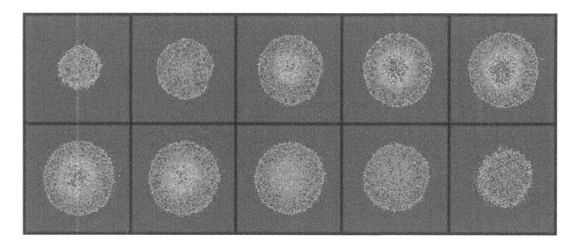

Figure 11.5
Motility contrast images of a multicellular tumor spheroid. NSD maps are shown at selected depths. The color scale is set to show highly variable pixels with red and less variable pixels with blue. The healthy shell surrounding the hypoxic or necrotic core appears in high contrast. (For interpretation of the references to color in this figure legend, the reader is referred to the web version of this book.)

Figure 11.6
Volumetric motility contrast image of a tumor spheroid that is 800 μm in diameter. Red denotes high motility and blue lower motility. (For interpretation of the references to color in this figure legend, the reader is referred to the web version of this book.)

by applying xenobiotics (pharmaceuticals). For instance, the motility contrast can be used to measure the effect of anticancer drugs that influence the cytoskeleton and prevent cell division. A series of motility contrast images were obtained for a tumor that responded to the antimitotic drug called nocodazole. The motility contrast, 3 min after the drug was applied, showed increased motion in the proliferating shell as a temporary response to the drug. However, at subsequent times, the motion in the proliferating shell is eventually quenched because many types of cellular motion are dependent on the cytoskeleton. Nocodazole causes the microtubules to depolymerize, and hence reduces the amount of microtubule-dependent motion. Time traces of the motility for increasing drug doses averaged over the proliferating shell are shown in Figure 11.7. The drug response is faster and stronger for high doses. Note that each curve in Figure 11.7 is from a different tumor spheroid. The monotonic dependence on drug dose demonstrates the high repeatability that is achievable with the controlled growth of tumor spheroids in a bioreactor.

11.4 Dynamic Light Scattering (DLS) Spectroscopy

The physical origin of dynamic speckle is from DLS. Although live tissues are far from being dilute, relationships can be established between the types of motion and the frequency dependence of the fluctuating light. DLS is caused by a change in the optical phase of scattered light as a particle moves [1]. The light scattered from a single moving particle is

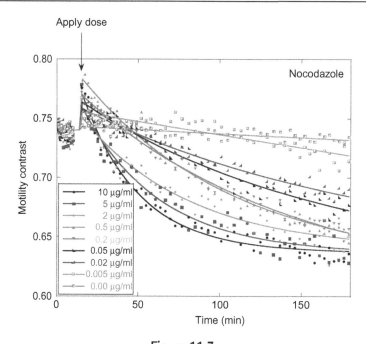

Figure 11.7
Dose–response curves of the NSD for increasing doses of nocodazole. Each curve is from a different tumor.

$$E(t) = E_s \exp(-i(\mathbf{k}_f - \mathbf{k}_i) \cdot \mathbf{r}(t)) \tag{11.8}$$

As the position $\mathbf{r}(t)$ of the particle changes, the phase of the scattered light changes. The difference in phase from one position to the next is

$$\begin{aligned}\Delta\phi &= -(\mathbf{k}_f - \mathbf{k}_i) \cdot \Delta\mathbf{r} \\ &= -\mathbf{q} \cdot \Delta\mathbf{r}\end{aligned} \tag{11.9}$$

where $\mathbf{q} = (\mathbf{k}_f - \mathbf{k}_i)$ is the scattering vector. The magnitude of $\mathbf{q}$ for a scattering angle θ is

$$\mathbf{q} = 2k \sin\frac{\theta}{2} = k\sqrt{2(1-\cos\theta)} \tag{11.10}$$

In field-based detection, such as holography, the mixing of the scattered field with a reference field produces the net field

$$E(t) = E_0 + \sum_{n=1}^{N} E_n \exp(-i\mathbf{q}x_n(0))\exp(-i\mathbf{q}\Delta x_n(t)) \tag{11.11}$$

where the coordinate x is defined parallel to the scattering vector $\mathbf{q}$. To construct the time autocorrelation function, the product of the field is taken with a delayed version of itself and in the limit of large N, the random initial phases average to zero to yield

$$E^*(0)E(t) = E_0^2 + \sum_{n=1}^{N} E_n^2 \exp(-i\mathbf{q}\Delta x_n(t)) \tag{11.12}$$

The autocorrelation is obtained by taking an ensemble average of this quantity. The ensemble average is provided by a stochastic sum that is evaluated using a probability distribution $P(\Delta x_n(t))$. The autocorrelation is

$$\langle E^*(0)E(t)\rangle = E_0^2 + \sum_{n=1}^{N} E_n^2 P(\Delta x_n(t)) \exp(-i\mathbf{q}\Delta x_n(t)) \tag{11.13}$$

and the stochastic sum is equivalent to an integral over the probability distribution

$$\begin{aligned}\langle E^*(0)E(t)\rangle &= E_0^2 + NI_s \int_{-\infty}^{\infty} P(\Delta x)\exp(-i\mathbf{q}\Delta x)\,dx \\ &= I_0 + NI_s FT(P(\Delta x))\end{aligned} \tag{11.14}$$

where the autocorrelation is proportional to the Fourier transform of the probability functional.

Different types of motion have different probability functionals. For diffusion, the diffusion autocorrelation function is

$$A_D^E(\tau) = \langle E^*(t)E(t+\tau)\rangle - I_0 = NI_s \exp(-\mathbf{q}^2 D\tau) \tag{11.15}$$

For a Gaussian distribution of velocities (Maxwellian), the autocorrelation function is

$$A_v^E(\tau) = NI_s \exp\left(\frac{-\mathbf{q}^2 v_{rms}^2 \tau^2}{2}\right) \tag{11.16}$$

The motion is coparallel to the scattering vector $\mathbf{q}$, which gives all these dynamic light scattering correlation function's one-dimensional character, even though the actual motion is in three dimensions.

Autocorrelation analysis is a natural way to analyze fluctuating time series. However, power spectra can provide a more direct way to capture a hierarchy of time scales when the fluctuations result from several distinct processes. The Wiener–Khinchin theorem makes a connection between autocorrelation functions and spectral power density. The Wiener–Khinchin theorem states:

$$S(\omega) = F(\omega)F^*(\omega) = FT\left(\int_{-\infty}^{\infty} f(t)f(t+\tau)dt\right) = FT(A(\tau)) \tag{11.17}$$

Therefore, any information contained within the autocorrelation function is also contained within the spectral power density. The autocorrelation and power spectra functions for diffusion and Maxwellian drift are given in Table 11.1. In our analysis of motility contrast and tissue dynamics, we choose to emphasize the power spectrum, and, in particular,

Table 11.1: Autocorrelations and Spectral Densities for DLS

	$A(\tau)$	$S(\omega)$
Diffusion	$N\exp(-\mathbf{q}^2 D\tau)$	$\dfrac{N}{\pi}\left(\dfrac{1}{1+(\omega/\mathbf{q}^2 D)^2}\right)$
Maxwell velocity	$N\exp(-\mathbf{q}^2 v_{rms}^2 \tau^2/2)$	$\dfrac{N}{2\pi}\exp(-\omega^2/2\mathbf{q}^2 v_{rms}^2)$

changes in the power spectrum caused by changing environmental conditions or by applied drugs on the live tissue.

As the optical thickness of a sample increases beyond one optical transport length, multiple scattering becomes significant, and the diffusive character of light begins to dominate [9–11]. To continue extracting meaningful dynamics from inside living tissues is a challenge. DLS information can still be extracted [12,13] using diffusing wave spectroscopy [14,15] or diffuse correlation spectroscopy [16–19]. Each scattering event along a path contributes stochastically to the phase shift of the photon following that path, and the stochastic phase modulations accumulate in the phase of the photon. Therefore, the multiple scattered wave can be expressed as a product of stochastic exponentials. All the possible paths that a photon can take must be considered in the total correlation function.

The correlation function for the field fluctuations is

$$g_k^{(1)}(\tau) = \exp(-i\omega t)\left\langle \prod_{n=1}^{N_k} |a_n|^2 \exp(i\bar{\mathbf{q}}_n \Delta \bar{r}_n(\tau)) \right\rangle$$
$$= \exp(-i\omega t)\langle |a_n|^2 \rangle \exp\left(-\langle \bar{\mathbf{q}}^2 \rangle \langle \Delta r^2(\tau) \rangle \frac{N_k}{6}\right) \quad (11.18)$$

where cross-terms in the correlation function vanish because scattering events are assumed to be uncorrelated. A mean value for the mean squared scattering vector is

$$\langle \bar{\mathbf{q}}^2 \rangle = 2k^2 \frac{\ell}{\ell^*} \quad (11.19)$$

where ℓ is the scattering mean free path, and ℓ^* is the transport mean free path. The number of scattering events for each path is $N_k = S_k/\ell$, where S_k is the total path length. The combined correlation function for a continuous distribution of possible paths s is

$$g^{(1)}(\tau) = \int_0^\infty \exp\left(-2\frac{\tau}{\tau_0}\frac{s}{\ell^*}\right)\rho(s)ds \quad (11.20)$$

where τ_0 is the characteristic time of the diffusive or Maxwellian processes and $\rho(s)$ is the probability density of possible paths that can be solved by solving the photon diffusion equation subject to the boundary geometry of the sample and the intensity distribution of the incident light [19]. The argument $2(\tau/\tau_0)(s/\ell^*)$ of the exponential is the product of the single-scattering rate multiplied by the average number of scattering events along the path (s/ℓ^*). Longer paths lead to faster decorrelation because more scattering events add to scramble the phase.

The multiple scattering analysis is important for fluctuation spectroscopy of dense tissue. The characteristic frequencies that appear in Table 11.1 are for quasi-elastic light scattering (QELS), but in multiple scattering, these frequencies must be multiplied by (s/ℓ^*). For coherence gating to a depth of 200–400 μm, this is a factor of 4–8. Therefore, the characteristic frequencies in MCI can be up to an order of magnitude higher than those obtained from single-scattering studies.

11.5 Tissue Dynamics Spectroscopy

MCI, described in Section 11.3, uses statistical properties to generate motility metrics and motility maps of live tissue samples. While these motility metrics, such as the NSD, capture the total motion, they do not provide information on the many different types of motion that occur inside the tissues. To access this more specific information, we perform fluctuation analysis of the fluctuating speckle originating from a coherence-gated depth inside the live tissue. Our emphasis for tissue dynamics spectroscopy (TDS) is on the spectral power density [20].

An example of the spectral power density from the proliferating shell of a tumor that is 500 μm in diameter is shown in Figure 11.8 at two temperatures. The spectrum spans three decades in frequency and in dynamic range. There is more spectral weight at higher frequencies at 37 °C than at 24 °C, and the knee frequency has shifted to higher rates as well.

To capture subtle effects of environmental or drug-induced changes in the power spectra, we define a normalized spectral difference as

$$D(\omega) = \frac{S_2(\omega) - S_1(\omega)}{S_1(\omega)} \quad (11.21)$$

where $S_1(\omega)$ is the baseline spectral power density, and $S_2(\omega)$ is the altered spectrum. An example for different osmolarities applied to four tumors is shown in Figure 11.9A. Hypotonic conditions enhance low frequencies and suppress high frequencies relative to hypertonic conditions. In the hypotonic case, the cells swell significantly to the bursting point and beyond. The enhanced low frequencies are likely to be related to gross shape changes of the membranes as they deform and form blebs. In the hypertonic case, the cells shrink and activate vesicle transport to attempt to compensate for the loss of water from the cytoplasm. The enhanced high frequencies may be related to enhanced

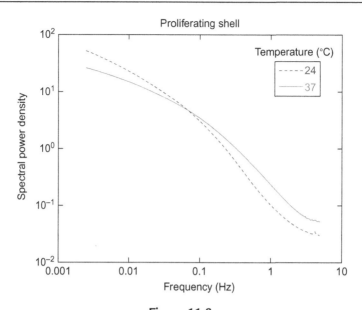

Figure 11.8
Spectral power density of the dynamic speckle from the proliferating shell at two temperatures.

cellular actions in response to the hypertonic insult. The effect of pH on the spectral content is very similar to osmolarity. Acidic conditions are known to induce cell swelling, and the spectral changes at low pH are consistent with hypotonic-induced cell swelling. The spectral density changes for pH = 5, 6, 8, and 9 are shown in Figure 11.9B, with many qualitative similarities to osmolarity-induced changes shown in Figure 11.9A.

To make a more detailed comparison of the effects of osmolarity and pH, we define a time-dependent differential spectral response as

$$D(\omega, t) = \frac{S(\omega, t) - S(\omega, 0)}{S(\omega, 0)}$$

where the time-dependent changes in the spectral power density are related to the baseline spectrum prior to the insult. The time-frequency data now represent a unique spectrogram, like a voice print, for each applied drug, dose, or condition. A comparison of the spectrograms subjected to osmolarity or pH is shown in Figure 11.10. Hypotonic osmolarity of 154 Osm is compared to pH 6, and hypertonic osmolarity of 428 Osm is compared to pH 8. The similarities are very strong for the hypertonic and alkaline conditions and nearly as strong for hypotonic and acidic conditions. Under application of 154 Osm, the spectrogram shows a clear transient volume-regulatory response that is entirely missing from the acidic pH spectrogram. This is consistent with the known physiology of hypotonic osmolarity and pH [21], in which pH-induced swelling does not

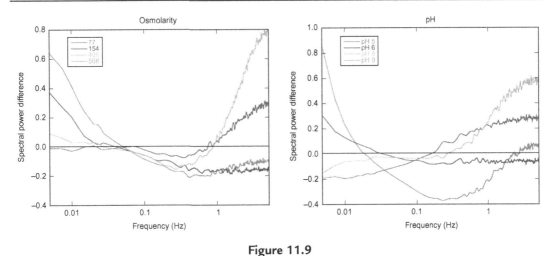

Figure 11.9
(A) Difference in spectral power density for postdose relative to predose spectra for a range of osmolarities. (B) Difference in spectral power density for postdose relative to predose spectra for a range of pH.

elicit a volume-regulatory response. On the other hand, for 428 Osm and pH 8, both spectrograms show a minor regulatory response.

One of the most important potential applications of TDS is for drug screening. TDS is label-free and captures the physiological changes induced by applied drugs. An example of two differential response spectrograms for colchicine and cytochalasin D is shown in Figure 11.11. Colchicine is an antimicrotubule drug that inhibits the polymerization of microtubules [22]. Microtubules are required for cellular mitosis, but they also play important roles in the cytoskeleton, in distributing stresses and in supporting organelle transport. When colchicine is applied, there is a rapid enhancement of low frequencies, followed by an enhancement of high frequencies approximately 3 h later. Cytochalasin D is an antiactin drug [23]. Actin is also a part of the cellular cytoskeleton and plays an important role in the stiffness of the cell membrane. Actin forms a cortex of interlinked filaments inside the cell membrane that helps give it rigidity. Cytochalasin D inhibits the polymerization of the actin filaments, and the cortex is degraded, leading to a more floppy cell membrane. The enhancement in midfrequencies for cytochalasin is likely due to enhanced membrane fluctuations and is shown in Figure 11.11. After approximately 3 h, the tissue undergoes a fast change to a new state with enhanced low and higher frequencies similar to the enhancements at long times observed for colchicine. These signatures may be related to cellular death under these high doses.

The time-frequency spectrograms in Figure 11.11 show distinctly different behaviors at different times and act like unique fingerprints of the drug effect on the tissue. By using

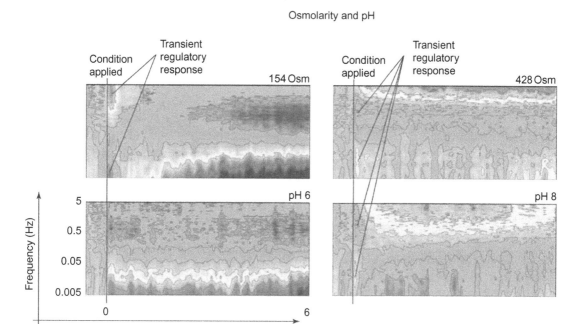

Figure 11.10
Differential spectrograms of hypo- and hypertonic conditions compared with acidic and alkaline conditions. The volumetric regulatory response in 154 Osm data is missing in the pH 6 data.

mathematical algorithms for feature extraction, these spectrogram fingerprints can be coded into unique feature vectors for different drugs, allowing a comparison of drug mechanism of action among known drugs and for comparison to new uncharacterized drug compounds. This approach to drug screening is novel and is compatible with high-content and high-throughput screening and is the subject of our ongoing research.

In conclusion, we have shown how to use motions inside live cells and tissues as endogenous contrast for functional imaging. The motions are captured through the effect on DLS that causes fluctuating speckle fields. Different types of motion have different characteristic time scales that allow the separation of different mechanisms into separate frequency bands. In general, small organelles have high-frequency signatures (1–10 Hz), membrane undulations have midfrequency signatures (0.1–1 Hz), and slow cell shape changes and tissue rearrangements have low-frequency signatures (0.01–0.1 Hz). The time development of the spectral power in the fluctuating speckle in response to applied environmental perturbations or drugs is captured in differential drug response spectrograms. Different drugs have different spectrogram signatures which might be useful for early drug discovery applications.

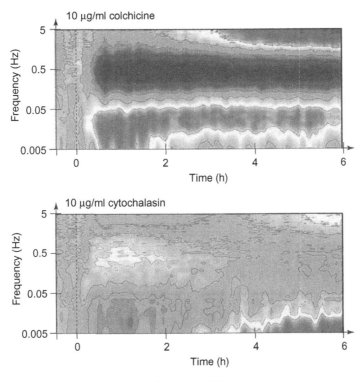

Figure 11.11
Differential response spectrograms for 10 µg/ml colchicine and cytochalasin D. The baseline is set prior to the application of the drugs.

Acknowledgment

This work was supported by the National Science Foundation Grant No. CBET-0756005.

References

[1] D.D. Nolte, Optical Interferometry for Biology and Medicine, Springer, New York, NY, 2011.
[2] L. Allen, D.G.C. Jones, Laser-produced speckle patterns, J. Opt. Soc. Am. 55 (1965) 1188.
[3] J.C. Dainty, Some statistical properties of random speckle patterns in coherent and partially coherent illumination, Opt. Acta 17 (1970) 761.
[4] J.W. Goodman, Some fundamental properties of speckle, J. Opt. Soc. Am. 66 (1976) 1145–1150.
[5] R. Hanbury-Brown, R.Q. Twiss, A test of a new type of stellar interferometer on Sirius, Nature 178 (1956) 1046–1048.
[6] K. Jeong, J.J. Turek, D.D. Nolte, Imaging motility contrast in digital holography of tissue response to cytoskeletal anti-cancer drugs, Opt. Express 15 (2007) 14057–14064.
[7] P. Yu, M. Mustata, P.M.W. French, J.J. Turek, M.R. Melloch, D.D. Nolte, Holographic optical coherence imaging of tumor spheroids, Appl. Phys. Lett. 83 (2003) 575–577.
[8] D. Huang, E.A. Swanson, C.P. Lin, J.S. Schuman, W.G. Stinson, W. Chang, et al., Optical coherence tomography, Science 254 (1991) 1178–1181.

[9] A. Wax, C.H. Yang, R.R. Dasari, M.S. Feld, Path-length-resolved dynamic light scattering: modeling the transition from single to diffusive scattering, Appl. Opt. 40 (2001) 4222–4227.

[10] R. Carminati, R. Elaloufi, J.J. Greffet, Beyond the diffusing-wave spectroscopy model for the temporal fluctuations of scattered light, Phys. Rev. Lett. 92 (2004) 213903-1–213903-4.

[11] P.A. Lemieux, M.U. Vera, D.J. Durian, Diffusing-light spectroscopies beyond the diffusion limit: the role of ballistic transport and anisotropic scattering, Phys. Rev. E 57 (1998) 4498–4515.

[12] G. Maret, P.E. Wolf, Multiple light-scattering from disordered media—the effect of Brownian motion of scatterers, Z. Phys. B: Condens. Matter 65 (1987) 409–413.

[13] M.J. Stephen, Temporal fluctuations in wave-propagation in random-media, Phys. Rev. B 37 (1988) 1–5.

[14] D.J. Pine, D.A. Weitz, P.M. Chaikin, E. Herbolzheimer, Diffusing-wave spectroscopy, Phys. Rev. Lett. 60 (1988) 1134–1137.

[15] G. Maret, Diffusing-wave spectroscopy, Curr. Opin. Colloid Interface Sci. 2 (1997) 251–257.

[16] B.J. Ackerson, R.L. Dougherty, N.M. Reguigui, U. Nobbmann, Correlation transfer—application of radiative-transfer solution methods to photon-correlation problems, J. Thermophys. Heat Transf. 6 (1992) 577–588.

[17] R.L. Dougherty, B.J. Ackerson, N.M. Reguigui, F. Dorrinowkoorani, U. Nobbmann, Correlation transfer—development and application, J. Quant. Spectrosc. Radiat. Transf. 52 (1994) 713–727.

[18] D.A. Boas, L.E. Campbell, A.G. Yodh, Scattering and imaging with diffusing temporal field correlations, Phys. Rev. Lett. 75 (1995) 1855–1858.

[19] T. Durduran, R. Choe, W.B. Baker, A.G. Yodh, Diffuse optics for tissue monitoring and tomography, Rep. Prog. Phys. 73 (2010) 076701, 1–43.

[20] D.D. Nolte, R. An, J.J. Turek, K. Jeong, Holographic tissue dynamics spectroscopy, J. Biomed. Opt. 16 (2011) 087004, 1–14.

[21] L.P. Sullivan, D.P. Wallace, R.L. Clancy, C. Lechene, J.J. Grantham, Cellular electrolyte and volume changes induced by acidosis in the rabbit proximal straight tubule, J. Am. Soc. Nephrol. 2 (1991) 1030–1040.

[22] R.B.G. Ravelli, B. Gigant, P.A. Curmi, I. Jourdain, S. Lachkar, A. Sobel, et al., Insight into tubulin regulation from a complex with colchicine and a stathmin-like domain, Nature 428 (2004) 198–202.

[23] J.A. Cooper, Effects of cytochalasin and phalloidin on actin, J. Cell Biol. 105 (1987) 1473–1478.

PART 3

Advanced Interferometric and Polarization Techniques

CHAPTER 12
Tomographic Phase Microscopy (TPM)

Wonshik Choi
Department of Physics, Korea University, Seoul, Korea

Editor: Natan T. Shaked

12.1 Introduction

Conventional interferometric microscopy techniques such as digital holographic microscopy and quantitative phase microscopy are classified as three-dimensional (3D) imaging techniques because a recorded complex field image can be numerically propagated to a different depth. Strictly speaking, however, a single complex field image contains only two-dimensional (2D) information on a specimen because the measured 2D image is only a subset of the 3D structure. In this chapter, we introduce tomographic phase microscopy (TPM) that experimentally implements optical diffraction tomography (ODT) suggested by Wolf in 1969 [1,2] for the 3D mapping of living specimens. The TPM enables us to record complex field images of a specimen at various angles of illumination. And the ODT makes it possible to reconstruct the 3D structure of a specimen from the multiple angle-dependent or wavelength-dependent complex field images. By mapping the acquired multiple independent 2D images onto the 3D Fourier space of the specimen, a tomographic map of the absorption coefficient and refractive index can be reconstructed. Since the refractive index is a measure of the molecular concentration, the tomographic map of the refractive index in a single intact living cell has drawn many interesting biomedical applications. In the latter part of this chapter, applications including quantitative assessment of disease progression in malaria-infected red blood cells (RBCs) are covered.

Refractive index serves as an important intrinsic contrast agent in visualizing nearly transparent living biological cells. Examples include phase contrast microscopy [3] and differential interference microscopy [4], which have been widely used in cell biology studies. In essence, both techniques make use of optical interferometry to enhance the contrast. Interferometry converts phase changes of the transmitted wave induced by the heterogeneous refractive index distribution within the cell into intensity variations.

But these techniques do not provide quantitative maps of phase change. Digital holographic microscopy and phase-shifting interferometric microscopy have been developed to overcome the qualitative limit of the phase contrast microscopy and differential interference contrast (DIC) microscopy. They can record the complex field (both amplitude and phase) of the wave generated by a specimen whose phase part provides a quantitative phase image of specimen-induced phase changes [5–9]. On the other hand, these phase microscopy techniques can either provide average refractive index of cells or cell thickness, but not detailed 3D structure. The imaging of a complex field is often considered to be 3D imaging due to the possibility of numerically propagating the recorded field to a different depth. However, the measured complex field image is only a 2D subset of the 3D object such that the information is significantly deficient. Depth-dependent information can be obtained only when there are additional constraints on the target object.

From the viewpoint of information capacity, multiple independent 2D images are to be acquired in order to obtain the 3D map of the object. In 1969, Wolf first proposed a theoretical framework known as ODT in which 2D complex field images of a specimen taken at various illumination angles fill the 3D Fourier space of the specimen [1,2]. Since then, various methods have been proposed to experimentally record multiple independent 2D complex field images, for example, either for various angles of illumination [10–13] or for different wavelengths of the light sources [14–15]. In the case of wavelength scanning, the bandwidth is often limited by the wavelength-dependent index change, i.e., dispersion, by the sample and optics. Therefore, spatial frequency coverage of the wavelength scanning is typically narrower than that of the angular scanning. In the case of the angular scanning method, there are two ways to change the relative angle of illumination with respect to the specimen. One is to rotate the sample with the illumination beam fixed, and the other is to rotate the illumination beam with the sample fixed. Rotating the sample makes it possible to cover the entire angular range, and thus ensures the same axial resolution as the transverse resolution. But it is difficult to fix the axis of rotation, and rotation inevitably perturbs the sample. In addition, data acquisition speed is limited due to the mechanical rotation of the sample. Therefore, the use of sample rotation is typically restricted to solid nonbiological objects such as optical fibers [10,16]. Special sample preparation is required for imaging biological cells [12,17].

On the other hand, the rotating-beam approach does not perturb the sample during data acquisition and is thus suitable for imaging live cells in their native state [11,13]. Data acquisition can be fast enough to study the dynamics of the live cells. Only small modifications are necessary for the instrument to fit into a conventional high NA (numerical aperture) optical microscope. A drawback of this method is caused by the lack of complete angular coverage due to the finite numerical aperture of an imaging system. In fact, conventional optical microscopes share the same limitation. Thus, the axial resolution is poorer than the transverse resolution. In an effort to attenuate the effect of this drawback, various algorithms have been developed to solve missing angle information with a prior knowledge of the specimen [18–20].

Along with the data acquisition in the experiment, the reconstruction algorithm is an important factor for the spatial resolution and accuracy of the complex refractive index measurement. The way of interpreting the experimentally measured complex field images determines the algorithm to be used. If the phase of the transmitted field is interpreted as a line integral of the refractive index along the propagation direction, then the filtered back-projection algorithm based on the inverse Radon transform can be applied [21]. For weakly scattering biological cells, this is often a good approximation [12,13] for points close to the plane of focus. However, since the effect of diffraction is ignored, there is loss of resolution for samples which are large compared to the depth of focus of the imaging system.

ODT is more of general approach in the sense that the effect of diffraction is taken into consideration. The Born approximation was first adopted by Wolf to make the relation linear between the complex refractive index of the object and the complex electric field (E-field). Several experimental studies have implemented diffraction tomography based on Born approximation in the optical regime [11,16,22]. But it turned out that the validity of the Born approximation was in question when the phase retardation by the specimen reached to $\pi/2$ [21] even if the attenuation of amplitude was negligible. This has led to the introduction of Rytov approximation in which the approximation is made on the complex phase of the scattered wave [2,23]. The Rytov approximation is more robust to perform imaging of the phase object than the Born approximation.

In this chapter, we introduce TPM [13,24,25] as an experimental method for quantitative 3D mapping of refractive index in live cells in their native state. TPM can collect angular images ranging from $-70°$ to $70°$ in as little as 1/30 s. The rotating-beam geometry was adopted to avoid perturbation of specimens during data acquisition, and filtered back-projection along with an iterative constraint algorithm was used at the original development for the 3D reconstruction. Later, the first experimental implementation of ODT was carried out based on the Rytov approximation to image live biological cells and to provide quantitative 3D refractive index maps. It was demonstrated that the Rytov approximation is valid for live cell imaging, while reconstruction based on the Born approximation leads to severe distortions. An iterative constraint algorithm is applied to minimize the effects of incomplete angular coverage. The acquired quantitative refractive index maps were used to quantify molecular concentrations without adding fluorescent agents [26]. They also provide a means of studying the light scattering of single cells [27], which helps the development of in vivo light scattering instruments for disease diagnosis.

12.2 Theory—Optical Diffraction Tomography

12.2.1 Inverse Radon Transform

For the illumination of plane waves on a thin sample with small index contrast, the phase of the transmitted wave is to a good approximation equal to the line integral of the refractive index along the path of beam propagation. Therefore, the phase image can simply be interpreted as the projection of refractive index, analogous to the projection of absorption in X-ray tomography. Then the Fourier slice theorem, also known as inverse Radon transform, can be used to reconstruct the 3D map. Since the theorem is relatively well documented and the application of the theorem to the experimental data is rather straightforward, we guide the readers to consult Ref. [21].

12.2.2 Optical Diffraction Tomography

It is relatively straightforward to implement a deconvolution algorithm for creating a 3D fluorescent image from a stack of full-field fluorescent images taken while scanning an objective lens. Each fluorescent particle acts as a point source, and there is negligible interference among the particles. The point spread function can be defined only by the imaging system. On the other hand, it is more complicated to implement 3D deconvolution for absorption and refractive index. Unlike fluorescent imaging, both amplitude and phase images of the transmitted field must be recorded since samples affect both amplitude and phase of the field. Moreover, interference among scatterers complicates the deconvolution process. To fully describe the effect of interference, the wave equation must be solved. This requires extensive computation time, and it is even more difficult to extract the structure of objects from the transmitted E-field images.

Approximations such as Born and Rytov have been employed in the past to make this problem relatively easy to solve [2,21]. According to these approximations, the relationship between the 3D scattering potential and the 2D measured field can be simplified by assuming that the scattered field is weak compared to the incident field. Using the Born approximation, Wolf derived a formulation that enables reconstruction of a 3D object from 2D measured E-fields [1]. For each illumination angle, the Fourier transform of the 2D measured E-field is mapped onto a spherical surface in the frequency domain of the 3D scattering potential. This spherical surface is called the Ewald sphere. In this section, we briefly introduce Wolf's original theory and Devaney's modification to adopt the first Rytov approximation [23].

With scalar field assumption, the propagation of light field, $U(\vec{R})$, through the medium can be described by the wave equation as follows:

$$\nabla^2 U(\vec{R}) + k_0^2 n(\vec{R})^2 U(\vec{R}) = 0 \qquad (12.1)$$

Here, $k_0 = 2\pi/\lambda_0$ is the wave number in the free space with λ_0 the wavelength in the free space, and $n(\vec{R})$ is the complex refractive index. If the field is decomposed into the incident field $U^{(I)}(\vec{R})$ and scattered field $U^{(S)}(\vec{R})$,

$$U(\vec{R}) = U^{(I)}(\vec{R}) + U^{(S)}(\vec{R}) \tag{12.2}$$

then, the wave equation becomes

$$(\nabla^2 + k_0^2 n_m^2) U^{(S)}(\vec{R}) = F(\vec{R}) U(\vec{R}) \tag{12.3}$$

with $F(\vec{R}) \equiv -(2\pi n_m/\lambda_0)^2((n(\vec{R})/n_m)^2 - 1)$, and n_m is the refractive index of the medium. The $F(\vec{R})$ is known as the object function. Based on Green's theorem, the formal solution to Eq. (12.3) can be written as:

$$U^{(S)}(\vec{R}) = -\int G(|\vec{R} - \vec{R}'|) F(\vec{R}') U(\vec{R}') d^3\vec{R}' \tag{12.4}$$

with $G(r) = \exp(i n_m k_0 r)/(4\pi r)$ the Green's function. Since the integrand contains the unknown variable, $U(\vec{R})$, we employ an approximation to obtain a closed form solution for $U^{(S)}(\vec{R})$. The first Born approximation is the simplest we can introduce when the scattered field is much weaker than the incident field ($U^{(S)} \ll U^{(I)}$), in which case the scattered field is given by the following equation:

$$U^{(S)}(\vec{R}) \approx -\int G(|\vec{R} - \vec{R}'|) F(\vec{R}') U^{(I)}(\vec{R}') d^3\vec{R}' \tag{12.5}$$

This approximation provides a linear relation between the object function and the scattered field $U^{(S)}(\vec{R})$. By taking the Fourier transform of both sides of Eq. (12.5), we obtain the following relation, known as the Fourier diffraction theorem [1]:

$$\hat{F}(K_x, K_y, K_z) = \frac{i k_z}{\pi} \hat{U}^{(S)}(k_x, k_y; z^+ = 0) \tag{12.6}$$

Here, $\hat{F}$ and $\hat{U}^{(S)}$ are the 3D and 2D Fourier transform of F and $U^{(S)}$, respectively; k_x and k_y are the spatial frequencies corresponding to the spatial coordinate x and y in the transverse image plane, respectively; $z^+ = 0$ is the axial coordinate of the detector plane, which is the plane of objective focus in the experiment. (K_x, K_y, K_z), the spatial frequencies in the object frame, define the spatial frequency vector of (k_x, k_y, k_z) relative to the spatial frequency vector of the incident beam (k_{x0}, k_{y0}, k_{z0}), and k_z is determined by the relation $k_z = \sqrt{(n_m k_0)^2 - k_x^2 - k_y^2}$. For each illumination angle, the incident wave vector changes, and so does (K_x, K_y, K_z). As a result, we can map different regions of the 3D frequency spectrum of the object function with various 2D angular complex E-field images. After completing the mapping, we can take the inverse Fourier transform of $\hat{F}$ to get the 3D distribution of the complex refractive index.

Numerical simulations have demonstrated that the Born approximation is valid when the total phase delay of the E-field induced by the specimen is less than $\pi/2$ [21]. The thickness of a single biological cell is typically about 10 μm, with index difference with respect to the medium about 0.03. Thus, the phase delay induced by typical cells is approximately π at a source wavelength of $\lambda = 633$ nm. Therefore, the Born approximation is not expected to be valid for imaging biological cells. In this respect, the Rytov approximation is more relevant than the Born approximation. It is not sensitive to the size of the sample or the total phase delay, but rather to the *gradient* of the refractive index. Specifically, the Rytov approximation is valid when the following condition is satisfied:

$$n_\delta \gg \left(\nabla \phi^{(S)} \frac{\lambda}{2\pi} \right)^2, \quad \text{with} \quad \phi^{(S)} = \ln\left(\frac{U(\vec{R})}{U^{(I)}(\vec{R})} \right) \qquad (12.7)$$

and n_δ is the index variation in the sample over the length scale of wavelength. This condition basically asserts that the Rytov approximation is independent of the specimen size and only limited by the phase gradient $\nabla \phi^{(S)}$. For a weakly scattering sample such as a biological cell, the phase change $\nabla \phi^{(S)}$ is linearly proportional to n_δ to a first approximation, such that the relation is valid when $n_\delta \ll 1$. The index variation n_δ is in the range of 0.03–0.04 for biological cells. Therefore, we can expect that the Rytov approximation is legitimate in imaging biological cells, while the Born approximation is subject to significant distortions in the reconstructed image.

As suggested by Devaney [23], the implementation of the Rytov approximation in the Fourier diffraction theorem requires a slightly different approach. Following Devaney's method, we introduce the complex phase, $\phi(\vec{R})$, defined by $U(\vec{R}) = e^{\phi(\vec{R})}$, and substitute this into the wave equation (Eq. (12.1)). After applying the approximation given by Eq. (12.7), we again obtain the Fourier diffraction theorem (Eq. (12.6)), but with $U^{(S)}$ replaced by $U^{(S)}_{\text{Rytov}}$ defined as:

$$U^{(S)}_{\text{Rytov}} = U^{(I)}(\vec{R}) \ln\left(\frac{U(\vec{R})}{U^{(I)}(\vec{R})} \right) \qquad (12.8)$$

The rest of the reconstruction is the same as described in the text following Eq. (12.6).

12.3 Experimental Implementation

12.3.1 Experimental Setup

ATPM instrument is shown in Figure 12.1. It is designed to record complex E-field images at various angles of illumination for the sample stationary at the sample stage of the microscope [13]. The heterodyne Mach–Zehnder interferometer [13] is used in the instrument to record amplitude images and phase images. An He–Ne laser beam

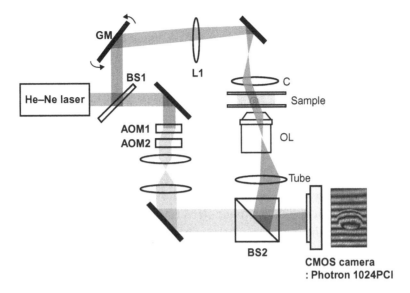

Figure 12.1
Tomographic phase microscope. GM, galvanometer scanning mirror; L1, lens ($f = 250$ mm); C, condenser lens (NA 1.4); OL, objective lens (NA 1.4); Tube, tube lens ($f = 200$ mm); BS1 and BS2, beamsplitters; and AOMs, acousto-optic modulators. The frequency-shifted reference laser beam is shown in blue [13]. (For interpretation of the references to color in this figure legend, the reader is referred to the web version of this book.)

(wavelength $\lambda = 633$ nm) was divided into sample and reference beams. The propagation direction of the sample beam is controlled by a galvanometer mirror, and the sample image of the transmitted beam is delivered to the camera by objective and tube lenses. Two acousto-optic modulators are used to shift the frequency of the reference beam by 1.25 kHz, and the frame rate of a CMOS (complementary metal-oxide semiconductor) camera (Photron 1024PCI) is adjusted to take images at 200 μs intervals. For each angle of illumination, four successive interferogram images are recorded in 800 μs, and phase-shifting interferometry is used to produce a pair of quantitative phase and amplitude images. To maximize the range of illumination angles, a high NA condenser (Nikon, 1.4 NA) and an objective lens (Olympus UPLSAPO, 1.4 NA) are used. The sample beam is rotated using a galvanometer mirror to cover from $-70°$ to $70°$ in $0.23°$ steps. It takes about 10 s to record a set of angular complex E-field images.

12.3.2 Data Processing by Inverse Radon Transform

To reconstruct a 3D refractive index tomogram from the projection phase images, a procedure based on the filtered back-projection method [21] is applied. A discrete inverse

Radon transform is applied to every $X-\theta$ slice in the beam rotation direction, with X the coordinate in the tilt direction. To compensate for the angle between imaging and illumination directions, the X values are divided by $\cos \theta$ (Figure 12.2).

To validate the instrument's measurements, refractive index tomograms are measured for 10 µm polystyrene beads (Polysciences #17136, $n = 1.588$ at $\lambda = 633$ nm) immersed in oil with a slightly smaller refractive index (Cargille #18095, $n = 1.559$ at $\lambda = 633$ nm). Tomograms showed a constant refractive index inside each bead, and the refractive index difference between the bead and its surroundings was $\Delta n = 0.0285 \pm 0.0005$, in agreement with the manufacturers' specifications for beads and oil ($\Delta n = 0.029$). Similar tests with a range of oil refractive indices from $n = 1.55$ to $n = 1.59$ also gave good agreement.
By measuring the full width at half maximum (FWHM) of the derivative of line profiles of refractive index normal to the boundary of the sphere, the spatial resolution of the

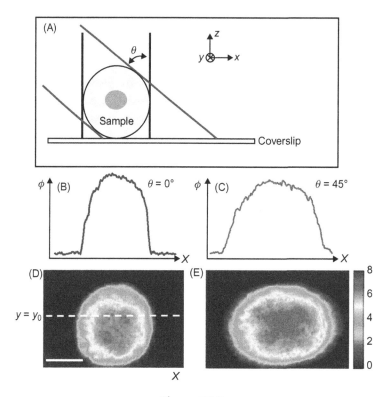

Figure 12.2
(A) Illumination geometry and projection on x–z plane. Blue lines represent the illumination at zero degrees and red lines at angle θ. (B and C) Projection phase line profiles of a HeLa cell at $\theta = 0°$ and $45°$, respectively, at fixed $y = y_0$. (D and E) Projection phase images corresponding to (B and C), respectively. The color bar indicates the phase in radian. Scale bar, 10 µm [13]. (For interpretation of the references to color in this figure legend, the reader is referred to the web version of this book.)

tomographic technique is measured to be approximately 0.5 μm in the transverse ($x-y$) directions and 0.75 μm in the longitudinal (z) direction.

A tomogram is measured for a single HeLa cell in culture medium. Cells were dissociated from culture dishes and allowed to partially attach to the coverslip substrate. A 3D index tomogram for a single cell (Figure 12.3A and B) and $x-y$ tomographic slices of the same cell at heights of $z = 12, 9.5, 8.5, 7.5, 6.5$, and 5.5 microns above the substrate (Figure 12.3C–H) show that the index of refraction is highly inhomogeneous, varying from 1.36 to 1.40. Brightfield images for objective focus corresponding to Figure 12.3E and F are shown in Figure 12.3I and J, respectively. There is a clear correspondence between the tomographic and brightfield images in terms of cell boundary, nuclear boundary, and size and shape of the nucleoli.

It is noteworthy that the refractive index of the nucleus ($n \approx 1.36$), apart from the nucleolus, is smaller than some parts of the cytoplasm ($n \approx 1.36-1.39$) and that the refractive index

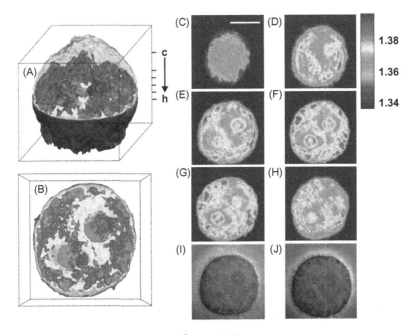

Figure 12.3
Refractive index tomogram of a HeLa cell. (A) 3D rendered image of a HeLa cell. The outermost layer of the upper hemisphere of the cell is omitted to visualize the inner structure. Nucleoli are colored green and parts of cytoplasm with refractive index higher than 1.36 are colored red. The dotted box is a cube of side 20 μm. (B) Top view of (A). (C–H) Slices of the tomogram at heights indicated in (A). Scale bar, 10 μm. The color bar indicates the refractive index at $\lambda = 633$ nm. (I and J) Brightfield images for objective focus corresponding to (E and F), respectively [13]. (For interpretation of the references to color in this figure legend, the reader is referred to the web version of this book.)

of the nucleoli, $n \approx 1.38$, is larger than that of the rest of the nucleus. This is contrary to the widely cited claims that the refractive index of the nucleus as a whole is higher than that of the rest of the cell [28]. Similar results were obtained for cultured HEK 293 cells, B35 neuroblastoma cells, and primary rat hippocampal neurons. All cells imaged contained many small cytoplasmic particles with high refractive index, which may be lipid droplets, lysosomes, vacuoles, or other organelles.

To demonstrate tomographic imaging of a multicellular organism, the nematode *Caenorhabditis elegans*, which is paralyzed with 10 mM sodium azide in nematode growth medium (NGM) buffer, is imaged in the same solution. Overlapping tomograms are created and the resulting data assembled into a mosaic (Figure 12.4B) and compared with its brightfield counterpart (Figure 12.4A). Several internal structures are visible, including a prominent pharynx and digestive tract.

12.3.3 Data Processing by the ODT

Using the set of phase and amplitude images taken at various angles of illumination, the diffraction tomography algorithm described in Section 12.2.2 can be applied for the 3D reconstruction. Given a quantitative phase image $\phi(x,y;\theta)$ and an amplitude image $A(x,y;\theta)$ taken at each illumination angle, θ, the complex E-field at the image plane is constructed by $U(x,y;\theta) = A(x,y;\theta)e^{i\phi(x,y;\theta)}$. The measured field image is composed of the phase change induced by the sample and the phase ramp introduced by the tilted illumination. A corresponding set of images $U_{bg}(x,y;\theta) = A_{bg}(x,y;\theta)e^{ik_{x0}x+ik_{y0}y}$ taken when no sample is present provides the background field, which can be considered as the incident fields.

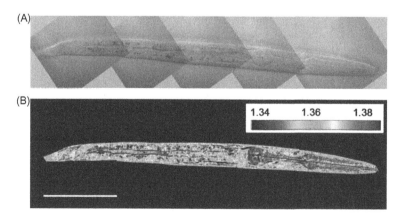

Figure 12.4
(A) Mosaic of brightfield images of a juvenile *C. elegans* nematode. (B) Mosaic of x–y slices of index tomograms through center of worm. The color bar indicates the refractive index at $\lambda = 633$ nm. Scale bar, 50 μm [13]. (For interpretation of the references to color in this figure legend, the reader is referred to the web version of this book.)

Figure 12.5A shows the phase image $\phi(x,y; \theta = 0)$ of a 6 μm polystyrene bead (Polysciences Inc.) taken at zero incidence angle. Figure 12.5B shows the typical amplitude image of $\hat{U}(k_x, k_y; \theta)$ on a logarithmic scale.

To apply the Fourier diffraction theorem (Eq. (12.6)), (k_x, k_y) on the right-hand side is converted into (K_x, K_y) as follows:

$$\hat{F}(K_x, K_y, K_z) = \frac{i(K_z + k_{z0})}{\pi} \hat{U}^{(S)}(K_x + k_{x0}, K_y + k_{y0}; z^+ = 0) \quad (12.9)$$

Then, either $\hat{U}^{(S)}(k_x, k_y; \theta)$ (Eq. (12.2)) or $\hat{U}^{(S)}_{\text{Rytov}}(k_x, k_y; \theta)$ (Eq. (12.8)) is calculated from measured complex fields. Shifting the spectrum by $(-k_{x0}, -k_{y0})$ in spatial frequency space

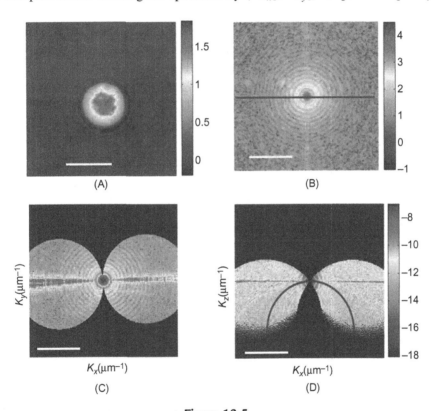

Figure 12.5

Mapping of the complex E-field onto the 3D Fourier space of the object function.
(A) Quantitative phase image of a 6 μm polystyrene bead at zero-degree illumination. The color bar indicates phase in radians. Scale bar, 5 μm. (B) Amplitude of the Fourier transform of the complex E-field image at zero-degree illumination on a logarithmic scale. (C) Amplitude distribution in K_x–K_y plane after mapping all the angular E-field images. (D) Amplitude distribution in the K_x–K_z plane. The color bar indicates base-10 logarithm of the amplitude of E-field. The scale bars in (B–D) indicate 2 μm^{-1} [24]. (For interpretation of the references to color in this figure legend, the reader is referred to the web version of this book.)

corresponds to the right-hand side of Eq. (12.9). In mapping the experimental data, this is equivalent to dividing the scattered field by the incident field $U_{bg}(x,y;\theta)$, which shifts $\hat{U}^{(S)}(k_x,k_y;\theta)$ in Fourier space. In other words, the scattered field used in the Fourier diffraction theorem in the experiment is as follows:

In the case of the first Born approximation:

$$\hat{U}^{(S)}(K_x+k_{x0}, K_y+k_{y0};\theta) = (U(x,y;\theta) - U_{bg}(x,y;\theta))/U_{bg}(x,y;\theta) \quad (12.10)$$

In case of the first Rytov approximation:

$$\hat{U}^{(S)}_{Rytov}(K_x+k_{x0}, K_y+k_{y0};\theta) = \ln(U(x,y;\theta)/U_{bg}(x,y;\theta)) \quad (12.11)$$

Figure 12.5C and D shows the results of this mapping on the $(K_x, K_y, K_z=0)$ and $(K_x, K_y=0, K_z)$ planes, respectively. The data along the blue line in Figure 12.5B are mapped onto the blue half circle on the (K_x, K_z) space of Figure 12.5D. Different angular images are mapped onto different spaces such that they eventually cover a significant portion of the (K_x, K_y, K_z) space of the object function $F(\vec{R})$. Looking at the frequency spectrum of Figure 12.5C and D, ring patterns are clearly visible after mapping various angular images, which is expected for the spherical shape of the sample. By taking the inverse Fourier transform of the entire 3D frequency spectrum, the 3D distribution of refractive index and absorption coefficient of the object can be obtained.

Even with the use of high NA object and condense lenses, an image within illumination angles of up to $\pm 70°$ can be measured. As a result, the entire region of frequency space cannot be filled as shown in Figures 12.5C and D. In other words, the inverse problem is underdetermined. In the first round of reconstruction, zero values can be put in the missing angle space. The reconstructed object function then exhibits negative bias around the sample and the refractive index is smaller than the actual value (Figure 12.6A). To minimize the artifact introduced by the missing angles, an iterative constraint algorithm [18,19] can be applied based on the prior knowledge that the object function is non-negative for the live cells. The index throughout the field of view, either inside or outside of the cell, is at least the same or higher than the medium. As a first step, zero values are filled with the missing space and then the inverse Fourier transform is taken to reconstruct a 3D map (Figure 12.6D). In the reconstructed image, there are pixels whose index values are smaller than the index of the medium (Figure 12.6A). These are forced to be the same as the index of the medium and then the modified map is taken Fourier transformed. The index values in the Fourier space in which zero values are assumed are no longer zero, and an approximate solution for the missing angles is acquired (Figure 12.6E). But, at the same time, the data in the space which contains measured data are now modified. Since the experimentally measured data are accurate, we replace the modified data with the measured data. This procedure can be iterated until the reconstructed object function converges (Figure 12.6C and F). Then, the negative bias is removed and the reconstructed image becomes more accurate. For the case of the

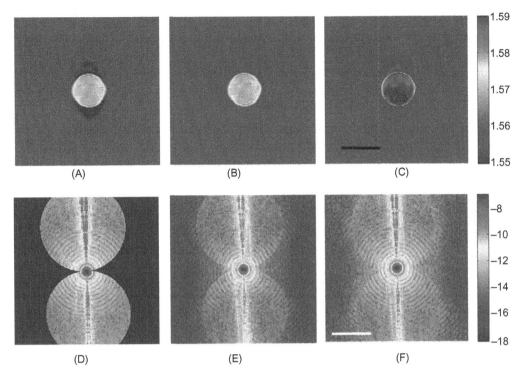

Figure. 12.6

Iterative constraint algorithm. (A) Slice image of a 6 μm bead before application of the constraint algorithm. (B) Same slice image as in (A) after application of the non-negative constraint. (C) Same slice image as in (B) after 100 iterations. The color bar indicates the refractive indices at 633 nm wavelength. Scale bar, 5 μm. (D) Amplitude distribution in K_x–K_y plane before application of the constraint algorithm. (E) 3D Fourier transform of tomogram after non-negative constraint. (F) 3D Fourier transform of tomogram after 100 iterations. The color bar indicates base-10 logarithm of E-field amplitude. Scale bar, 2 μm^{-1} [24]. (For interpretation of the references to color in this figure legend, the reader is referred to the web version of this book.)

polystyrene bead, we estimate that the accuracy of the measured refractive index is close to 0.001 after application of the iterative constraint algorithm. When the Fourier maps before (Figure 12.6D) and after (Figure 12.6F) iterations are compared, the ring patterns are generated in the missing angle regions (Figure 12.6F). This indicates that iterative constraint algorithm can generate reasonably accurate solutions for the missing angle regions.

To compare the performance of the inverse Radon transform and ODT, two sets of angular E-field images are acquired for 6 μm polystyrene beads (Polysciences) immersed in oil (Cargille, $n = 1.56$) at two different foci, one in the middle of the bead and the other 4 μm above the center (Figure 12.7). When the inverse Radon transform is applied, the slice image at the middle of the bead is uniform when the objective focus is set to the middle of the bead (Figure 12.7A). However, when the focus is above center, the slice image in the middle of the bead presented ring patterns (Figure 12.7B), which are due to diffraction of the propagating beam. This suggests that the inverse Radon transform be subject to the error in the reconstruction due to its inability to take the effect of diffraction into account. On the other hand, ODT based on the Rytov approximation is free from this artifact. Slice images of tomograms reconstructed by ODT are shown in Figure 12.7C and D at the objective focus in the middle of the bead and 4 μm above the center of the bead, respectively. Both images show clear boundaries of the bead with uniform index distributions. This indicates that the diffraction tomography properly accounts for the effects of diffraction. Note that the index of the bead relative to that of the oil is set to 0.03. This difference is very close to the relative index of the cell to the culture medium. Hence, the Rytov approximation is expected to be applicable to imaging of single cells.

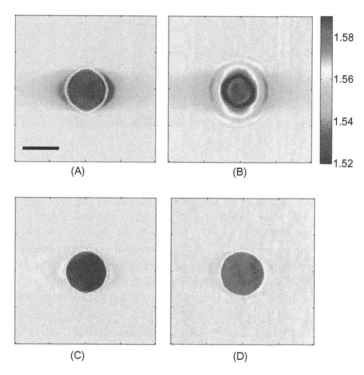

Figure 12.7

Comparison between filtered back-projection algorithm and diffraction tomography with the Rytov approximation. (A and B) Slice images of tomogram in the middle of a 6 μm bead reconstructed by the filtered back-projection algorithm when the objective focus is in the middle of the bead (A) and 4 μm above the center of the bead (B). (C and D) Same slice images as (A and B) after reconstructed by the diffraction tomography based on the Rytov approximation at objective focus in the middle of the bead (C) and 4 μm above the center of the bead (D). The color bar indicates refractive indices at 633 nm wavelength. Scale bar, 5 μm [24]. (For interpretation of the references to color in this figure legend, the reader is referred to the web version of this book.)

Next, the performance of the Rytov and Born approximations is compared. Figure 12.8A and C show the slice images of a 6 μm bead reconstructed from the Born and Rytov approximations, respectively. The index in the middle of the bead is lower for the Born approximation while it is relatively uniform for the Rytov approximation. For a 10 μm bead, distortion inside the bead is even more pronounced for the Born approximation (Figure 12.8B), whereas the refractive index of the slice image of the Rytov approximation is still uniform. This demonstrates that the validity of the Born approximation is highly dependent on the size of the object and is not suitable at all for the 10 μm polystyrene bead in oil, whose phase delay is close to 3 rad. This suggests that the Born approximation cannot be used for imaging a biological cell which typically induces similar phase delay as a 10 μm bead. On the other hand, the Rytov approximation is less affected by the object size and valid for the index difference of 0.03. Thus, the Rytov approximation will be appropriate for imaging biological cells.

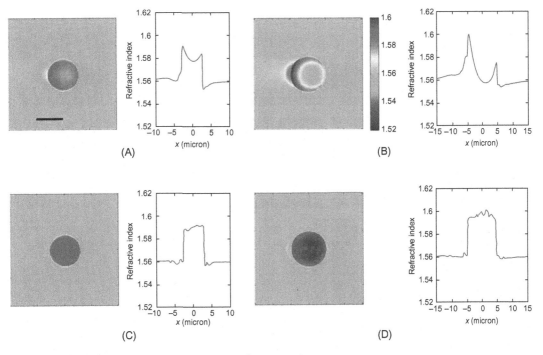

Figure 12.8

Comparison of the Born and Rytov approximations in diffraction tomography. (A and C) Slice images of a 6 μm bead reconstructed based on the Born and Rytov approximations, respectively. Line profiles across the center of the bead are presented next to the slice images. (B and D) Same as (A and C) for a 10 μm bead. The color bar indicates refractive indices at 633 nm wavelength [24]. (For interpretation of the references to color in this figure legend, the reader is referred to the web version of this book.)

246 Chapter 12

ODT is now applied for imaging live HT29 cells, a human colon adenocarcinoma cell line. Cells were prepared in an imaging chamber specially designed for the imaging of a live cell. It is composed of two coverslips separated by a 125 μm thick plastic spacer. Cells were incubated at 37 °C for 12 h before the measurements such that they become fully attached to the coverslip surface. For a fixed objective focus, a set of angular complex E-field images are taken and applied to both filtered back-projection algorithm and diffraction tomography based on the Rytov approximation.

Figure 12.9F–J are x–y slices of tomogram images processed by the inverse Radon transform. Figure 12.9K–N are slice images of a tomogram reconstructed by diffraction tomography based on the Rytov approximation, corresponding to Figure 12.9F–K, respectively. Figure 12.9H is the slice corresponding to the objective focus plane, and its counterpart for ODT is not shown since it is very similar to Figure 12.9H. Figure 12.9G and L is the slice image 1.7 μm above the focus, Figure 12.9F and K is the slice image 2.9 μm above the focus, and Figure 12.9I and M 2.9 μm below the focus. The brightfield images

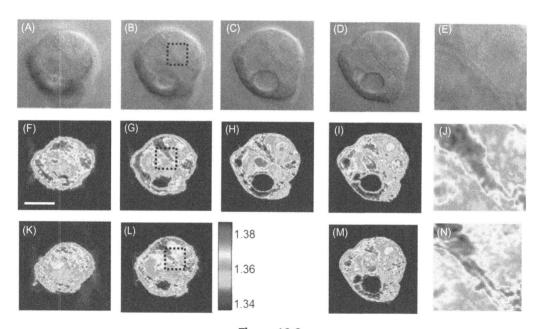

Figure 12.9
Brightfield images (A–E), and 3D tomogram of HT29 cells reconstructed by the filtered back-projection algorithm (F–J), and diffraction tomography based on the Rytov approximation (K–N). (H) is the slice image corresponding to the objective focus. (G and L) Slice images 1.7 μm above the original focus, and (F and K) 2.9 μm above the focus. (G and K) Slice images 2.9 μm below the focus. (A–E) Brightfield images at the same foci as (F–J). (E, J, and N) Zoom-in images of the rectangular boxes in (B, G, and L), respectively. The color bar indicates refractive indices at 633 nm wavelength. Scale bar, 10 μm. (For interpretation of the references to color in this figure legend, the reader is referred to the web version of this book.)

(Figure 12.9A–E) were taken for comparison by moving the objective lens at the same height as Figure 12.9F–J, respectively. It is clear that the structures in both of the refractive index tomograms are well matched to the brightfield images. However, if we compare the details of the images at the tomogram slices 1.7 μm above the focus, the difference between tomograms can be seen. Figure 12.9E, J, and N is the zoom-in images of the black rectangles in Figure 12.9B, G, and L, respectively. Compared with the brightfield image, tomograms processed using the inverse Radon transform show blurring of fine structures in the cytoplasm. In contrast, in the tomogram processed by diffraction tomography, the two lines that separate two nuclei are clearly resolved (Figure 12.9N). This demonstrates that the Rytov approximation is valid for taking the effect of diffraction into account in reconstructing 3D refractive index maps of live biological cells. As a result, we could clearly image the details of the 3D structures of a single live cell throughout its entire volume as well as quantify the refractive index of subcellular organelles.

12.4 Video-Rate TPM

TPM is beneficial in 3D imaging with minimal assumptions. But its data acquisition is slow because large numbers of independent 2D images are to be recorded to fill the 3D object spectrum. This fundamentally impedes studying fast dynamics in biological cells. A single TPM tomogram requires about 10 s of data acquisition [13]. Improving the speed of tomographic imaging will open up new possibilities in imaging rapidly changing, moving, or flowing cells. The acquisition time of TPM has been limited by two factors. First, for optimum image quality, phase images must be acquired at approximately 100 illumination angles; each phase image requires the capture of four raw frames due to the use of phase-shifting interferometry, for a total of 400 images per tomogram. Second, the galvanometer controlling sample illumination angle must be held constant during the acquisition of the four frames, requiring a settling time of approximately 100 ms after each change in angle. In order to overcome these two limitations, a spatial fringe pattern demodulation technique [29,30] can be applied for individual phase recording in TPM. The method uses only 150 raw images per tomogram and does not require galvanometer settling time. As a result, full 3D tomograms can be acquired at a rate of 30 Hz [25].

12.4.1 Experimental Scheme

The setup (Figure 12.10) resembles the TPM system described in Section 12.3 without acousto-optic frequency shifting in the reference arm. A helium–neon laser beam ($\lambda = 632.8$ nm) is divided into sample and reference arm paths by a beamsplitter. In the sample arm, the beam reflected from a galvanometer-mounted mirror (HS-15, Nutfield Technology). A lens (L1, $f = 250$ mm) is used to focus the beam at the back focal plane of an oil-immersion condenser lens (Nikon 1.4 NA), which recollimates the beam to a diameter of approximately

600 μm. Light passing through the sample is collected by an oil-immersion objective lens (Olympus UPLSAPO 100XO, 1.4 NA), and an achromatic doublet tube lens ($f = 200$ mm) was used to focus an image of the sample onto the camera with magnification M = 250. The reference laser beam is enlarged by a 10× beam expander (L2, L3) and combined with the sample beam through a beamsplitter. The resulting interference pattern is captured at 10-bit resolution by a high-speed CMOS camera (Photron Fastcam APX RS, 512 × 512 pixels). A mercury arc lamp, LED illuminator, dichroic mirror, optical filters, and CCD camera (Roper CoolSnap HQ) are also integrated into the setup for correlated brightfield and fluorescent imaging. The galvanometer is driven by a symmetric triangle wave with amplitude corresponding to ±60° at the sample and frequency of 15 Hz. A total of 150 images are acquired during each galvanometer sweep. Irradiances at the detector plane are ~10 μW/cm² for both the sample and reference fields; camera exposure times were typically ~20 μs.

To obtain angle-dependent quantitative phase images, a fringe pattern demodulation technique [29,30] is used. First, the Fourier transform of the raw image is calculated; it contains peaks centered at 0 and $\pm\vec{q}_\theta$, where $\vec{q}_\theta$ is the spatial frequency of the fringe pattern, equal to the difference between sample and reference wave vectors at the image plane. The Fourier components are then shifted by $-\vec{q}_\theta$ such that the $+\vec{q}_\theta$ peak is translated to 0. A 2D Hanning low-pass filter is applied to select only this central component. Applying the inverse Fourier transform then gives a complex-valued function $Z_\theta(x,y)$, from which the phase image is calculated by $\phi_\theta(x,y) = \mathrm{Arg}\, Z_\theta(x,y)$.

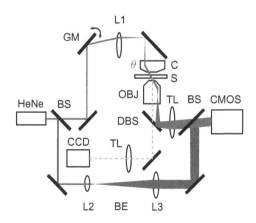

Figure 12.10
Spatial modulation tomography setup; HeNe, helium−neon laser. In sample path (red)—BS, beamsplitter; GM, galvanometer-controlled mirror; C, condenser lens; θ, beam tilt angle; S, sample; OBJ, objective lens; and DBS, dichroic beamsplitter. In reference path (blue)—BE, beam expander; TL, tube lenses; CMOS, camera; L1−L3, lenses; CCD, camera for brightfield and fluorescent imaging (light path shown by dotted line); and not shown, illuminators and filters for brightfield and fluorescent imaging [25]. (For interpretation of the references to color in this figure legend, the reader is referred to the web version of this book.)

To achieve phase images with optimum spatial resolution, two conditions need to be met. First, the period of spatial fringes should be no larger than the diffraction-limited spot, which corresponds to approximately 0.3 μm at the sample. Second, for the adequate sampling of the fringe, the pixel resolution should be fine enough to have at least 3 pixels per fringe. Typically, it is optimal to set 4 pixels per fringe. For our camera pixel size of 17 μm, the magnification is set to be 250 such that the 4 pixels correspond to 272 nm, thus satisfying both conditions.

In the original TPM experiment [13], due to the rotation of the sample beam, the fringe spatial frequency varied in magnitude from 0 to its maximum value $k|\theta_{max}|/M$, where $k = 2\pi/\lambda$. This large variation in spatial frequency impedes the maintenance of an optimal spatial frequency of the interference fringe. To avoid this problem, a fixed tilt of the reference beam is introduced in a direction normal to the sample beam tilt, with an angle such that in the absence of sample beam tilt, there are 4 pixels per fringe in the y-direction, as illustrated in Figure 12.11. The fringe period is fixed along the y-direction as the sample angle is varied from $-\theta_{max}$ to $+\theta_{max}$ (Figure 12.11B−D). To calculate quantitative phase images we applied the demodulation process only along y-direction.

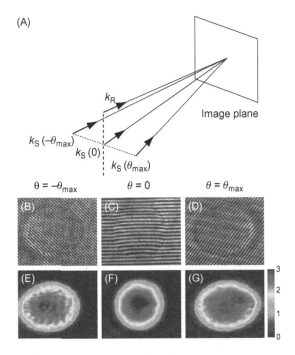

Figure 12.11

(A) Sample and reference beam geometry incident on image plane. k_R is reference beam wave vector and $k_S(\theta)$ is sample beam wave vector. (B−D) Detail of raw images of a 10 μm polystyrene bead for $\theta = -\theta_{max}$, 0, and θ_{max}. Scale bar, 5 μm. (E−G) Corresponding phase images. Color bar, phase in radians [25]. (For interpretation of the references to color in this figure legend, the reader is referred to the web version of this book.)

A set of angle-dependent background phase images is acquired with no sample present and subtracted from the sample phase images to reduce fixed-pattern noise from dust, optical aberrations, and imperfect optical alignment. The background-subtracted phase images are used to reconstruct the 3D refractive index of the sample using a diffraction tomography algorithm based on the Rytov approximation introduced in the previous section.

For cell imaging, cells are dissociated from culture dishes and allowed to attach to coverslips in normal culture medium (Mediatech DMEM + 10% fetal calf serum) for about 6 h at 37 °C before imaging at room temperature. Coverslips are placed inside a flow chamber (custom made or Bioptechs FCS2). The culture medium is injected into the chamber using either a manual syringe or a syringe pump (Harvard Apparatus PHD 22/2000). A valve is used to switch the input to the flow chamber to either a culture medium containing 0.5% acetic acid (for acetic acid experiments) or a hyperosmolar phosphate-buffered saline solution (for osmolarity experiments). Using the syringe pump, the hyperosmotic solution is injected into the chamber at a rate of 1.5 ml/min. Tomograms are continuously acquired while the new medium is added. Data acquisition is performed using custom software written in MATLAB (MathWorks, Natick, MA). The 3D diffraction tomography reconstruction algorithm is performed by custom software written in C. The rest of the data analysis is performed by custom software written in MATLAB. Using a computer running Windows XP 64-bit edition with an Intel Core 2 6600 processor running at 2.4 GHz and 2.93 GB of RAM, the computation time required to construct a single tomogram from 150 interferogram images is approximately 5 min.

12.4.2 Results

To demonstrate the instrument's capabilities, changes in structure of a single cell are monitored during exposure to acetic acid. Acetic acid is widely used during colposcopy to identify suspicious sites on the cervix, due to its whitening effect in precancerous lesions [31]. Previously, it was shown that acetic acid causes an increase in refractive index inhomogeneity throughout a cell and increases the index of the nucleolus [13]. However, the time course of these changes remains unclear due to a limited temporal resolution.

To examine these changes in refractive index in detail, tomograms of a HeLa cell are recorded while the cell is exposed to a new medium containing acetic acid. Almost all changes in the cell structure were found to occur within a 2.75 s interval. Figure 12.12A–C shows $x-y$ slices through the center of the cell at the start, midpoint, and end of this interval. An increase in refractive index heterogeneity is observed throughout the cell and the index of the nucleolus increases dramatically. To assay the effects of acetic acid on different components in the cell, we partitioned the $x-y$ slices into three distinct regions of interest (ROIs) approximately corresponding to (1) the region between the cell boundary and nuclear boundary, (2) the region enclosed by the nuclear boundary but not including the nucleolus, and (3) the nucleolus. Boundaries between ROIs (Figure 12.12D) are drawn manually based on correlations between index tomograms, brightfield images, and widefield fluorescent images using the nucleic acid stain SYTO (Invitrogen).

Figure 12.12E shows the time dependence of the average refractive index of the three ROIs. In the nucleolus (ROI 3), a steady increase is observed in average index, which reaches a stable value about 1.386 within about 2 s. The remainder of the nucleus (ROI 2) exhibits an average index with similar time course but in the opposite direction, decreasing from about 1.364 to 1.359. It is found that the average refractive index of the nucleus, apart from nucleoli, to be smaller than that of the cytoplasm. The time dependence of the average index in nucleus and nucleolus suggests a condensation of nuclear proteins into the nucleoli. The average refractive index of parts of the cell outside the nucleus (ROI 1) is largely unaffected by the addition of acetic acid.

Optical scattering properties of a cell are largely determined by spatial variations in refractive index. To characterize these variations, the standard deviation σ_n of refractive index in the three ROIs (Figure 12.12F) is calculated as a function of time. All three ROIs display a marked increase in refractive index heterogeneity. Remarkably, the three ROIs converge to similar large values for postacetic acid σ_n, despite a difference in preacetic acid values of about 40%. The more than twofold greater increase in σ_n for the nucleus and nucleolus

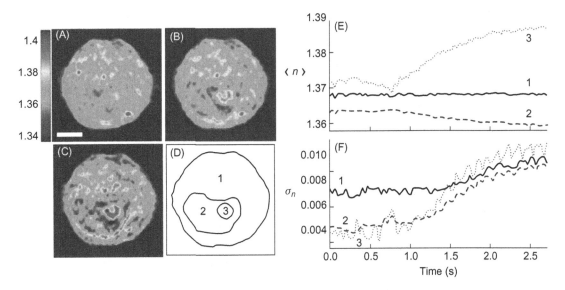

Figure 12.12

Refractive index tomograms (cross-section in x−y plane) of a HeLa cell during exposure to acetic acid solution at (A) $t = 0.0$ s, (B) $t = 1.3$ s, and (C) $t = 2.6$ s. Scale bar, 5 μm. Image sequence recorded at 30 frames per second. (D) ROIs described in text. (E) Average refractive index of each ROI. (F) Standard deviation σ_n of refractive index for each ROI. Solid line: ROI 1. Dashed line: ROI 2. Dotted line: ROI 3 [25].

compared with the rest of the cell suggests that increased whitening of precancerous cells may reflect the greater nuclear-to-cytoplasmic volume ratio in such cells [32].

A similar analysis is performed to monitor changes in shape and structure of a single cell during exposure to a hypertonic-buffered saline solution. Figure 12.13 shows an HT29 (human colonic adenocarcinoma) cell, during a change in solution osmolarity from 300 mosm/l to 975 mosm/l. To determine the changes in index of refraction of different components of the cell, a 2D mask is drawn around a section of the cytoplasm, nucleolus, and nucleus regions. Because the boundary of the cytoplasm and other organelles vary slightly over the course of the video, the masks are drawn to maintain validity throughout the video. The average index of refraction inside each mask is calculated over the 15.4 s of recording time. After exposure to the hyperosmolar solution, the cell shrinks, and the average nuclear and cytoplasmic refractive indices exhibit a roughly linear increase of approximately 1.6×10^{-3}/s and 1.7×10^{-3}/s, respectively. The nucleolar refractive index increases only slightly.

In summary of this section, the use of spatial fringe pattern demodulation enables the acquisition of tomograms about 300 times faster than with the original TPM based on phase-shifting technique. The improved system is used to measure region-specific temporal dynamics of refractive index upon changes in acidity and osmolarity. Video-rate acquisition will also make it possible to acquire tomograms of flowing cells, with applications to studies of cell structure using flow cytometry or microfluidic chambers [33].

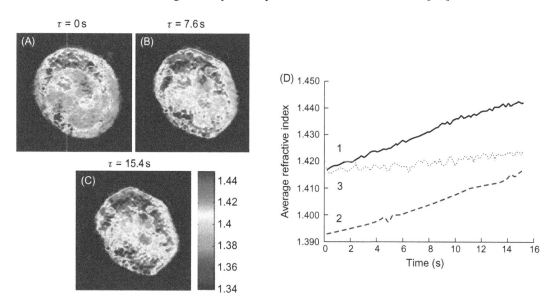

Figure 12.13
Refractive index tomograms (x–y slice) of an HT29 cell during exposure to hyperosmolar solution at time (A) 0.0 s, (B) 7.6 s, and (C) 15.4 s (movie available in Supplemental Movie S2). (D) Time-dependent average refractive index of three ROIs as described in text [25].

12.5 Biological Applications

12.5.1 Study Malaria-Infected Red Blood Cells

TPM is applied to studying the disease states of malaria-infected RBCs [26]. TPM can quantitatively and noninvasively elucidate the consequences on cell biomechanics of *Plasmodium falciparum* malaria by mapping 3D distributions of refractive index. The refractive index maps of *Pf*-RBCs show the morphological alterations of host RBCs and the structures of vacuoles of parasites. In addition, the refractive index is translated into quantitative information about Hb content of individual *Pf*-RBCs. During the intraerythrocytic stages of *P. falciparum*, we show the decrease of both the total amount and the concentration of Hb in the cytoplasm of *Pf*-RBCs.

TPM quantitatively provides the 3D distribution of refractive index, $n(x,y,z)$. As shown in Figure 12.14, the refractive index maps of *Pf*-RBCs is measured during all intraerythrocytic stages: healthy RBC (Figure 12.14A), ring (Figure 12.14B), trophozoite (Figure 12.14C), and schizont stage (Figure 12.14D). Images in the horizontal rows show refractive index

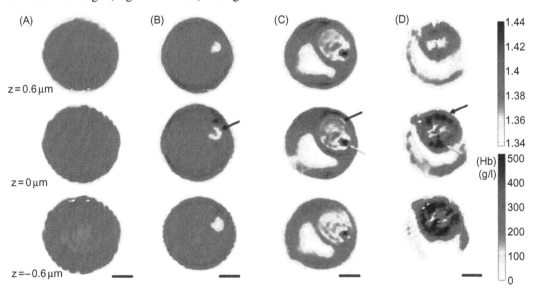

Figure 12.14

3D refractive index maps of *Pf*-RBCs reveal the structural modifications and the hemoglobin concentration of cytoplasm. (A) Healthy RBC. (B) Ring stage. (C) Trophozoite stage. (D) Schizont stage. Images in row show three different cross-sections: 0.6 μm above the focused plane (top), at the focused plane (middle), and 0.6 μm below the focused plane (bottom). Black arrows indicate the location of *P. falciparum*, and the gray arrows the location of hemozoin. Color maps show the refractive index (*n*) (top right) and the Hb concentration (bottom right). Scale bar, 1.5 μm [26]. (For interpretation of the references to color in this figure legend, the reader is referred to the web version of this book.)

maps at three different cross-sections: 0.6 μm above the focused plane (top), at the focused plane (middle), and 0.6 μm below the focused plane (bottom). Although healthy RBCs show homogeneous distribution of refractive index, *Pf*-RBCs are not optically homogeneous. Many factors contribute to refractive index change: the vacuole of parasite occupies a fraction of volume in cytoplasm of RBC; Hb is metabolized and converted into hemozoin crystal in the parasite membrane; and various parasite proteins are exported from parasite into cytoplasm of *Pf*-RBCs. Regions of low refractive index indicate the vacuole of *P. falciparum* (Figure 12.14B–D, black arrows) and regions of high refractive index suggest the position of hemozoin (Figure 12.14C and D, gray arrows).

Refractive index maps measured by TPM can be used to quantitatively investigate the Hb content in *Pf*-RBCs. The refractive index is averaged over the cytoplasmic volume of *Pf*-RBC for 15 cells per group (Figure 12.15A). Their mean values are 1.399 ± 0.006, 1.395 ± 0.005, 1.383 ± 0.005, and 1.373 ± 0.006 for healthy RBCs, ring, trophozoite,

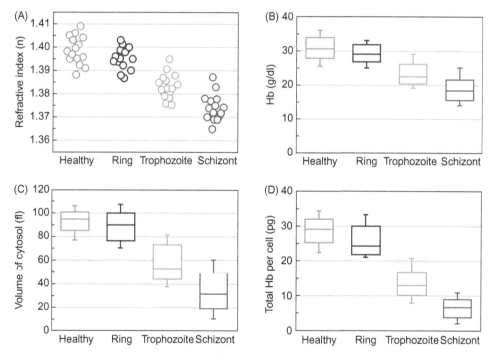

Figure 12.15

Host RBC Hb concentrations decrease as *P. falciparum* matures. (A) Refractive index of healthy RBC and *Pf*-RBCs at the indicated stages. (B) Mean corpuscular Hb concentration (MCHC) of healthy and *Pf*-RBCs at the indicated stages. (C) Cytoplasm volume of *Pf*-RBCs at the indicated stages. (D) Hb content in healthy and *Pf*-RBCs at the indicated stages. Each point in (A) represents average refractive index for one cell. Graphs in (B–D) show the median (central horizontal line), standard deviations (box), and minimum and maximum values (vertical lines). For each condition, 15 samples were tested [26].

and schizont stage, respectively. Given that the cytoplasm of RBCs consists mainly of Hb, it is likely that the refractive index is mostly due to the content of Hb. Hb concentration for individual *Pf*-RBCs can thus be calculated by calibrating from the refractive index of Hb solutions reported [34]. The results are shown in Figure 12.15B. The mean values of Hb concentration are 30.9 ± 3.1, 29.3 ± 2.4, 23.3 ± 2.7, and 18.7 ± 2.9 g/dl for healthy RBCs, ring, trophozoite, and schizont stage, respectively. Because TPM also provides 3D structural information, cytoplasmic volume of *Pf*-RBCs can be obtained by subtracting the volume of parasites' vacuole from the volume of whole RBC. The results are shown in Figure 12.15C. Their mean values are 93.1 ± 7.9, 88.5 ± 11.8, 57.5 ± 13.8, and 34.2 ± 15.1 fl for healthy RBCs, ring, trophozoite, and schizont stage, respectively. In addition, total Hb content is calculated per each *Pf*-RBC. The total amount of Hb in cytoplasmic volume is given by multiplying Hb concentration and cytoplasmic volume. The results are shown in Figure 12.15D. Their mean values are 28.8 ± 1.2, 25.9 ± 4.2, 13.4 ± 3.4, and 6.3 ± 2.5 pg for healthy RBCs, ring, trophozoite, and schizont stage, respectively.

12.5.2 Assessing Light Scattering of Intracellular Organelles in Single Intact Living Cells

The measurement of a refractive index map by TPM makes it possible to assess the contribution of individual organelles to the light scattering spectrum. This is because the refractive index map is an object function that determines the scattering of light. One can therefore apply TPM for a model-independent method of assessing contributions to the light scattering from individual organelles in single intact cells. After measuring the 3D index map of a living cell, it can be modified in such a way so as to eliminate contrast due to a particular intracellular organelle. By calculating and comparing the light scattering distributions calculated from the original and modified index maps using the Rytov approximation, the light scattering contribution from the particular organelle of interest can be directly calculated. The relative contributions of the nucleus and nucleolus to the scattering of the entire cell were determined in this way, and the applicability of the homogeneous spherical model to nonspherical and heterogeneous organelles in forward scattering was evaluated.

The method is first applied for extracting the scattering distribution of the nucleolus. The region associated with nucleolus (2) was replaced with refractive index values of the rest of the nucleus, drawn according to uniform distribution on a pixel-by-pixel basis (Figure 12.16B). This procedure was repeated at every section of different heights in which the nucleolus border could be clearly determined, leading to generation of a new 3D tomogram in which the nucleolus is absent. The difference between scattering fields calculated from the original tomogram and the nucleolus-free tomogram thus provides the scattering distribution contributed solely by the nucleolus. Figure 12.16E shows the scattering distribution of nucleolus when the incident beam is parallel to the optical axis. Oscillations in angular distribution are much coarser than those of the whole cell, and the asymmetry of the light scattering distribution was as expected from the asymmetry of the shape. This is the first time in which light scattering of a single nucleolus in an intact living cell has been assessed.

A similar procedure was applied to assess the scattering of the nucleus. In the nucleolus-free tomogram (Figure 12.16B), there are three major compartments causing light scattering: cell boundary, particles in the cytoplasm, and the nucleus. The cell boundary and particles in the cytoplasm were eliminated at every section by setting the index of the surrounding area of the nucleus as the average index of cytoplasm (Figure 12.16C). This modified tomogram contains only the scattering contribution of the nucleus. Using the Rytov approximation, nuclear scattering was directly calculated from this modified tomogram (Figure 12.16F). The oscillations in angular distribution are finer than those of the nucleolus, as expected.

Next, we studied the relative strengths of the scattering from the different organelles. In order to determine the morphology of the specific organelles such as nuclei from the light scattering measurement of the entire cell, the measurement must be sufficiently sensitive to account for baseline scattering from the whole cell. The knowledge of the

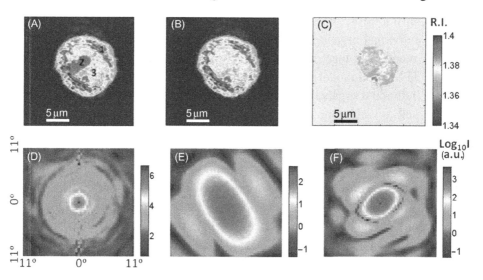

Figure 12.16

Cross-section of 3D refractive index tomograms and 2D angular scattering maps. (A) Original tomogram section of HT29 cell. (B) Tomogram section with nucleolus replaced with random nuclear index values. (C) Tomogram of the cell nucleus. The color bar indicates refractive index at wavelength of 633 nm. Scale bars, 5 μm. (D) Angular scattering intensity of the original cell tomogram. (E) Scattering intensity of the nucleolus. (F) Angular scattering of the nucleus. The color bar indicates intensity in base-10 logarithm with arbitrary units [39]. (For interpretation of the references to color in this figure legend, the reader is referred to the web version of this book.)

relative strengths of light scattering among various intracellular compartments can serve as an important criterion in designing light scattering spectroscopy systems. For this purpose, we compared scattering from three major categories: whole cell, entire intracellular particles, and nucleus. The original tomogram (Figure 12.17A), the same tomogram with the index of the medium set to be the same as the average index of the cell (Figure 12.17B), and the nuclear tomogram (Figure 12.17C) were prepared for each of the three categories. Figure 12.17A represents a cell in suspension, while Figure 12.17B simulates an intact cell in cell monolayer or tissues, in which scattering at the cell boundary is attenuated. Angular scattering spectra were first calculated and then averaged azimuthally to sum up all the scattering signals (Figure 12.17D). The angular scattering of the nucleus is about two orders of magnitude smaller than the scattering of the whole cell in the culture media. This agrees well with the expectation from the relative index contrast of the nucleus to the cytoplasm, about 0.996, and the relative index contrast of the cell to the medium of 1.021. Notably, the shape of the nuclear scattering spectrum (solid red in Figure 12.17D) does not exhibit obvious oscillatory features associated with the cross-sectional diameters

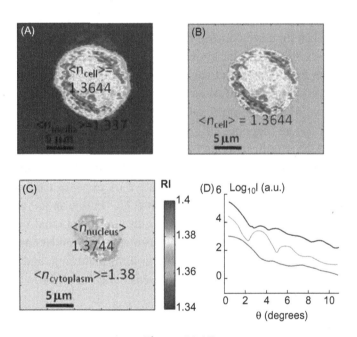

Figure 12.17
Comparison of relative scattering strengths among the whole cell, the entire intracellular organelles, and the nucleus. (A) Tomogram of the HT29 cell in culture medium. (B) Tomogram of HT29 cell with index in the media matched to an average index of the whole cell. (C) Nuclear tomogram surrounded by the average index of the cytoplasm. Scale bars, 5 μm. (D) Angular scattering spectrum from the whole cell in the culture medium (blue), index-matched cell (green), and nucleus (red) [39]. (For interpretation of the references to color in this figure legend, the reader is referred to the web version of this book.)

of the nucleus observed in Figure 12.17B and C, which is due to the azimuthal averaging of the spectrum, equivalent to the averaging of the various cross-sectional diameters.

As shown in this example, the implication of finding the object function, which is the map of 3D refractive index and absorption coefficient, is beyond the identification of a structure. It enables us to estimate the light scattering by individual compartment constituting the object. This will in return lead to developing the method to find the structure of interest with minimal amount of data acquisition, which is the key concept of light scattering spectroscopy.

12.6 Conclusion

Throughout this chapter, the TPM was introduced that can experimentally record angle-dependent complex field images of a sample. Detailed data processing methods were explained for the application of ODT for the 3D reconstruction from the acquired 2D images. It was found that the first-order Rytov approximation enabled accurate imaging of biological cells, whereas the first-order Born approximation caused distortion in the reconstructed images. The iterative constraint algorithm helped to reduce the effect of missing angles. But the employed prior knowledge, non-negative constraint, is a rather weak constraint. With a better constraint such as the support constraint using cell boundary, the accuracy of reconstruction can be further improved, especially in the axial direction. Theoretically, the spatial resolution of the ODT can be better than twice the diffraction limit due to the Ewald sphere mapping [11], which was experimentally demonstrated for nonbiological objects [35]. But in imaging biological cells, weak contrast due to small index differences poses practical limits to the resolution beyond diffraction limit.

For biological and biomedical applications, quantitative index maps can provide molecular concentrations without the need of fluorescent agents [23]. Furthermore, since the refractive index is an intrinsic quantity, the dynamics of molecules can be studied without such artifacts as photobleaching. Refractive index maps can also help understand the way individual organelles in single live cells contribute to light scattering, and thus help in the design of in vivo light scattering instruments for disease diagnosis [24].

In addition to the mapping of the structure, refractive index data can be used to characterize sample-induced aberrations in microscopy. Characterization and correction of such aberrations may be particularly important for modern super-resolution techniques such as STED [36] and structured illumination [37]. The TPM system can also be used to measure the transmission matrix of a turbid medium such as biological tissue, which deterministically characterizes the input–output response of the medium. Using the matrix, it was recently shown that the image can be delivered through turbid medium, and the turbid medium can be used to break the diffraction limit posed by the conventional lens [38].

References

[1] E. Wolf, Three-dimensional structure determination of semi-transparent objects from holographic data, Opt. Commun. 1(4) (1969) 153–156.

[2] M. Born, E. Wolf, A.B. Bhatia, Principles of Optics: Electromagnetic Theory of Propagation, Interference and Diffraction of Light, Cambridge University Press, Cambridge [England], New York, 1999.

[3] F. Zernike, Phase-contrast, a new method for microscopic observation of transparent objects. Part I, Physica 9 (1942) 686–698.

[4] G. Nomarski, Microinterféromètre différentiel à ondes polarisées, J. Phys. Radium 16 (1955) 9S–11S.

[5] Y.K. Park, G. Popescu, K. Badizadegan, R.R. Dasari, M.S. Feld, Diffraction phase and fluorescence microscopy, Opt. Express 14(18) (2006) 8263–8268.

[6] G. Popescu, K. Badizadegan, R.R. Dasari, M.S. Feld, Observation of dynamic subdomains in red blood cells, J. Biomed. Opt. 11(4) (2006) 040503.

[7] C. Fang-Yen, S. Oh, Y. Park, W. Choi, S. Song, H.S. Seung, et al., Imaging voltage-dependent cell motions with heterodyne Mach-Zehnder phase microscopy, Opt. Lett. 32(11) (2007) 1572–1574.

[8] P. Marquet, B. Rappaz, P.J. Magistretti, E. Cuche, Y. Emery, T. Colomb, et al., Digital holographic microscopy: a noninvasive contrast imaging technique allowing quantitative visualization of living cells with subwavelength axial accuracy, Opt. Lett. 30(5) (2005) 468–470.

[9] A. Barty, K.A. Nugent, D. Paganin, A. Roberts, Quantitative optical phase microscopy, Opt. Lett. 23(11) (1998) 817–819.

[10] A. Barty, K.A. Nugent, A. Roberts, D. Paganin, Quantitative phase tomography, Opt. Commun. 175(4–6) (2000) 329–336.

[11] V. Lauer, New approach to optical diffraction tomography yielding a vector equation of diffraction tomography and a novel tomographic microscope, J. Microsc. 205(Pt 2) (2002) 165–176.

[12] F. Charriere, N. Pavillon, T. Colomb, C. Depeursinge, T.J. Heger, E.A.D. Mitchell, et al., Living specimen tomography by digital holographic microscopy: morphometry of testate amoeba, Opt. Express 14(16) (2006) 7005–7013.

[13] W. Choi, C. Fang-Yen, K. Badizadegan, S. Oh, N. Lue, R.R. Dasari, et al., Tomographic phase microscopy, Nat. Methods 4(9) (2007) 717–719.

[14] F. Montfort, T. Colomb, F. Charrière, J. Kühn, P. Marquet, E. Cuche, et al., Submicrometer optical tomography by multiple-wavelength digital holographic microscopy, Appl. Opt. 45(32) (2006) 8209–8217.

[15] L.F. Yu, M.K. Kim, Wavelength-scanning digital interference holography for tomographic three-dimensional imaging by use of the angular spectrum method, Opt. Lett. 30(16) (2005) 2092–2094.

[16] W. Gorski, W. Osten, Tomographic imaging of photonic crystal fibers, Opt. Lett. 32(14) (2007) 1977–1979.

[17] F. Charriere, A. Marian, F. Montfort, J. Kuehn, T. Colomb, E. Cuche, et al., Cell refractive index tomography by digital holographic microscopy, Opt. Lett. 31(2) (2006) 178–180.

[18] K.C. Tam, V. Perezmendez, Tomographical imaging with limited-angle input, J. Opt. Soc. Am. 71(5) (1981) 582–592.

[19] B.P. Medoff, W.R. Brody, M. Nassi, A. Macovski, Iterative convolution backprojection algorithms for image-reconstruction from limited data, J. Opt. Soc. Am. 73(11) (1983) 1493–1500.

[20] Y. Sung, R.R. Dasari, Deterministic regularization of three-dimensional optical diffraction tomography, J. Opt. Soc. Am. A 28(8) (2011) 1554–1561.

[21] A.C. Kak, M. Slaney, Principles of Computerized Tomographic Imaging, Academic Press, New York, 1999.

[22] M. Debailleul, B. Simon, V. Georges, O. Haeberle, V. Lauer, Holographic microscopy and diffractive microtomography of transparent samples, Meas. Sci. Technol. 19(7) (2008) 074009.

[23] A.J. Devaney, Inverse-scattering theory within the Rytov approximation, Opt. Lett. 6(8) (1981) 374–376.

[24] Y.J. Sung, W. Choi, C. Fang-Yen, K. Badizadegan, R.R. Dasari, M.S. Feld, Optical diffraction tomography for high resolution live cell imaging, Opt. Express 17(1) (2009) 266–277.

[25] C. Fang-Yen, W. Choi, Y.J. Sung, C.J. Holbrow, R.R. Dasari, M.S. Feld, Video-rate tomographic phase microscopy, J. Biomed. Opt. 16(1) (2011) 011005.

[26] Y.K. Park, M. Diez-Silva, G. Popescu, G. Lykotrafitis, W.S. Choi, M.S. Feld, et al., Refractive index maps and membrane dynamics of human red blood cells parasitized by Plasmodium falciparum, Proc. Natl. Acad. Sci. USA 105(37) (2008) 13730–13735.

[27] W. Choi, C.C. Yu, C. Fang-Yen, K. Badizadegan, R.R. Dasari, M.S. Feld, Field-based angle-resolved light-scattering study of single live cells, Opt. Lett. 33(14) (2008) 1596–1598.

[28] A. Brunsting, P.F. Mullaney, Differential light scattering from spherical mammalian cells, Biophys. J. 14(6) (1974) 439–453.

[29] M. Takeda, H. Ina, S. Kobayashi, Fourier-transform method of fringe-pattern analysis for computer-based topography and interferometry, J. Opt. Soc. Am. 72(1) (1982) 156–160.

[30] T. Ikeda, G. Popescu, R.R. Dasari, M.S. Feld, Hilbert phase microscopy for investigating fast dynamics in transparent systems, Opt. Lett. 30(10) (2005) 1165–1167.

[31] E. Burghardt, Colposcopy – retrospective and prospective views, Arch. Gynecol. Obstet. 242(1–4) (1987) 240–244.

[32] D.C. Walker, B.H. Brown, A.D. Blackett, J. Tidy, R.H. Smallwood, A study of the morphological parameters of cervical squamous epithelium, Physiol. Meas. 24(1) (2003) 121–135.

[33] N. Lue, W. Choi, G. Popescu, K. Badizadegan, R.R. Dasari, M.S. Feld, Synthetic aperture tomographic phase microscopy for 3D imaging of live cells in translational motion, Opt. Express 16(20) (2008) 16240–16246.

[34] M. Friebel, M. Meinke, Model function to calculate the refractive index of native hemoglobin in the wavelength range of 250–1100 nm dependent on concentration, Appl. Opt. 45(12) (2006) 2838–2842.

[35] M. Debailleul, V. Georges, B. Simon, R. Morin, O. Haeberlé, High-resolution three-dimensional tomographic diffractive microscopy of transparent inorganic and biological samples, Opt. Lett. 34(1) (2009) 79–81.

[36] S.W. Hell, Toward fluorescence nanoscopy, Nat. Biotechnol. 21(11) (2003) 1347–1355.

[37] M.G. Gustafsson, Nonlinear structured-illumination microscopy: wide-field fluorescence imaging with theoretically unlimited resolution, Proc. Natl. Acad. Sci USA 102(37) (2005) 13081–13086.

[38] Y. Choi, T.D. Yang, C. Fang-Yen, P. Kang, K.J. Lee, R.R. Dasari, et al., Overcoming the diffraction limit using multiple light scattering in a highly disordered medium, Phys. Rev. Lett. 107(2) (2011) 023902.

[39] M. Kalashnikov, W. Choi, C.C. Yu, Y.J. Sung, R.R. Dasari, K. Badizadegan, et al., Assessing light scattering of intracellular organelles in single intact living cells, Opt. Express 17 (2009) 19674.

CHAPTER 13

Phase-Sensitive Optical Coherence Microscopy (OCM)

Itay Shock and Natan T. Shaked
Department of Biomedical Engineering, Faculty of Engineering, Tel Aviv University, Tel Aviv, Israel

Editor: Zeev Zalevsky

13.1 Introduction

As previously presented in this book, developments in quantitative phase-imaging techniques have given insights to dynamic mechanisms in living, unstained biological cells with fast acquisition rates and unprecedented accuracies. When measuring fluctuations in time, these techniques can give us knowledge on physiological processes as well as pathological conditions with subnanometric optical path delay (OPD) or optical thickness accuracy.

Optical coherence microscopy (OCM) is an imaging technique that is capable of recording the amplitude and phase information of light that has interacted with live biological tissues for line or point measurement [1]. By analyzing the phase information retrieved by a standard OCM setup, it is possible to obtain quantitative OPD or optical thickness accuracy, when measuring accuracy of several tens of picometers, at least an order of magnitude better compared to wide-field interferometric methods such as digital holographic microscopy [2], presented in the second part of the book.

In standard OCM, the amplitude of the interference pattern is analyzed. Since the phase oscillates 2π rad at every shift of half a wavelength of the light source, high-sensitivity phase measurements of the fringes provide an ultrahigh-accuracy measurement of the change in the OPD. Because of its high sensitivity to phase changes, several system designs have been implemented to increase phase stability and reduce fluctuations due to ground vibrations, air turbulences, acoustic noise, temperature change, and other ambient noises.

Section 13.2 gives a brief overview of OCM principles which is the basis of phase OCM. After giving the basics of OCM, the section then describes how phase OCM can be derived from the OCM data. Section 13.4 describes several phase OCM application for biomedical research.

13.2 OCM Principles

A typical OCM system records the interference pattern between light that has interacted with the sample and a reference wave, where the light source used is a low-coherence source. By analyzing the interference pattern, a reflectivity profile along the axial dimension can be constructed, similar to A-scan in ultrasound.

The performance of an OCM system is mainly determined by its longitudinal (axial) and transverse resolutions, data acquisition specifications, including digitization resolution and speed [3].

OCM achieves high axial image resolution independent of focusing conditions because the axial and transverse resolutions are determined independently. The transverse resolution as well as the axial field of view (FOV) is governed by the focal spot size as in conventional microscopy. However, the axial resolution is determined by the coherence length of the light source.

The transverse resolution, $\Delta x, y$, is determined by the focused beam spot size which, according to the common Rayleigh criterion, is given by:

$$\Delta x, y = \frac{0.61 \lambda}{\text{NA}} \qquad (13.1)$$

where λ is the central wavelength of the source and NA is the numerical aperture of the objective lens.

The axial FOV, assuming a detector with infinitely small pixel size, is given by:

$$\text{FOV} = \frac{\pi \lambda}{2 \text{NA}^2} \qquad (13.2)$$

From Eqs. (13.1) and (13.2), it can be deduced that increasing the NA of the objective will result in higher transverse resolution ($\Delta x, y$ are smaller) but with the tradeoff of decreased axial FOV.

The axial resolution is inversely proportional to the bandwidth of the light source (assuming Gaussian-shaped spectrum profile)[1]:

$$\Delta z = \frac{2 \ln(2) \lambda^2}{\pi \Delta \lambda} \approx 0.44 \frac{\lambda^2}{\Delta \lambda} \qquad (13.3)$$

where $\Delta \lambda$ is the full-width-half-maximum (FWHM) spectral bandwidth of the light source.

---

[1] In FD-OCM with tunable source, the axial resolution is determined by the scanning spectral range, and the bandwidth determines the working range.

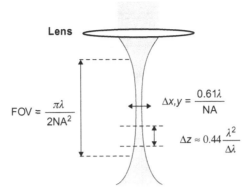

Figure 13.1
Transverse and axial resolutions and the FOV in OCM.

An illustration of the transverse and axial resolutions and the FOV is shown in Figure 13.1.

For example, using a setup with a light source with a central wavelength $\lambda = 840$ nm, a bandwidth of $\Delta\lambda = 50$ nm, and an objective lens with NA = 0.6, one can achieve a transverse resolution $\Delta x, y \approx 1$ μm, FOV > 2 μm and an axial resolution $\Delta z = 6$ μm. The latter value means that in order to resolve two reflecting layers in the axial axis using an OCM system, the reflecting layers must be separated by a distance of at least 6 μm.

Other factors governing the system performance include the light source penetration depth and the sample properties such as scattering and absorption, as well as the detector characteristics such as data acquisition rate, resolution, and sensitivity. Because OCM is often used for live cells and tissues and even in vivo, there are further limitations on the maximum irradiance levels allowed in order to avoid damage to the sample.

The signal to noise ratio (SNR) performance of the system can be derived from the optical communication field and is given by:

$$\text{SNR} = \frac{\xi P}{E_p \text{NEP}} \quad (13.4)$$

where P is the power that reaches the detector after being reflected back from the sample multiplied by the detector efficiency ξ, E_p is the photon energy, and NEP is the noise equivalent power. The top expression is the total power that is detected divided by the electronic bandwidth or data acquisition rate.

In OCM imaging, two approaches are used to obtain the image: time-domain OCM (TD-OCM) and Fourier-domain OCM (FD-OCM). Both systems can be implemented using a Michelson interferometer setup. In TD-OCM, the reference arm path length is scanned

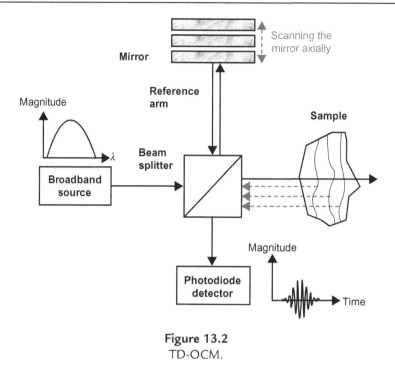

Figure 13.2
TD-OCM.

using a scanning mirror and an internal reflectivity profile of the sample is constructed as illustrated in Figure 13.2.

In FD-OCM, a stationary mirror is used, the spectral density of the interference pattern is captured, and a Fourier analysis is used to reconstruct the internal sample reflectivity profile. There are two types of implementations of FD-OCM systems: spectral-domain OCM (SD-OCM) and swept-source OCM (SS-OCM). In SD-OCM, a broadband light source with constant spectrum is used in combination with a spectrometer allowing for all the depth information to be acquired in a single exposure. SS-OCM uses a swept light source with adjustable wavelength which is scanned over time and is detected using a photodiode, allowing higher control of the wavelength intervals. Implementations of the two systems are shown in Figures 13.3 and 13.4.

The major advantage of TD-OCM over FD-OCM is that the length of the reference arm can be synchronized with the position of the sample arm giving a high axial resolution along the penetration depth. However, because TD-OCM implements a mechanically moving mirror in the separate reference arm, these systems cannot measure fast dynamic processes and suffer from high phase noise. On the other hand, FD-OCM can have no mechanical moving parts, which decreases the phase noise and as so, it is more adequate for phase measurements of dynamic objects. Both of these concepts can be used to obtain either

Phase-Sensitive Optical Coherence Microscopy (OCM) 265

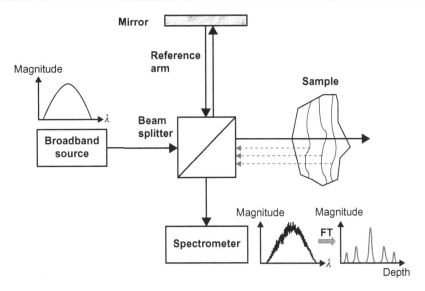

Figure 13.3
SD-OCM.

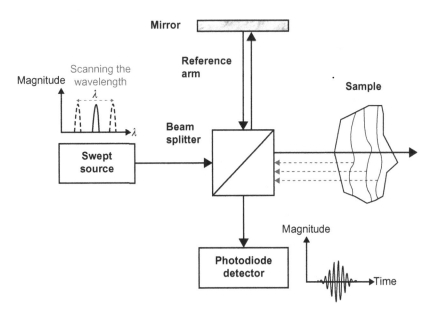

Figure 13.4
SS-OCM.

single-point imaging, similar to A-scan in ultrasound, line-scan, or full-field imaging. When applying the techniques for full-field imaging, the FOV, which is governed by the objective NA, plays a crucial role, as high lateral resolution means lower FOV, so part of the full-field image is defocused. Various methods to extend the FOV in this case have been proposed in the literature [4].

13.3 Phase OCM

The interference spectral data that are collected by the OCM system contains both the amplitude and the phase information of the light beam. When using Fourier transform on the interference signal, a complex expression is resulted. OCM estimates the axial position of the internal reflective structures in the sample by taking the magnitude of the Fourier function in the space domain. Phase OCM goes one step further and uses the phase information of the obtained Fourier transform as well. The phase information allows measuring very small time-dependent variations in the OPD. This information is already contained in the OCM image and does not require additional hardware setup.

A description will be given here for the SD phase-sensitive OCM technique that was proposed by Izatt and Choma [1] and later enhanced [5–7] to allow increased OPD range measurements.

The intensity of the interference pattern on the spectrometer can be expressed as follows:

$$I(k) = S(k)R_R + S(k)R_S + 2S(k)\sqrt{R_R R_S}\cos(2k\Delta d + \theta) \tag{13.5}$$

where k is the wavenumber, $S(k)$ is the spectral density of the light source, R_R and R_S are the reflectivities from the reference surface and the sample surface, respectively, Δd is the OPD between the sample and the reference signals, and θ is a constant phase shift.

The first two expressions in the equation are constant intensities for a given setup and sample, so the intensity is modulated by the cosine expression as a function of the OPD. The challenge is to determine the phase expression and extract the depth information from it. The phase is defined as follows:

$$\varphi(k) = 2k\Delta d + \theta \tag{13.6}$$

The system sensitivity in time depends on the stability of this phase expression. If the system has a phase stability or phase jitter of $\Delta\varphi$, the system SNR can be expressed as:

$$\text{SNR} = \frac{2}{\pi^2 \langle \Delta\varphi^2 \rangle} \tag{13.7}$$

To determine the OPD, the phase function needs to be unwrapped. Then, the slope of the phase function is used and combined with the phase value to increase depth precision.

To retrieve the phase information, several single-frame techniques have been proposed for the evaluation of the phase [8]. These include Fourier transform [9], Hilbert transform [10], spatial phase shifting [11], windowed Fourier transform [12], and wavelet transform [13]. The phase reconstructions are more or less identical except for spatial phase shifting, indicating that the reconstruction of the profiles does not suffer from large errors. Temporal phase shifting [14] is a multiframe method for reconstructing the phase information. This method can obtain accurate results; however, it typically requires five frames.

As an example of the full steps required for retrieving the OPD, the use of the Fourier transform method is described next. First, a Fourier transform is performed on the spectral data giving the depth space. A Gaussian bandpass filter is then applied on the positive term of the complex depth function to remove noise and unwanted reflections. Then, an inverse Fourier transform is applied returning to wavenumber space and giving the complex expression:

$$\tilde{I}(k) = 2S(k)\sqrt{R_R R_S}\exp[j(2k\Delta d + \theta)] \tag{13.8}$$

The phase term can be retrieved as:

$$\varphi(k) = 2k\Delta d + \theta \tag{13.9}$$

where

$$\varphi(k) = \arctan\left(\frac{\mathrm{Im}(\tilde{I})}{\mathrm{Re}(\tilde{I})}\right) \tag{13.10}$$

To determine the OPD, the phase function needs to be unwrapped. First, a least-squares algorithm is used to determine the slope of the phase as a function of the wavelength. Then, the slope of the phase function is used as the reference to remove 2π ambiguity, and one phase component for wavenumber k_i is chosen to give the absolute OPD as:

$$\Delta d = \frac{\varphi(k_i)}{2k_i} + \frac{\pi}{k_i}\left[\mathrm{floor}\left(\frac{\varphi'}{2\pi}\right)\right] \tag{13.11}$$

where φ' is the phase that was retrieved from the slope of the phase function. An illustration of the process is shown in Figure 13.5.

13.4 Phase OCM Applications

In this section, we review several biomedical applications of phase-sensitive OCM. These applications take advantage of dynamic measuring capabilities and high sensitivity for small changes in living cells that can be obtained by phase OCM. The high stability of these systems allows observing biological processes over an extended period of time. In addition, the imaging technique can be combined with labeling techniques such as using gold particles, to achieve even greater specificity of the cellular processes.

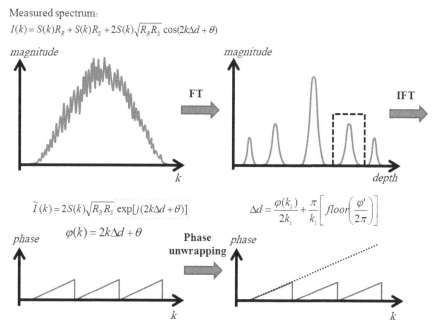

Figure 13.5
Possible digital processing for phase OCM.

13.4.1 Imaging of the Outer Retina

The cone photoreceptor outer segment (OS) changes over time in response to visible stimuli and due to the daily course of renewal and shedding. This results in submicron changes in the OPD, which are too small to be studied using standard OCM. Ref. [15] shows that it is possible to measure phase information from the outer retina in vivo. It does so by quantifying of phase differences within the retina, which improves the sensitivity to OS length change by more than an order of magnitude, down to 45 nm, slightly thicker than a single OS disk. This technique was applied for measuring changes in the OS for hours in hundreds of cones over time.

In this method, the phase difference between different reflective layers in the retina is quantified, which, unlike absolute phase, is immune to decorrelation due to axial motion. Critical to the technique is the ability to measure phase differences between reflective layers located at different retinal depths. A number of preliminary analytical steps are required, the most important of which is the selection of one dimensional scans or A-lines for analysis and spatial unwrapping of phase within the regions of interest.

The setup consists of an ultrahigh-resolution SD-OCM system, combined with a woofer−tweeter adaptive optics system, which provides cellular-level lateral resolution.

Together they achieved a three-dimensional (3D) resolution of 3 μm × 3 μm × 3 μm. The OCM system included a spectrometer, consisting of a transmissive grating and two-line CMOS detector. The light source used is a bandpass-filtered ultrafast Ti:sapphire laser, with 160 nm bandwidth and central wavelength of 800 nm. The line acquisition rate was 167 kHz.

13.4.2 Determining Elastic Properties of Skin

The mechanical properties of the skin are an important tissue parameter used in the study of skin pathophysiology. Most skin pathologies result in changes to their elastic properties and/or thickness. Therefore, the evaluation of the elastic properties of skin tissues is important for the early diagnosis and the treatment of many skin diseases. One such use of this parameter is in the detection of skin cancer, which has a different stiffness than the surrounding healthy skin.

The method described in Ref. [16] combines the use of an impulse-stimulated surface wave with a phase-sensitive OCM system to evaluate the mechanical properties of skin. In this study, an inexpensive homemade shaker was used to generate physical impulse stimulations, which produced surface waves. The shaker head needs to be in contact with the sample surface in order to generate the surface waves. The Young's modulus can be deduced from measuring the behavior of these waves using the phase OCM system. Measurements were performed in tissue-mimicking phantoms and human skin in vivo. Phantoms were made out of different concentrations of agar solution and with different layer thickness to simulate similar elastic properties of skin. Dispersion phase-velocity curves were calculated to obtain the elastic properties from layers with different mechanical properties.

The shaker applied a tissue displacement of approximately 100 nm in the axial direction and produced frequencies of approximately 10 kHz.

The phase OCM imaging system was based on a Michelson interferometer and used a superluminescent diode with 1310 nm central wavelength and 46 nm bandwidth. It provided an axial resolution of approximately 15 mm in air and 10 mm within the skin (assuming the refractive index (RI) is approximately 1.4). The focal length of the objective lens was approximately 50 mm and provided a transverse resolution of approximately 18 mm. The interference pattern recorded with a spectrometer with a maximum acquisition rate of approximately 47 kHz.

For the measurements, seven single and double-layer agar–agar phantoms were used to simulate soft tissues. In addition, in vivo experiments were carried out on five healthy human volunteers with an age span between 25 and 45 years old. Measurements were obtained from two skin sites, the forearm and the palm.

Phase and amplitude measurements results from an agar sample are shown in Figure 13.6. To determine the system noise, data were collected from the phase surface with no sample

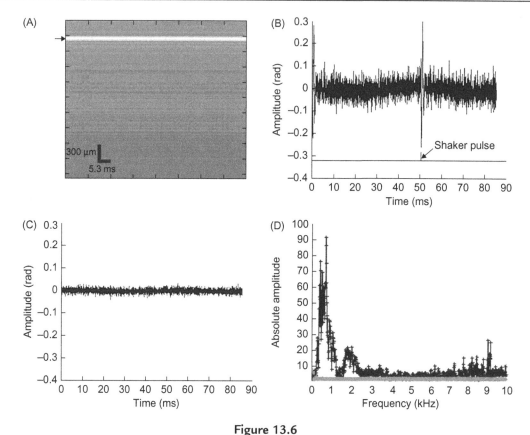

Figure 13.6

Surface waves due to shaker pulse in agar sample. (A) The amplitude data of the phase OCM image at the detection point over time (black arrow points to the surface data selected to analyze the phase change). (B) The phase change of the surface wave signal and waveform of the shaker pulses on a 1% agar phantom. (C) The phase change of the detected system noise. (D) The frequency contribution of the system noise and the detected surface wave signal. Source: Taken with permission from Ref. [16]. Copyright of Optical Society of America.

present. From the phase-velocity graph, the elastic properties of different layers were evaluated and shown to be in good agreement with the literature. This system has the potential for clinical use in diagnosing dermis layer diseases and in other types of tissue pathologies that change the elastic property of the tissue.

13.4.3 Motion Correction of Rodent Cerebral Cortex

Generally, during dynamic imaging in vivo, heart beat and breathing result in regular movements, which are the main cause of image quality degradation. As the phase of the SD-OCM signal is very sensitive to sample movements, these cardiac and respiratory motions of the animal cause especially high noise in dynamic SD-OCM imaging. These motions make long-time laps measurements difficult. Therefore, motion correction processing is required for such types of long-term dynamic imaging. Lee et al. [17] proposed a motion correction method that is especially suitable for phase-resolved dynamic OCM imaging for recording temporal responses of the brain functionality.

Respiratory and cardiac motions contribute differently to image noise. Respiratory motions cause bulk image shifts (BISs) and cardiac motions cause global phase fluctuations (GPFs). A cross-correlation maximization-based shift correction algorithm was effective in suppressing BISs, while GPFs were significantly reduced by removing axial and lateral global phase variations. GPFs were suppressed by removing phase variations that are global in either the axial or the lateral direction. By using a combination of BIS and GPF correction algorithms, the motion artifacts were significantly reduced. To demonstrate the use of the algorithm, dynamic imaging data from the rodent cerebral cortex were acquired using an SD-OCM system. In addition, a nonorigin-centered GPF correction algorithm was examined. Several combinations of these algorithms were tested to find an optimized approach. It was demonstrated that the image stability was improved from 0.5 to 0.8 units in terms of the cross-correlation over 4 s of dynamic imaging, with a reduction of phase noise by two orders of magnitude in $\sim 8\%$ voxels.

The light source consisted of two superluminescent diodes which yielded a combined 170-nm bandwidth centered at 1310 nm, giving an axial resolution of 3.5 μm in tissue. The spectrum of interfered light was measured with a 1024 pixel InGaAs line-scan camera at 47,000 spectra/s. A 5× objective was used, giving a transverse resolution of 7 μm in tissue. The surface of the cortex was illuminated by another light source, with a wavelength of 570 ± 5 nm. This light was used for imaging of the cortical surface by a CCD camera.

13.4.4 Visualizing the Microvasculature within the Retina

Phase-resolved Doppler OCM measures the phase difference between adjacent A-scans, which have to be recorded at overlapping positions within the sample. This phase difference is directly proportional to the velocity of the moving particle. An extension of this idea was proposed in Ref. [18]. The sample beams of two identical SD-OCM setups are scanned over the object at different lateral positions. This generates two tomograms which are slightly separated in time. During the post processing, both data sets are merged and an extended algorithm is applied to extract a 3D capillary network tomogram of the retina. The separation between both sample beams can be adjusted arbitrarily, and hence the velocity measurement range can be freely chosen. By using this method, it was possible to contrast the major vessels in the human retina in vivo as well as display the microvasculature.

13.4.5 Refractive Index Measurements

In Ref. [19], the use of common-path phase OCM was demonstrated for the measurement of the refractive index (RI) of a biomimetic material (glucose solution in water having intralipid as the scattering medium) and a single biological cell (keratinocyte). The RI of glucose solutions was measured with a precision of ~ 0.00015, which corresponds to a precision of ~ 2 nm in the OPD measurement using this setup. The precision obtained in the measurement of the RI of a single keratinocyte was ~ 0.0004. By taking advantage of the phase stability and robustness of the common-path geometry, the enhanced repeatability of the results was demonstrated.

To measure the RI (n) of glucose solution, the OPDs were measured first with an empty chamber with a geometrical thickness t, followed by OPD measurements in a liquid-filled chamber. This was used to estimate the optical thickness, that is, $n \times t$. The observed changes in RI for different glucose concentrations in water and in two different concentrations of intralipid (0.5% and 1%) were measured. A linear increase in RI was observed with an increase in concentrations of glucose, in agreement with the literature.

To measure the RI of a single cell, the setup was coupled with a microscope. A CCD camera was placed in the viewing port of the microscope, and it displayed the bright field image from the same objective lens with which the OCM signal was collected. The cell diameter (D) of keratinocyte cells was measured using the bright field image. This cell diameter was then used to estimate the cell thickness by assuming that the cells are spherical in shape. In order to measure the RI, two OPD measurements are taken, that of the sample chamber when the light passes through the cell L_c and that of the outside of the cell L_s. From these measurements the RI of the cell can be derived as:

$$n_c = n_s + \frac{L_c - L_s}{D}$$

where n_c is the RI of the cell, and n_s is the RI of the cell media.

13.4.6 Nerve Displacement during Action Potential

Nerve fibers undergo rapid outward lateral surface displacements during propagation of action potentials. This swelling phenomenon, which is generally attributed to water influx into axons, has been observed in a number of invertebrate and vertebrate preparations. Measurement of nerve displacements requires a system that is capable of measuring nanometer-scale displacements in surfaces with low reflectivity and at a rate of ~ 2 kHz bandwidth. Techniques such as laser Doppler vibrometry are sensitive to small displacements but are susceptible to external perturbations and may be difficult to apply to weakly reflecting objects.

Reference [20] presents a phase OCM setup in which phase noise is canceled by measurement of motions relative to a nearby surface. An unequal-arm Michelson interferometer compensates for the path delay between sample and reference reflections, and a differential measurement scheme that uses a passive reference gap compensates for phase noise in the Michelson interferometer.

To assess the noise performance of the interferometer, a glass coverslip was used such that the light reflected from the bottom surface of the glass interfered with light from the top surface and the interference pattern was recorded. A test measurement at 1 kHz bandwidth gave phase noise of 0.069 mrad, corresponding to displacement of 8.5 pm. To measure the fiber activity, a walking leg nerve with 1-mm diameter and 50-mm length from an American lobster (*Homarus americanus*) was dissected and placed in an acrylic nerve chamber. The chamber contained five wells filled with a saline solution. Between wells the nerve was surrounded by an insulating layer of petroleum jelly to maximize interwell resistance. A compound action potential was generated by a current pulse from a stimulus isolator and was detected at the other end by an amplifier with a gain of 10^4. In the central well, the nerve rested upon a small glass platform such that it was not submerged in the saline solution.

The optical signal showed a peak height of 5 nm and FWHM duration of 10 ms, with a direction corresponding to an upward displacement. The noise of the displacement measurement was ∼0.25 nm for 1-kHz bandwidth. The displacements were observed in approximately half of the nerve preparations and varied in amplitude from 0 to 8 nm for 5-mA, 1-ms stimulation.

Figure 13.7 shows the nerve displacement and electrical potential that were measured using the system. Positive displacements correspond to an increase in the height of the nerve surface.

Other studies on giant axon of the squid using phase OCM with and without dyes can be found in Ref. [21].

13.4.7 Measurements of Red Blood Cells

In Ref. [22], we used a low-coherence spectral domain phase microscopy (SDPM) system for accurate quantitative phase measurements in red blood cells (RBCs) for the prognosis and monitoring of disease conditions that manifest in mechanical and structural changes in RBCs. Using the system, a comparison was done on the cellular dynamics of healthy RBCs and glutaraldehyde-treated RBCs that have lower amplitude of vibrations, and the membrane vibrational fluctuations were measured in time to reflect their membrane stiffness. Due to its common-path geometry, the OPD stability of SDPM is less than

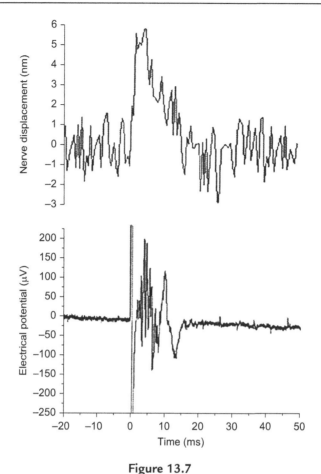

Figure 13.7

Nerve displacement and electrical potential measured using the phase OCM system. Source: *Taken with permission from Ref. [20]. Copyright of Optical Society of America.*

0.3 nm, at least three times better compared to the commonly used holographic phase microscopy system under the same conditions. In single-point interferometric measurements of live cells, the light is focused on a small area and therefore higher irradiance allows for ultrafast acquisition times with compact and reasonable-price detectors. Furthermore, single-point measurements also allow for an easy-to-implement common-path geometry and a compact and portable fiber optics configuration, and thus can work even without using floating optical tables, which has a significantly higher potential for clinical applications.

When measuring RBCs with quantitative phase microcopy, a constant RI assumption can be taken, so that the phase measurement is proportional to the thickness of the RBC. Using the time-dependent thickness profile of the RBC, cell stiffness properties can be calculated.

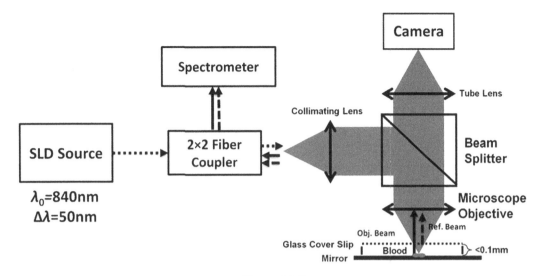

Figure 13.8
SD-OCM system for measuring OPD of RBCs. Source: *Taken with permission from Ref. [22] Copyright of SPIE.*

The system is illustrated in Figure 13.8. The blood sample was placed in a closed chamber such that the reflection from the upper surface of the chamber is used as the reference beam and the light passing through the media and the RBC is reflected from the bottom surface and is used as the sample beam. A compact digital camera arm was added to the SD-OCM system. The camera displays the light source beam and the sample bright-field image simultaneously and allows the correct positioning of the light source beam over the sample.

To demonstrate the system ability to measure live-cell fluctuations, glutaraldehyde-treated RBCs at various concentration levels were used. Glutaraldehyde causes cross-linking of proteins and destroys the lipid membrane, leading to an altered structure and thus affecting the RBC membrane elasticity characteristics. A comparison between the SDPM and a wide-field digital interferometry (WFDI) system (holographic phase microscopy) is shown in Figure 13.9.

By inducing low-membrane fluctuations, it was demonstrated that the system is more sensitive than commonly used quantitative phase measurement systems, such as ones that are based on WFDI.

13.4.8 Detection of Gold Nanoparticles

Gold nanoparticles are used as contrast agents to image disease states by targeting to biochemical makers unique to specific diseases, such as surface membrane proteins in cancer cells. This in vivo approach currently is being developed as a diagnostic tool and

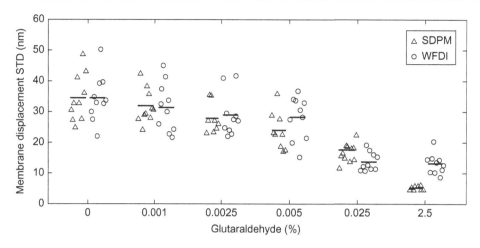

Figure 13.9

Membrane displacement values of 10 RBCs from single-point phase OCM (SDPM) and wide-field phase interferometry (WFDI) for different glutaraldehyde conditions. Each shape represents a single cell membrane displacement standard deviation (STD) over a 30-s time period. The horizontal line is the average value of the displacement in each concentrations.
Source: *Taken with permission from Ref. [22] Copyright of SPIE.*

also to monitor disease progression. Gold nanoshells consist of an inner silica core surrounded by a thin gold shell. They can be engineered with different dimensions of the core and shell. The optical resonance frequency of the particles can vary from ultraviolet to near infrared wavelengths according to their dimensions. This allows customized tailoring of the optical scattering and absorption properties of the particles to suit the needs of the specific application. Gold nanoshells are highly biocompatible, water-soluble, and commercially available.

The detection of a gold nanoparticle contrast agent is demonstrated using a photothermal modulation technique and phase-sensitive OCM [23].

An active contrast agent detection technique was demonstrated for high-speed OCM imaging based on photothermal modulation. The technique uses gold nanoshells designed to have high absorption at 808 nm whereas tissue absorption at this wavelength is inherently low. A multimode laser diode operating at 808 nm is used to heat up the gold particles, causing small-scale, localized temperature gradients, in places where the gold particles are present. These temperature variations cause a local change in the OPD of sample. These changes were detected using an SS-/FD-OCM phase microscopy system built using a double-buffered FD mode-locked laser operating at 1315 nm and a sweep rate of 240,000 sweeps/s (240 kHz). By modulating the 808 nm laser diode at a known frequency and observing variations in the OPD that occur only at the frequency, the contrast agent

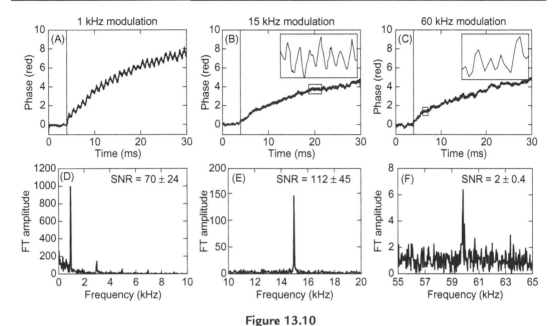

Figure 13.10

(A–C) Phase of the gold nanoparticles measured by photothermal OCM. (D–F) Fourier transforms of the measured phases at various 808 nm laser modulation frequencies. The vertical lines in (A–C) show the time points when the 808 nm laser was activated. Insets in (B and C) show enlarged views of the measured phase. The 808 nm laser modulation frequencies were 1 kHz (A and D), 15 kHz (B and E), and 60 kHz (C and F). Source: *Taken with permission from Ref. [23]. Copyright of J. R. Soc. Interface.*

can be detected in a way that significantly reduces background noise. The SNR was defined as the ratio of the signal peak to the noise value. The phantom used in the photothermal experiments described achieves mean SNRs of 2–131. The high SNR suggests that lower concentrations of nanoparticles could be detectable using this method compared to other methods, allowing for in vivo applications where the marking agent is administered systemically and accumulates at lower levels in diseased tissue. Figure 13.10 shows the measured phases and their Fourier transforms. A similar photothermal OCM method was used in Ref. [24].

13.5 Conclusions

In recent years, use of phase-sensitive OCM in biomedical research has grown widely. Technological improvements resulted in the utilization of this method variety of in this field of research. Phase OCM has the combined advantages of in vivo imaging, noninvasive, subnanometer OPD sensitivity for internal structures in the sample, and high acquisition speeds, allowing study of dynamic biologic processes. Further advantages of the method include robustness, system stability due to common-path design, transportability, and the minor radiation exposure, which give this technology great potential for clinical applications.

SD-OCM and SS-OCM methods are much more common in phase OCM applications since TD-OCM is more likely to have higher phase noise. In addition, the SD-OCM can achieve higher acquisition rates than TD-OCM, an advantage which is vital in studying dynamic biological processes. With the development of fast scanning-wavelength lasers, SS-OCM can be also used to study cell activity with high phase stability and high SNR. Single-point imaging inherently has a higher SNR than the full-field or line-scan systems, and so it is ideal in measuring very small phase changes, in cases where the two-dimensional spatial changes are less of interest than the temporal changes in the point measured. Determining the central wavelength to be used also is important as using longer wavelengths means obtaining reduced axial and lateral resolution but, on the other hand, usually means obtaining higher penetration depth.

References

[1] J.A. Izatt, M.A. Choma, Theory of optical coherence tomography, in: W. Drexler (Ed.), Optical Coherence Tomography: Technology and Applications, Springer, London, UK, 2008, pp. 47–72 (Chapter 2).
[2] M.A. Choma, A.K. Ellerbee, C. Yang, T.L. Creazzo, J.A. Izatt, Spectral-domain phase microscopy, Opt. Lett. 30 (2005) 1162–1164.
[3] W. Drexler, Ultrahigh-resolution optical coherence tomography, J. Biomed. Opt. 9(1) (2004) 47.
[4] A. Zlotnik, Y. Abraham, L. Liraz, I. Abdulhalim, Z. Zalevsky, Improved extended depth of focus full field spectral domain optical coherence tomography, Opt. Commun. 283(24) (2010) 4963–4968.
[5] D. Parshall, M.K. Kim, Digital holographic microscopy with dual-wavelength phase unwrapping, Appl. Opt. 45(3) (2006) 451–459.
[6] H.C. Hendargo, M. Zhao, N. Shepherd, J.A. Izatt, Synthetic wavelength based phase unwrapping in spectral domain optical coherence tomography, Opt. Express 17 (2009) 5039–5051.
[7] J. Zhang, B. Rao, L. Yu, Z. Chen, High-dynamic-range quantitative phase imaging with spectral domain phase microscopy, Opt. Lett. 34 (2009) 3442–3444.
[8] S.K. Debnatha, M.P. Kothiyalb, S.-W. Kim, Evaluation of spectral phase in spectrally resolved white-light interferometry: comparative study of single-frame techniques, Opt. Lasers Eng. 47 (2009) 1125–1130.
[9] M. Takeda, H. Ina, S. Kobayashi, Fourier-transform method of fringe-pattern analysis for computer-based topography and interferometry, JOSA 72 (1983) 156–160.
[10] D.A. Zweig, R.E. Hufnagel, A Hilbert transform algorithm for fringe-pattern analysis (advanced optical manufacturing and testing), Proc. SPIE 1333 (1990) 295–302.
[11] P. Sandoz, G. Tribillon, H. Perrin, High resolution profilometry by using phase calculation algorithms for spectroscopic analysis of white light interferograms, J. Mod. Opt. 43 (1996) 701–708.
[12] Q. Kemao, Windowed Fourier transform for fringe pattern analysis, Appl. Opt. 43 (2004) 2695–2702.
[13] Y. Deng, Z. Wu, L. Chai, C. Wang, K. Yamane, R. Morita, et al., Wavelet transform analysis of spectral shearing interferometry for phase reconstruction of femtosecond optical pulses, Opt. Express 13 (2005) 2120–2126.
[14] P. Hariharan, B.F. Oreb, T. Eiju, Digital phase-shifting interferometer: a simple error-compensating phase calculation algorithm, Appl. Opt. 26 (1987) 2504–2506.

[15] R.S. Jonnal, O.P. Kocaoglu, Q. Wang, S. Lee, D.T. Miller, Phase-sensitive imaging of the outer retina using optical coherence tomography and adaptive optics, Biomed. Opt. Express 3 (2012) 104–124.

[16] C. Li, G. Guan, R. Reif1, Z. Huang, R.K. Wang, Determining elastic properties of skin by measuring surface waves from an impulse mechanical stimulus using phase-sensitive optical coherence tomography, J. R. Soc. Interface 9 (2012) 831–841.

[17] J. Lee, V. Srinivasan, H. Radhakrishnan, D.A. Boas, Motion correction for phase-resolved dynamic optical coherence tomography imaging of rodent cerebral cortex, Opt. Express 19 (2011) 21258–21270.

[18] S. Zotter, M. Pircher, T. Torzicky, M. Bonesi, E. Götzinger, R.A. Leitgeb, et al., Visualization of microvasculature by dual-beam phase-resolved Doppler optical coherence tomography, Opt. Express 19 (2011) 1217–1227.

[19] Y. Verma, P. Nandi, K.D. Rao, M. Sharma, P.K. Gupta, Use of common path phase sensitive spectral domain optical coherence tomography for refractive index measurements, Appl. Opt. 50 (2011) E7–E12.

[20] C.F. Yen, M.C. Chu, H.S. Seung, R.R. Dasari, M.S. Feld, Noncontact measurement of nerve displacement during action potential with a dual-beam low-coherence interferometer, Opt. Lett. 29 (2004) 2028–2030.

[21] T. Akkin, D. Landowne, A. Sivaprakasam, Detection of neural action potentials using optical coherence tomography: intensity and phase measurements with and without dyes, Front. Neuroenergetics 2 (2010) 22.

[22] I. Shock, A. Barbul, P. Girshovitz, U. Nevo, R. Korenstein, N.T. Shaked, Optical phase measurements in red blood cells using low-coherence spectroscopy *J. Biomed. Opt.* 17 (2012) 101509:1–5.

[23] D.C. Adler, S.W. Huang, R. Huber, J.G. Fujimoto, Photothermal detection of gold nanoparticles using phase-sensitive optical coherence tomography, Opt. Express 16 (2008) 4376–4393.

[24] M.C. Skala, M.J. Crow, A. Wax, J.A. Izatt, Photothermal optical coherence tomography of epidermal growth factor receptor in live cells using immunotargeted gold nanospheres, Nano Lett. 8 (2008) 3461–3467.

CHAPTER 14

Polarization and Spectral Interferometric Techniques for Quantitative Phase Microscopy

Yizheng Zhu[1], Matthew T. Rinehart[1], Francisco E. Robles[1,2], and Adam Wax[1]
[1]Department of Biomedical Engineering
[2]Medical Physics and the Fitzpatrick Institute for Photonics,
Duke University, Durham, NC

Editor: Natan T. Shaked

14.1 Introduction

Imaging the internal structure of biological cells is an essential component of many clinical and laboratory investigations. Histopathology, the main method for clinical diagnosis of disease, relies on scoring altered cell structure by studying fixed, stained, and sectioned cells using a light microscope. In the laboratory, examination of structural changes with this approach reveals the origins of disease as well as mechanisms that govern cellular function. Unfortunately, this widely used method has a number of limitations.

Investigations that are based on the analysis of histological samples must contend with the artifacts that arise due to sample preparation. Fixatives, staining agents, and cell sectioning alter the structure of the cells in nontrivial ways. In addition, such analysis can only glimpse a snapshot in the life of an individual cell, relying on the study of large ensembles of cells to develop a picture of their temporal evolution. Quantitative phase-imaging techniques offer an alternative to these approaches for studying the structure of living cells in situ, by providing a highly sensitive, noninvasive means of measuring cellular structure. In addition, since optical imaging does not perturb the structure or function of cells, techniques based on phase imaging permit investigation of the development, formation, and function of cellular structures through examination of the properties of the *same* cells over time.

As presented in Part 2 of the book, there have been many recent efforts to develop techniques for measuring the structure and dynamics of cells and subcellular components using quantitative phase measurements. They all share the common aspect of providing measurements on the nanometer scale, achieved by interferometric techniques which measure the phase of light. Here, we focus on the extensions of these techniques which provide additional capabilities and unique information by exploiting other aspects of light, including its polarization and spectral dependence.

This chapter presents a suite of novel quantitative phase measurement tools designed to probe the structure and dynamics of living cells by exploiting either the polarization or spectral dependence of light to increase the sensitivity and stability of these measurements. These tools share the use of interferometric schemes that measure phase changes in light to learn about biological cells and tissues. Here, we show how polarization can be used to improve imaging throughput by providing two phase-shifted images in a single exposure. We then present a method that uses polarization to generate two spatially displaced beams, as done in Nomarski microscopy, but the interference that is generated is detected in the spectral domain, taking advantage of recent instrumental advances in the field of optical coherence microscopy. Finally, we exploit the rich source of information in the optical spectrum to obtain unique measurement capabilities, including the simultaneous use of multiple wavelengths to aid in phase unwrapping and the measurement of dispersion properties to provide access to spectroscopic features in phase measurements.

14.2 Dual-Interference Channel Quantitative Phase Microscopy

Imaging live cells requires a system that is able to visualize mostly transparent three-dimensional objects with very little inherent absorption, imparting almost no change to the light amplitude reflected from or transmitted through them. Conversely, the phase of light transmitted through these transparent objects can provide information about cell structure by recording subtle changes in density as variations in optical path delay (OPD). Phase imaging can be performed by conventional microscopic techniques, such as phase-contrast microscopy and differential interference contrast microscopy [1]. However, these techniques do not yield quantitative phase measurements. In addition, they suffer from various artifacts that make it hard to quantitatively interpret the resulting phase images in terms of OPDs.

Digital holography, on the other hand, yields quantitative measurement of the phase distribution across a field of view [2–6]. Therefore, it is possible to manipulate the complex wavefront to conduct quantitative analysis. Digital holography, however, requires interferometric setups, typically yielding phase images that are vulnerable to phase noise. Elimination of most phase noise, which can arise from perturbations in the interferometer arms, can be accomplished by acquiring two or more phase-shifted interferograms of the same sample [5,6]. However, this approach can be limiting, as certain biological processes, such as cell membrane fluctuations and neuron activity, occur faster than the acquisition rates of most optical wide-field interferometric imaging systems.

In this section, we introduce a new optical polarization technique for obtaining dynamic quantitative phase measurements with high precision and a low degree of phase noise. The method, termed dual-interference channel quantitative phase microscopy (DQPM), is based on a novel dual-polarization channel interferometric setup that is capable of simultaneous acquisition of two phase-shifted interferograms of the same sample. These interferograms are then digitally processed to produce the phase profile of the sample. Since the two interferograms are acquired at the same time, this approach eliminates most of the common phase noise in the final phase image. Additionally, due to the simultaneous nature of the acquisition, one can capture rapid cell phenomena with subnanometer temporal resolution, limited only by the speed of the digital camera used.

Ikeda et al. [7] have suggested an alternative dynamic QPM method which is based on the acquisition of a single interferogram. This method is able to avoid the temporal phase noise between successive frames, but since only one interferogram is acquired, common phase noise due to the sample is not eliminated. Various attempts for simultaneous phase imaging have been reported in the optical testing field [8–12]. However, these methods have not been implemented for biologically relevant applications.

The DQPM system is shown in Figure 14.1, which includes a Mach–Zehnder interferometer-based off-axis digital holographic microscopy setup, followed by an image-splitting system. A HeNe laser, linearly polarized at 45°, is split into sample and reference beams by beamsplitter BS_1. The sample arm contains the sample, and microscope objective (MO) forms a 4 f configuration with lens L_2, as does lens L_1 in the reference arm. Beamsplitter BS_2 combines the sample and reference beams, and the interference pattern of interest appears one focal length from lens L_2. This interference pattern is spatially restricted by an aperture to half of the digital camera sensor size and imaged through the 4 f image-splitting system shown in Figure 14.1B, onto the digital camera. This 4 f image-splitting system includes two identical lenses, L_3 and L_4, creating a 1:1 image of the aperture plane on the camera. A Wollaston prism is positioned between these two lenses,

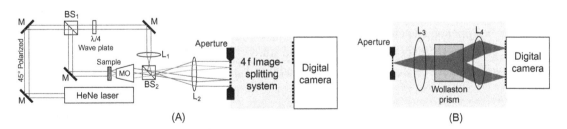

Figure 14.1
The DQPM system: dual-channel, single exposure interferometer for obtaining phase profiles of live dynamic biological cells. (A) The entire interferometer. (B) The 4 f image-splitting system. *Source: From Ref. [13].*

splitting the horizontal and vertical polarization components to create two spatially separated and phase-shifted interferograms on the camera.

This setup introduces a phase shift between the two parallel interferograms, which can be explained as follows. In the reference arm, a quarter ($\lambda/4$) waveplate is oriented along the horizontal axis, so that the light transmitted through it has a 90° phase difference between the horizontal and vertical components. On the other hand, in the sample arm, the light remains linearly polarized at 45°, so there is no phase difference between the horizontal and vertical components. When the beams are combined by BS_2, there is a $\alpha = 90°$ shift between the interference patterns formed by the horizontal and the vertical components. The Wollaston prism outputs two perpendicularly polarized beams, separating the horizontal or vertical polarization components for each of the interferograms, yielding a phase shift of α between the interferograms.

The two interferograms, I_1 and I_2, acquired by the digital camera, can be mathematically written as follows:

$$I_1 = I_R + I_S + 2\sqrt{I_S I_R}[\cos(\phi_{OBJ} + \phi_C)] \\ I_2 = I_R + I_S + 2\sqrt{I_S I_R}[\cos(\phi_{OBJ} + \phi_C + \alpha)] \quad (14.1)$$

where I_R and I_S are the reference and sample intensity distributions, respectively; ϕ_{OBJ} is the spatially varying phase associated with the object; and ϕ_C is the spatially varying phase of the interferometer without the object present. Note that ϕ_C and α can be digitally measured, by fitting the background interference signal (interference pattern without sample) in each interferogram to a sine wave. The wrapped object phase ϕ_{OBJ} is computed as [6]:

$$F = \frac{\exp(-j\phi_C)}{1 - \exp(j\alpha)}[I_1 - I_2 + jHT\{I_1 - I_2\}]; \quad \phi_{OBJ} = \arctan\left\{\frac{\operatorname{Im} F}{\operatorname{Re} F}\right\} \quad (14.2)$$

where HT denotes a Hilbert transform. The final object phase is then obtained using an unwrapping algorithm. Note that this process removes most common noise and background elements thus producing highly stable phase measurements. In addition, since the interferograms I_1 and I_2 are acquired in a single exposure, without the need for raster scanning, fast phenomena can be visualized. Using Eq. (14.1) and recalling that $HT\{\cos(\phi)\} = \sin(\phi)$, $\cos(\phi) = 0.5[\exp(j\phi) - \exp(-j\phi)]$, and $\sin(\phi) = 0.5j[\exp(-j\phi) - \exp(j\phi)]$, one can easily show that $F = 2\sqrt{I_S I_R}\exp(j\phi_{OBJ})$, and thus Eq. (14.2) produces the desired signal.

To demonstrate the utility of the DQPM system (Figure 14.1), we imaged live unstained MDA-MB-468 human breast cancer cells in standard growth medium which contained phenol red growth medium. Figure 14.2A shows a typical light microscopic image of the sample through the optical system, demonstrating the very low visibility achieved with simple brightfield imaging of this sample. The specific setup incorporated a HeNe laser

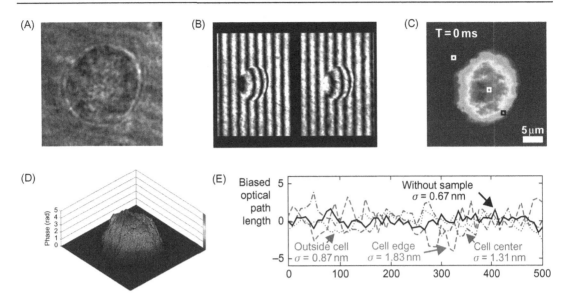

Figure 14.2
MDA-MB-468 human breast cancer cell. (A) Transmission brightfield microscopic imaging through the system; (B) the two phase-shifted interferograms in a single camera exposure; (C) final unwrapped phase profile; (D) surface plot of the phase profile shown in (C); (E) temporal phase stability without the sample and with the sample in the three points marked in (C). *Source: Adapted from Ref. [13].*

(5 mW, 633 nm) as the light source. In both interferometer arms, 40×, 0.65 NA achromatic objective lenses were used, each creating 33× magnification when paired with lens L_2 (15 cm focal length). The resulting complex amplitude was spatially restricted by a 3.2 mm × 2.3 mm aperture. The 4 f image-splitting system, as shown in Figure 14.1B, consists of two lenses L_3 and L_4 of similar type, each with a focal length of 7.5 cm. Between these lenses, the Wollaston prism (crystal quartz, 2° of angular separation) splits the pattern from the aperture plane so that the charge-coupled device (CCD) camera (AVT Pike F032-B, 640 × 480 pixels) captures the two phase-shifted interferograms I_1 and I_2 shown in Figure 14.2B in one frame but with a spatial separation of about 2.4 mm. To determine the coregistration between the two phase-shifted interferograms recorded in the single CCD image, images of a test target (United State Air Force (USAF) resolution chart) were used as a calibration standard. The frequency q and the actual phase shift α of the interferometric fringes were extracted by fitting each of the interferograms to a sine wave. The background interference fringes are oriented vertically in each of the interferograms, yielding $\phi_C = qx$, where x is the horizontal coordinate in each interferogram frame. Figure 14.2C and D shows the final unwrapped phase profile of the sample, obtained by applying Eq. (14.2), and a phase unwrapping algorithm. The same cell was imaged every 8.3 ms (120 frames per second (fps)) for a duration of 500 ms.

We quantified the temporal phase stability by recording the millisecond-scale temporal pathlength fluctuations originating within a diffraction-limited spot, as defined by the imaging optics. The stability was first measured without the presence of a sample and then with the sample at three representative locations: on the cell center, on the cell edge, and at the background (through the medium only), as shown in Figure 14.2C. For these four situations, the standard deviations of the measured phase fluctuations were determined to be 0.67, 1.31, 1.83, and 0.87 nm, respectively, as shown in Figure 14.2E. The first result represents the fundamental temporal stability of the entire optical system, demonstrating subnanometer path-length sensitivity on the millisecond time scale. The slightly degraded stability in the three latter cases may be attributed to small vibrations occurring inside the entire sample.

To illustrate the utility of DQPM for imaging dynamic cellular phenomena, we recorded a 2-s movie at 120 fps that shows the phase profiles of a beating rat neonatal cardiomyocyte (myocardial cell isolated from the heart of a neonatal rat) in growth medium.

Here, we have demonstrated a new polarization-based method for imaging fast biological phenomena with high accuracy. Experimental results show good spatial resolution for dynamic cell phase imaging and excellent temporal phase stability. Due to its simultaneous acquisition, the system's ability to observe fast cellular phenomena is only limited by the full frame rate of the camera, and thus this method provides a powerful tool for dynamic studies of biological cells.

14.3 Spectral-Domain Differential Interference Contrast Microscopy

In Section 14.2, an improvement in system stability was realized by utilizing light polarized in two orthogonal directions. Now, we discuss another polarization-based system for quantitative measurement of phase gradient. More importantly, this technique, termed spectral-domain differential interference contrast microscopy (SD-DIC), integrates spectral techniques to greatly improve the sensitivity of these measurements.

First invented in the 1950s, DIC microscopy has since become a standard imaging modality in modern optical microscopes, and today it sees widespread use for enhanced contrast in imaging unstained and transparent biological specimens [14]. Similar to phase-contrast microscopy, DIC takes advantage of optical phase variations across an object to achieve better visualization, with the help of interferometry. In a typical DIC microscope, the interference occurs between the two orthogonally polarized components of the input light, which are split in space to pass the sample with a slight lateral shear and hence experience a differential optical phase/path-length change between them. When subsequently recombined, they interfere to produce drastically improved contrast compared with brightfield images. DIC microscopy hence measures the phase/path-length *gradient* of a sample and generates images with a shadow-casting effect.

Although widely used for sample characterization, conventional DIC microscopy is inherently qualitative due to the nonlinear relationship between the image intensity and the optical path-length (OPL) gradient. Such a difficulty is further complicated by intrinsic intensity variations due to scattering or absorption within the sample. Recently, quantitative DIC microscopy has received growing attention, with a number of implementations proposed to isolate and quantify the OPL gradient from the conventional DIC intensity images [15–17]. From these gradient measurements, quantitative OPL or phase maps of the sample can be reconstructed and analyzed [16,18]. In most of these systems, partially coherent light sources were used in order to reduce speckle noise, but the broadband nature of these sources has not been fully exploited.

On the other hand, broadband light sources have found key applications in spectral-domain low-coherence interferometry (SD-LCI), perhaps best exemplified by SD-optical coherence tomography (SD-OCT). SD-LCI has been shown to have a sensitivity advantage over its time-domain counterparts [19]. Furthermore, when combined with common-path interferometry, it can produce superior sensitivity when measuring path length or path-length gradient [20–22].

In Section 14.3.1, we demonstrate SD-DIC microscopy as a method that combines the common-path nature of DIC microscopy with the high sensitivity of SD-LCI to produce high-resolution, quantitative measurement of the OPL gradient across a sample. We introduce its principles, demonstrate imaging of both reflective objects (a USAF resolution target) and transparent objects (live neonatal cardiomyocytes), and study the dynamics of cardiomyocyte contraction at selected cellular locations.

14.3.1 Principles of Spectral Domain-DIC Microscopy

Figure 14.3A shows a first generation SD-DIC system. The broadband light from a single-mode superluminescent diode (Superlum, Inc.; $\lambda = 840$ nm, $\Delta\lambda = 40$ nm) is split equally in power into slow and fast axes of a 32.5 cm long polarization-maintaining (PM) fiber using a polarization controller. The two copropagating orthogonal polarization components are then passed through a Nomarski prism, which splits them into spatially separated o- and e-waves. The axes of the PM fiber are aligned with the o- and e-polarizations of the Nomarski prism; therefore the o- and e-waves experience a differential optical delay through the PM fiber. An objective lens (Carl Zeiss, Inc.; 40×, 0.75) focuses both beams onto the sample with a slight lateral separation, which is smaller than the size of diffraction-limited spot. The sample is mechanically scanned in $x-y$ using motorized actuators for two-dimensional imaging. A CCD is used to monitor the sample and the beam location.

In this particular SD-DIC implementation, the sample could be configured into two different modes. For surface profiling applications shown in Figure 14.3B, photons directly

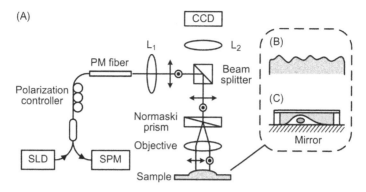

Figure 14.3
(A) SD-DIC system; (B and C) possible sample configurations for surface profiling and live cell imaging, respectively. SLD, superluminescence diode; SPM, spectrometer; L_1 and L_2, lenses.
Source: From Ref. [23].

reflected from the sample surface are collected. When imaging thin, transparent samples, such as cells, they are plated on a reflective surface to enable double-pass transmission measurement, as shown in Figure 14.3C. A small glass chamber is typically used to keep cells alive in culture media. In either mode, reflected o- and e-waves return through the same optical path to be mixed in the fiber coupler to interfere, before entering a spectrometer (OceanOptics: HR4000) for detection. A polarization controller may be used in the detection arm to maximize the fringe visibility.

Although the system shown in Figure 14.3A resembles a conventional reflected-light DIC microscope, the fundamental differences are the PM fiber and the spectral-domain detection, which enable the high-resolution operation. The PM fiber functions as a phase retarder, which adds a large bias, approximately 215 μm (single pass) in this system, to the OPL gradient between the o- and e-waves. Hence, the sample signal intensity received by the spectrometer can be written as:

$$I_t(k;x,y) \propto r_o^2(x,y) + r_e^2(x,y) + 2\eta r_o(x,y)r_e(x,y)\cos[2k(\Delta L_{PR} + \delta L_{NP} + \delta L_{DIC}(x,y))] \quad (14.3)$$

where r_o and r_e represent the reflectivities of a surface or the transmissivities of a cell at position (x,y) for the o- and e-waves, respectively; k is the wavenumber; η is the interference visibility; ΔL_{PR}, δL_{NP}, and δL_{DIC} are single-pass OPL differences generated by the phase retarder, the Nomarski prism, and the sample gradient, respectively.

Equation (14.3) shows that the phase retarder functions similarly to the frequency shifter used to generate a carrier frequency in a heterodyne system. Without the phase bias, the submicron/nanometer magnitudes of δL_{NP} and δL_{DIC} will generate only low-frequency interferometric fringes in the spectrum. Such modulation makes it difficult to accurately determine the phase signal in the presence of a DC background and various low-frequency noises. In contrast, these undesired signals can be effectively avoided by introducing a large phase bias ΔL_{PR}, which shifts the interferometric signal to higher spectral frequency.

Once the high-frequency interferogram is acquired, it is processed to extract the phase term $(\Delta L_{PR} + \delta L_{NP} + \delta L_{DIC})$ [20,22], where $(\Delta L_{PR} + \delta L_{NP})$ constitutes a background constant, and δL_{DIC} represents the quantitative OPL gradient of the sample. Meanwhile, the process also produces a signal intensity image, obtained by summing $I_t(k; x,y)$ across the entire spectral bandwidth at each point on the sample. The image can be interpreted as a very close approximation of the conventional brightfield (intensity) image, representing the reflectivity or transmission distribution across the sample.

14.3.2 Characterization of USAF Resolution Target

To demonstrate the performance of the SD-DIC system for surface profiling, we imaged the chromium (Cr) pattern of a positive USAF resolution target. Figure 14.2A shows the δL_{DIC} image of Group 7, Element 1, with a clear shadow-cast appearance typical to DIC images. The DIC contrast is provided by the surface step between the Cr coating and the uncoated glass substrate. For comparison, the intensity image is shown in Figure 14.2B. Figure 14.4C and D shows the cross-sectional profiles from both the DIC and intensity images, respectively. In the DIC curve, the rising and falling edges are clearly indicated by the positive and negative peaks. The unequal magnitudes suggest an oblique incident angle possibly due to slight sample misalignment. The axial resolution of δL_{DIC} measurement on the coated surface is 32 pm. The transverse resolution is 0.95 µm, measured as the 10–90% edge response in the intensity curve.

It is worth noting that the bars appear wider in the DIC image (Figure 14.4A) than in the intensity image (Figure 14.4B). This is further shown in Figure 14.4C and D, where the DIC peaks are not precisely located at the middle of the intensity slopes where the differential signal is expected to be the largest; rather, they shift toward the slope bottom. Therefore, the separation between the positive and negative DIC peaks is 5.0 µm, greater than the 3.8 µm full width at half maximum of the intensity peak (actual bar width 3.9 µm). We note that the 1.2 µm discrepancy is a consequence of the drastic contrast in reflectivity between the glass substrate and the Cr coating. As shown in the inset of Figure 14.4C, where the incident beam covers more glass than coating, the reflected light and its phase can be dominated by photons reflected from the coating. In terms of phase measurement, this is equivalent to the situation where the entire incident beam is on the coated area, hence extending the apparent size of the bar. It should be noted that this dimensional artifact is a consequence of the decoupling of phase and intensity and therefore is present in perhaps all quantitative DIC techniques, but will manifest itself only where there is steep contrast in reflectance or transmittance. Under these conditions, the apparent transverse size

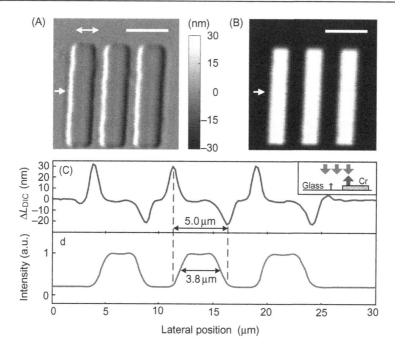

Figure 14.4
SD-DIC images of a USAF resolution target. (A and B) OPL gradient and brightfield intensity images of Group 7, Element 1; double arrow indicates the shear direction; scale bar is 10 μm. (C and D) Cross-sectional profiles of (A and B) at positions indicated by the single arrows; inset explains the increased bar width in (A). Source: *From Ref. [23].*

in the phase image of an object may not accurately reflect its true size. Despite this artifact, the intensity image will still carry size information faithfully and may be used together with the phase image to better characterize the object. In addition, this phenomenon is expected to have minimal effect for most biological samples since they do not usually exhibit such steep intensity changes.

14.3.3 Live Cell Imaging with Rat Cardiomyocytes

We also performed live cell imaging with the SD-DIC system, recording fast cellular dynamics at selected sites. For sample preparation, ventricular cardiomyocytes were isolated from 2-day-old Sprague Dawley rat neonates using sequential trypsin and collagenase digests, depleted of fibroblasts by differential attachment, and then cultured for 24 h on a reflective surface coated with human fibronectin before imaging in low-serum differentiation medium at 22 °C [24].

Figure 14.5A and B shows the OPL gradient and intensity images of an isolated cardiomyocyte. The intensity image has an artifact, appearing as a bright spot on the left side, possibly caused by the disturbance to the fiber during scanning. Yet it does not affect the DIC image, demonstrating the excellent decoupling of gradient signal from intensity using the spectral-domain demodulation approach and signifying one of the important advantages of the proposed system.

Live cell dynamics were monitored by recording the DIC signal at the circled spot (Figure 14.5A) for 120 s with a sample rate of 83 Hz, which is limited by spectrometer data acquisition speed. The top curves in Figure 14.5C show that two pulsatile motion events were observed, each with a duration of 0.5 s. For comparison, a background signal was recorded outside the cell, showing a resolution of 36 pm, which is similar to that obtained

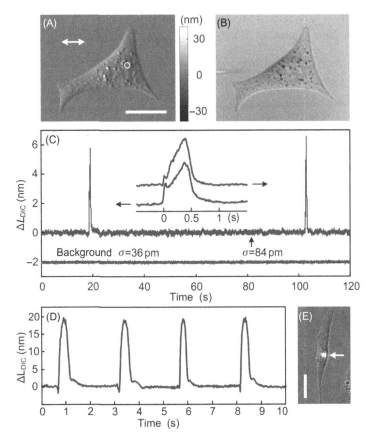

Figure 14.5

Live cardiomyocyte measurements. (A) OPL gradient image of an isolated cardiomyocyte; double arrow indicates shear direction. (B) Corresponding brightfield image. (C) Upper: stochastic beating events at the circle in (A) for 2 min; inset shows enlarged view of the two beating events; Lower: background signal taken at the upper-left corner of the image. (D) Regular beating sequence of another cardiomyocyte shown in (E), where the bright spot indicates illuminated location. Scale bars are 20 μm. Source: From Ref. [23].

previously with the resolution target, although the beam now passes through the liquid culture medium, which in general is a source of instability in temporal measurement of phase through cell samples. This is evidence of superior suppression of ambient fluctuations by the common-path geometry of DIC microscopy. The dynamics of a second cardiomyocyte are shown in Figure 14.5D, registering four periodic beating events in 10 s with an average duration of 0.4 s.

The relatively rare pulsatile events seen for the first cardiomyocyte are consistent with stochastic contractions often observed in freshly isolated ventricular cardiomyocytes, whereas the regular events in the second cardiomyocyte are of the same frequency as contraction cycles observed in "pacing cells," which is a type of specialized cardiomyocyte readily identified in culture [25]. The observation of similar durations of the events in Figure 14.5C (0.5 s) and Figure 14.5D (0.4 s) are consistent with this explanation.

In this section, we have demonstrated that polarization diversity can be used to provide highly stable measurements of cell samples. A fiber-optic SD-DIC microscope for imaging both reflective surfaces and live cells was presented and applied to monitoring dynamics at selected sample sites. SD-DIC offers a high-sensitivity means to quantitatively decouple OPL gradient and intensity in a single measurement. As a point-scanning system, the SD-DIC achieves enhanced OPL gradient resolution at the expense of acquisition speed. Using existing technologies, the system can be adapted to achieve sub-Hz frame rates for two-dimensional imaging and greater than 50 kHz acquisition rates for single-point measurements. Hence, wide-field imaging using SD-DIC is most suited for high-resolution, low-speed characterization, such as surface profiling or observation of relatively slow dynamics (on the order of a few seconds) of live biological specimens, while fast dynamics below 0.1 ms can be monitored for selected locations.

14.4 Spectral Multiplexing by RGB Color Channels

As described earlier, light sources with continuous, broad bandwidths can be exploited to provide drastic improvement of phase measurement resolution. The concept of exploiting the spectral domain can also be leveraged with sources that have multiple discrete bandwidths. We now show how color cameras can be incorporated to facilitate spectral multiplexing that introduces new functionalities to conventional quantitative phase measurements.

The spectral channels of commercial RGB cameras can be used not only for color imaging applications but also for multiplexing of independent information through the use of spectral encoding. With careful consideration of illumination and detection schemes, the three color channels of an RGB camera can be used to multiplex both phase and fluorescence information about biological samples with minimal crosstalk between color channels. Here, we describe applications for multiple-wavelength phase unwrapping and also coregistered phase and fluorescence imaging to provide morphological information in conjunction with molecular specificity. The primary advantage of using a commercial RGB camera for multimodality imaging is that all information is captured simultaneously through different color channels and thus is immune to temporal motion artifacts.

14.4.1 RGB Camera Multiplexing for Phase Unwrapping

QPM is useful for investigating mostly transparent samples. By interferometrically detecting a transmitted wavefront in weakly diffracting samples, such as cells, microfluidic chambers, and other protein-based microstructures [26–28], it is possible to quantify the OPDs across a field of view. However, any structure delaying the light by more than one wavelength introduces phase ambiguities, which arise as a direct result of mathematically calculating phase by an arctangent operation. The two-argument variant of arctangent is limited to an unambiguous range of $\pm\pi$; therefore, any relative phase changes across a field of view exceeding 2π require additional processing to remove the incorrect discontinuities.

There have been many algorithms developed for removing 2π ambiguities from quantitative phase measurements, including two-dimensional spatial algorithms with robustness against noisy data [29–32] and also temporal algorithms [33]. However, these algorithms all involve minimizing the total phase gradient, which enforces the presumption of a smooth phase that does not vary by more than π between two adjacent points. Samples that have high aspect ratios or sharp edges violate this condition and therefore require alternative approaches for accurate phase unwrapping.

An alternative approach for digital unwrapping relies on measuring the OPDs at a specific location using more than one wavelength. The OPDs through a sample are related to the optical phase, ϕ_m, measured modulo 2π at a given wavelength, λ_m, as:

$$\text{OPD}_m(x, y) = \frac{\phi_m(x, y)\lambda_m}{2\pi} \quad (14.4)$$

Several methods of two-wavelength unwrapping extend the range of measurable OPDs by first creating a synthetic phase map, $\phi_{12} = (\phi_1 - \phi_2) + 2\pi[(\phi_1 - \phi_2) < 0]$, that is unambiguous over the range of the "beat" wavelength, defined as $\Lambda_{12} = (\lambda_1\lambda_2/|\lambda_1 - \lambda_2|)$ [34–36]. While the resulting phase map has a larger range of measurement, the phase noise in ϕ_{12} is also larger than either ϕ_1 or ϕ_2 [34]. Gass et al. [34] reported that if the phase noise of ϕ_m is defined as $2\pi\varepsilon_m$, the phase noise in ϕ_{12} becomes $2\pi\varepsilon_{12} = 2\pi(\varepsilon_1 + \varepsilon_2)$. For transparent samples, the noise in OPD_m is $\varepsilon_m\lambda_m$; therefore, the noise in OPD_{12} is amplified by a factor of $\sim 2\Lambda_{12}/\lambda_m$. Because of this amplification inherent to the mathematical process, the beat wavelength OPD map is only used as a guide for adding multiples of λ_m

to OPD_m and not for direct imaging. As long as the amplified noise does not exceed $\lambda_m/2$, this process remains valid and the single-wavelength noise level ε_m can be maintained while the unambiguous measurement range can be increased to Λ_{12}. Since two-wavelength unwrapping relies on a mathematical analytical solution rather than the gradient minimization strategies, this unwrapping method is particularly useful for structures with sharp phase discontinuities.

There is an inherent trade-off between measurement range and noise amplification in two-wavelength phase unwrapping. As the separation between λ_1 and λ_2 is decreased, both the synthetic wavelength, Λ_{12}, and the total measurement range increase. With unlimited signal-to-noise ratio and perfect system stability, the measurement range could be extended arbitrarily using two laser sources with an infinitesimally small wavelength spacing; however, practical noise limits require careful instrument design and illumination selection in order to extend the measurement range while avoiding overamplification of the phase noise. To address this limitation of noise amplification, Mann et al. [37] proposed a hierarchical method of phase unwrapping that uses three wavelengths and intermediate synthetic wavelength profiles.

In order to use multiple-wavelength phase unwrapping with dynamic samples, detection and illumination schemes that allow the acquisition of multiple measurements in a single snapshot are required. Previously, reflection-geometry holographic microscopy systems have used spatial frequency multiplexing [37,38] or camera color channels with matched illumination wavelengths [39,40] to simultaneously acquire separate complex wavefront information. However, transmission-geometry phase measurements for semitransparent samples had only been demonstrated with sequential illumination schemes that are intended for measuring static samples [41,42].

In order to adapt multiple-wavelength unwrapping to imaging of dynamic semitransparent samples, such as cell cultures and other biological phenomena in microfluidic devices, we proposed the use of an RGB color camera with transmission-geometry QPM [43]. The illumination of the optical system shown in Figure 14.6A consisted of two lasers ($\lambda = 532$ and 633 nm) that were chosen to match the peak spectral responses of the *red* and *green* color channels of the Bayer pattern color camera (Roper Scientific, CoolSNAP *cf*) as seen in Figure 14.6B. Light from the two interferometer arms was collected by matched MO and imaged onto the camera by common tube lens L_1. A slight lateral shift in the reference arm MO created a linear off-axis interference fringe for each illumination wavelength.

Off-axis interferograms were obtained simultaneously for each wavelength. Because the color channels have low-intensity crosstalk, 4.3% and 5.4% for the green and red channels, respectively, the interferograms were easily isolated. After separating the color channels, the complex information in the Fourier domain was spatially filtered and recentered according to the methods of Cuche et al. [44]. This operation removed the zero-order

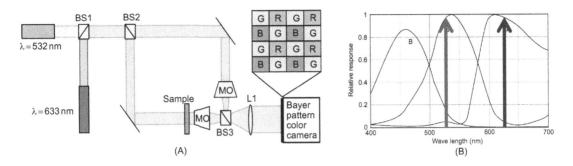

Figure 14.6
(A) Simultaneous two-wavelength transmission QPM optical system [18]; (B) spectral response of the camera color channels overlaid with the illumination wavelengths. *Source: Adapted from Ref. [43].*

autocorrelation and complex conjugate terms. Wrapped phase maps at $\lambda_1 = 532$ nm and $\lambda_2 = 633$ nm were recovered by inverse Fourier transforming and taking the arctangent of each pixel's resulting complex value.

To demonstrate the sensitivity of this system, we measured National Institute of Standards and Technology (NIST)-certified polystyrene microspheres ($n = 1.56$, $d = 19.99 \pm 0.20$ μm) that were immersed in index-matching oil ($n = 1.515$) to create a relative refractive index (RI) difference of $\Delta n = 0.045$. Two-wavelength phase unwrapping resulted in a profile with a maximum measured bead thickness of 20.105 μm and an RMS deviation from an ideal sphere of 17.312 nm.

We also demonstrated the ability of simultaneous two-wavelength transmission QPM for unwrapping high-aspect-ratio objects by imaging transparent 10–15 μm tall microstructures (Figure 14.7A–D). These microstructures were printed in UV-cure epoxy using a maskless holographic lithography process [45]. This procedure was also used to create a pattern of bovine serum albumin conjugated with fluorescein isothiocyanate on substrates for use as biosensor islands. Additionally, we demonstrated the utility of single shot two-wavelength phase unwrapping with biological samples by imaging and unwrapping A431 human skin cancer cells in culture (Figure 14.7E).

Multiple-wavelength phase unwrapping has technical drawbacks to consider. Since the process amplifies the noise, the initial phase noise levels must be sufficiently low to keep the amplified noise below $\lambda_m/2$ for successful implementation of this unwrapping procedure. Wider spacing between the wavelengths, as what we have employed here, will also help control the amplified noise to acceptable levels. Furthermore, highly dispersive samples may suffer from wavelength-dependent RI differences when multiple-wavelength phase unwrapping is applied to transmission-geometry systems. Nevertheless, the initial results presented here demonstrate that this approach has the potential to extend the

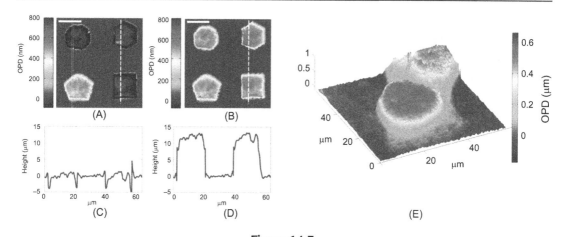

Figure 14.7
Microstructure OPD and height maps after quality-map guided phase unwrapping (A and C), compared to unwrapping performed with two-wavelength unwrapping (B and D). OPD map of A431 human skin cancer cells unwrapped using two-wavelength unwrapping (E). *Source: Adapted from Ref. [43].*

applicability of digital holography and quantitative phase measurements to structures such as sharp slopes and edges with greater than 2π wraps.

14.4.2 Spectral Multiplexing of Quantitative Phase and Fluorescence Biomarkers

Quantitative phase measurements are useful for investigating dynamic morphological changes in individual cells without the use of exogenous optical labels, and optical phase measurement has even been used to directly define quantitative parameters for characterizing cardiomyocyte [24] and red blood cell (RBC) [46] dynamics. However, while imaging with endogenous contrast removes the complexity of bioconjugating fluorescent species and enables to label cell features, there is an overall loss of molecular specificity. Therefore, the full value of QPM as a cell imaging modality can be realized when paired with traditional fluorescence techniques.

Previous efforts to combine phase and fluorescence imaging include coregistration of diffraction phase and fluorescence microscopy to image both the quantitative phase delays of kidney cells while simultaneously locating their nuclei [47]. Molecular labeling was achieved by introducing DNA-binding Hoescht dye. Coregistered images were acquired in a single snapshot in this study, which points to the utility of this method for imaging dynamic systems. A more recent study integrated multiple-wavelength phase unwrapping with widefield epi-illumination fluorescence microscopy [48]. However, phase maps and fluorescence images were captured sequentially with shutters used to gate the corresponding illumination lasers. Without simultaneous acquisition, dynamic changes can take place in a biological sample between frames, which introduces registration errors, and image throughput is lowered, an important parameter when imaging cell dynamics.

In order to simultaneously capture both fluorescence and quantitative phase images, we used a Bayer-mosaic-pattern color CCD to separate fluorescence and phase information by color channel. While there is some crosstalk between color channels ($\sim 5\%$, see earlier), the illumination wavelengths and fluorophores can be carefully chosen to minimize these effects. The system as shown in Figure 14.8 consists of a transmission-geometry QPM illuminated by a 532 nm laser source. Two additional laser sources at 632.8 and 376 nm are incorporated at the unused port of the recombining beamsplitter and illuminate the entire field of view on the sample after passing through the condenser lens and sample arm MO. Fluorescence emission is then imaged using the same optical pathway as the transmitted QPM light onto the CCD by the MO and tube lens. Excitation light from the 376 and 632.8 nm sources is suppressed by a band-pass filter that transmits all other visible light (including the QPM illumination).

In our experiments, we examined rat ventricular cardiomyocytes and recorded their contractions at 5 fps. The sample chamber was filled with growth media to keep the cells alive and topped by a coverslip to reduce phase noise from media surface fluctuations. The cell nuclei were labeled with Hoescht 34580 dye (Invitrogen, Carlsbad, CA, USA), which was excited by the 376 nm laser and had an emission peak at 440 nm, which is closely

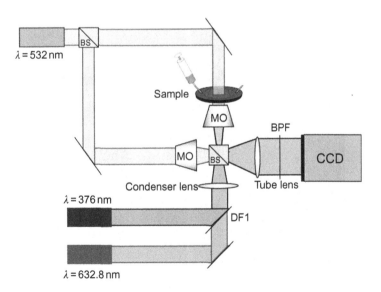

Figure 14.8
Quantitative phase and fluorescence microscope using a color camera to multiplex the imaging modalities. Source: *Adapted from Ref. [24].*

matched to the blue channel of the CCD. The mitochondria were labeled with Mitotracker Deep Red (Invitrogen), which was excited by the 632.8 nm laser and had an emission peak at 670 nm, which closely matched the red channel of the CCD. Since fluorophore emission is not a single wavelength but rather a broad spectrum, there was some concern that the nuclear stain could be partially visible in the green color channel, which was used for QPM. However, since QPM employs off-axis holography, any fluorescence crosstalk that is concentrated in the low-spatial frequencies is filtered and removed, and therefore does not significantly affect the phase measurements. The three color channels from each frame were then processed individually to produce fluorescence maps with molecular specificity and a quantitative phase image to visualize the overall cellular morphology. Figure 14.9A shows the mitochondrial distribution (red) and nuclear positions (blue) of cardiomyocytes in a single frame of the sequence, and Figure 14.9B shows the quantitative phase map.

There are some unique considerations that arise from the use of an RGB color camera to multiplex-independent imaging modalities. The mosaic color filter effectively reduces the number of pixels and therefore the spatial resolution of each independent color channel. To compensate for this loss, the optical system must be designed to further magnify the object, which in turn reduces the field of view. The trade-off between resolution and field of view can be adjusted based on the application and may be alleviated by using a higher pixel count sensor. A further consideration for separating fluorescence and quantitative phase information is that commercial digital cameras often perform on-chip image demosaicing that mixes data from all color channels before outputting the processed image to the user

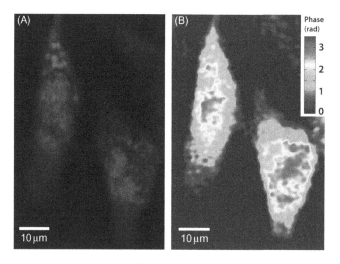

Figure 14.9
Rat ventricular cardiomyocytes with (A) fluorescently labeled mitochondria (red) and nuclei (blue), and (B) quantitative phase delays. (For interpretation of the references to color in this figure legend, the reader is referred to the web version of this book.) Source: *Adapted from Ref. [24]*.

[49,50]. Since the multiplexing methods presented here collect independent information in the spectral channels, it is important that the raw pixel intensities are read out from the sensor and processed separately.

The primary advantages of using an RGB sensor to multiplex phase and fluorescence information are the simplicity of the optical system designs and the nature of simultaneous acquisition. A single sensor removes the need for complex detection schemes that can require synchronization of multiple cameras, shutters, and lasers. Finally, since the quantitative phase and fluorescence modalities are acquired simultaneously on the RGB sensor, these systems minimize motion artifacts and theoretically allow frame rates to be increased to the limit of the camera electronics.

14.5 Nonlinear Phase Dispersion Spectroscopy

We now introduce an interferometric method for profiling the spectral properties of cell samples. The approach is based on spectral-domain phase microscopy (SDPM), which is an extension of OCT that provides quantitative OPD information with nano- to picometer sensitivity [20]. This method is similar to quantitative digital holographic techniques [51,52] in that SDPM is realized by interfering a sample wave with a reference wave. However, SDPM retrieves the phase information from spectrally resolved oscillations rather than spatial fringes. Consequently, spectroscopic information is more readily accessible in SDPM. In addition, SDPM traditionally uses a focused beam that is swept across the sample rather than a wide-field imaging approach. Likewise, spectroscopic OCT (SOCT) [53] is another extension of OCT that takes advantage of the spectral dimension to provide functional information regarding the wavelength-dependent total attenuation coefficient due to a sample [54–57]. Similar to SOCT, spectroscopic phase techniques have been developed to probe the wavelength-dependent real part of the RI, thereby granting access to the dispersion-inducing biochemical properties of samples [51,58,59]. However, to date, such methods have required multiple light sources or narrow band-pass filters, thus limiting the spectral information to a few narrow spectral regions.

In this section, we introduce a spectroscopic phase technique, termed nonlinear phase dispersion spectroscopy (NLDS), which provides broadband, high-resolution spectral information of the real part of the RI. In the meantime, complementary information is obtained from the same sampled data using SDPM and SOCT processing to yield quantitative topographical maps and spectral profiles of the total attenuation coefficient of samples, respectively. We present results from proof-of-concept experiments using fluorescent and nonfluorescent polystyrene beads, and RBCs, and also discuss the technique's potential applications [60].

To understand how the dispersive properties (i.e., real parts of the RI) may be obtained, consider a Michelson interferometer with a reference field, $E_r(\omega)$, and a sample field, $E_s(\omega)$, described as:

$$E_r(\omega) = s(\omega) \cdot \exp(i(\omega/c_0)2z_r) \tag{14.5a}$$

$$E_s(\omega) = \sqrt{I_s(\omega)} \cdot \exp(i(\omega/c_0)2z_d)\exp(i(\omega/c_0)n(\omega)2(z_s - z_d)) \tag{14.5b}$$

where z_r, z_d, and z_s are the distances from the beamsplitter to the reference mirror, dispersive medium, and scatterer, respectively; $S(\omega)$ is the spectrum of the source field, $I_s(\omega)$ is the sample field intensity, c_0 is the speed of light in vacuum, and $n(\omega)$ is the real part of the RI of the sample [61]. The sample can be described by bulk absorption and scattering coefficients μ_a and μ_s, respectively, such that the sample field intensity may be written as $I_s(\omega) = |S(\omega)|^2 \exp(-(\mu_a + \mu_s) 2(z_s - z_d))$. Thus, the interferometric signal, after elimination of DC background terms, may be expressed as:

$$\begin{aligned}\tilde{I}(\omega) &= 2\sqrt{I_s(\omega)I_r(\omega)} \cdot \exp(i(\omega/c_0)2(z' - dn(\omega))) \\ &= 2I_r(\omega)\exp(-\mu_{tot}(\omega)d) \cdot \exp(i(\omega/c_0)2(z' - dn(\omega)))\end{aligned} \tag{14.6}$$

where $I_r = |S|^2$, $\mu_{tot} = \mu_a + \mu_s$, $z' = z_r - z_d$, and $d = z_s - z_d$ is the sample thickness. To analyze the information contained in the dispersion of the signal, we employ a Taylor series expansion of $n(\omega)$,

$$n(\omega) = n(\omega_0) + \left.\frac{dn(\omega)}{d\omega}\right|_{\omega_0}\left(\frac{\omega - \omega_0}{c_0}\right) + \frac{1}{2!}\left.\frac{d^2n(\omega)}{d\omega^2}\right|_{\omega_0}\left(\frac{\omega - \omega_0}{c_0}\right)^2 + \cdots = n(\omega_0) + \Delta n(\omega) \tag{14.7}$$

where $n(\omega_0)$ may be evaluated at an arbitrary frequency or wavelength, and Δn incorporates all the high-order ω-dependent terms from the Taylor series expansion. Thus, Eq. 14.6 may be rewritten as follows:

$$\tilde{I}(\omega) = 2I_r(\omega)\exp(-\mu_{tot}(\omega)d) \cdot \exp(i(\omega/c_0) \cdot 2(z' - dn(\omega_0))) \cdot \exp(-i(\omega/c_0) \cdot 2d\Delta n(\omega)) \tag{14.8}$$

Equation 14.8 clearly shows that the measured signal contains three distinct parts. The first part attenuates the signal intensity, providing access to spectroscopic measurement of μ_{tot}. This term may be analyzed using SOCT techniques such as the dual-window method [62], or in the case of a thin sample, the complex signal $I(\omega)$ may simply be demodulated. The second term of $I(\omega)$ describes the linear phase and can be obtained using SDPM. Let us consider that the sample thickness, d, may be expanded as multiples of the source's coherence length, l_c, and a subcoherence length deviation, δ, given as $d = ml_c + \delta l_c$ with m being an integer. Thus, a fast Fourier transform (FFT) of Eq. 14.8, ignoring the third term for now, reveals a peak located at $2n_0ml_c$, corresponding to the sample thickness. Additionally, the phase at the same point gives $\phi = (2\pi/\lambda_0) \cdot 2n_0\delta l_c$, with λ_0 being the center wavelength. Thus, with a reasonable assumption regarding the average RI, the linear phase term yields the value d and provides quantitative structural information. Lastly, the third term contains the variations in the RI with respect to wavelength, $\Delta n(\omega)$, and is identified in Eqs. (14.7) and (14.8) by its nonlinear phase. This nonlinear phase term is the source of information for NLDS, which grants access to the dispersion-causing biochemical properties of samples. It is important to note that even though multiple parameters are embedded in the sampled signal—specifically, μ_{tot} in the intensity modulated term, d, in the linear phase term, and Δn in the nonlinear phase term—each is present in only one of the three parts of Eq. (14.8) and hence can be retrieved independently.

We demonstrate the NDLS technique using a super continuum laser source (Fianium, MA, USA) and a 4 f Michelson interferometer. A similar setup was described previously in Ref. [63], which used collimated light to probe the sample. To match beam dispersion between the sample and reference arms of the interferometer, identical MO (40×, 0.66 NA, infinity-corrected, L-40×, Newport Corp., CA, USA) are used in each arm, producing a lateral resolution of 0.46 μm. The samples are plated on a silvered surface such that input light passes through the cell media and the sample and is reflected by the silver coating and then imaged onto the entrance slit of an imaging spectrograph. The use of the imaging spectrograph allows for acquisition of multiple spectral interferograms across one lateral dimension. To obtain sample information in the other dimension, the sample is then translated using a motorized actuator. The imaging spectrograph spans the visible spectrum of the super continuum source, with 575 nm central wavelength, 240 nm bandwidth, and 0.2 nm spectral resolution.

14.5.1 Fluorescent and Nonfluorescent Polystyrene Beads

To verify the concept of NDLS, we measured fluorescent (Invitrogen F8833) and nonfluorescent (Thermo Scientific), 10 μm polystyrene beads. Samples were prepared by drying bead solutions on a silver-coated coverglass followed by immersion with glycerol, mimicking the environment of cell culture. To prepare the raw interferograms for further processing, they were resampled from the wavelength domain to the wavenumber domain with linearly spaced sampling points.

The linear phase is processed using SDPM to acquire structural information. To achieve this, an FFT of the interferograms needs to be computed, but dispersion effects must be eliminated first. This can be accomplished by first unwrapping the phase of each interferogram and then fitting it to a line of the form of $\phi_{lin} = (\omega/c_0)L - 2\pi q$ [64], where L is the best estimate of $2n_0 m l_c$, and q is an integer describing the initial phase. Figure 14.10A shows the unwrapped phase of $I(\omega)$, along with the corresponding linear fit, ϕ_{lin}, for a point located at the center of the fluorescent bead. The residual phase, $\Delta\phi(\omega)$, which is the difference between the unwrapped phase and its linear fit, is removed from each interferogram and stored for

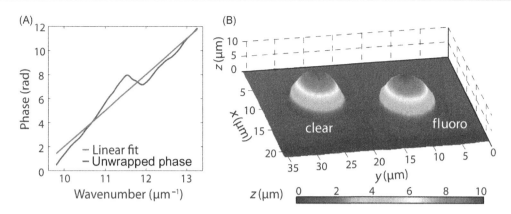

Figure 14.10
(A) Unwrapped phase of an interferogram acquired from the center of the fluorescent bead and the line of best fit used to correct for dispersion effects. (B) Topological map of nonfluorescent (clear) and fluorescent (fluoro) beads. Source: From Ref. [60].

analysis using NLDS, as discussed later. The removal of this residual phase yields a dispersion-compensated interferogram. Finally, an FFT of the compensated signal may be computed, and the location of the peak and the angle (phase) at the same point are used to determine the optical path length, $n(\omega_0)d$, with nano- to picometer sensitivity. For this bead sample, an average RI of $n_0 = n_{bead} - n_{glycerol} = 1.59 - 1.47 = 0.12$ is used to convert the phase information to a topological map of the sample. The resulting maps for both a nonfluorescent and a fluorescent bead are shown in Figure 14.10B. The sensitivity of the system is 5 nm as determined by the standard deviation of fluctuations within a background region.

Now the absorption spectra are analyzed. Returning to Eq. (14.8), now the dispersion-compensated interferograms contain only the first two terms (intensity and linear phase). Consequently, the absorption spectrum, which is the first term of Eq. (14.8), can be computed by taking the absolute value of the signal. Figure 14.11A shows two curves corresponding to the negative log of the spectrally resolved intensity, i.e., $\mu_{tot}(\omega)d$, for both the fluorescent and nonfluorescent beads. The figure also shows the ideal extinction coefficient of the fluorescent bead, as provided by the manufacturer, which is in good agreement with the experimental results, exhibiting an extinction maxima located at 540 nm. For each point in the sample, we repeat this procedure. The obtained absorption spectra are then normalized and divided into red, green, and blue channels using the Commission Internationale d'Eclairage (CIE) color functions, providing a true color representation of the samples, as shown in Figure 14.11B, where the hue and saturation information is superposed on the topological map. As expected, the background and clear bead are relatively colorless, in contrast to a pink hue shown by the fluorescent bead. These colors are consistent with the appearance as seen by the naked eye or under a microscope.

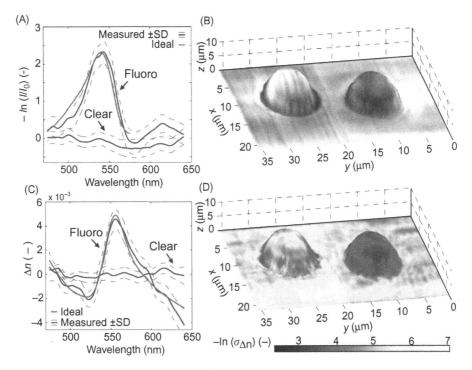

Figure 14.11
(A) Negative log of the normalized spectrum from the center of each bead. (B) True color representation of the sample superposed on the topological map. (C) Changes in the RI for a point at the center of each bead. (D) Negative log of the standard deviation of the changes in the real part of the RI superposed with the topological map. (For interpretation of the references to color in this figure legend, the reader is referred to the web version of this book.) *Source: From Ref. [60].*

The last step is to investigate the dispersive properties of these samples using NLDS. For this analysis, recall that we have already obtained the residual phase, $\Delta\phi(\omega)$, which contains the cumulative dispersion from the system, background medium, and sample. In order to isolate the sample-induced dispersion only, the average residual phase of a background region is used as a reference to remove all dispersive contributions from outside of the sample of interest. The resulting phase contribution is related to the changes in the sample's RI by $\Delta\phi(\omega) = (\omega/c_0)2d\Delta n(\omega)$ [55]. Figure 14.11C shows Δn for locations corresponding to the middle of the clear and fluorescent beads. For comparison, Figure 14.11C also shows the ideal dispersion for the fluorescent bead, which is calculated using a subtractive Kramers–Kronig relation [65], and agrees well with the measured spectra. Additionally, Figure 14.11C indicates that the fluorescent bead exhibits much greater amount of dispersion than the clear bead, a result anticipated by the principles of causality.

This NLDS procedure is repeated to obtain dispersion measures for all points on the sample. The results are shown in Figure 14.11D, where the negative log of the standard deviation of Δn is superposed on the topological map. Note that the dark red hue of the fluorescent bead indicates higher dispersion compared to the background and clear bead, which appear in a light yellow hue. These dispersion results provide complementary information to the absorption map in Figure 14.11B.

14.5.2 Red Blood Cells

The second experiment conducted to demonstrate proof of the NLDS principle examines normal, fully oxygenated RBCs. A sample of whole blood was obtained from a healthy donor and processed as described previously [52]. The RBCs were plated on a silvered coverglass and imaged at ambient room temperature and pressure. All procedures were adhered to approved Institutional Review Board protocols.

We followed the same signal-processing method described earlier to analyze an isolated RBC. Again, the topological map was obtained from the linear phase (SDPM) analysis, as shown in Figure 14.12A, illustrating the well-known round biconcave shape of a healthy RBC. The attenuation-based spectral information is shown in Figure 14.12B as the true color spectroscopic map. Here, the RBC appears colorless, which is in fact in good agreement with typical appearance of isolated RBCs when observed with a phase-contrast microscope. These results indicate that the absorption incurred by a single RBC is not sufficient to impart a visible spectral modulation. The lack of spectroscopic information hinders accurate quantification of important biomarkers, namely hemoglobin (Hb) concentration (C_{Hb}) and oxygen saturation (SO_2). This observation is also reflected in previous findings, where quantification of C_{Hb} and SO_2 of individual RBCs have shown large uncertainties [56,66]. In contrast, the NLDS technique is capable of providing higher sensitivity as shown in Figure 14.12C by the good contrast between the RBC and the background. In fact, based on the current system's phase sensitivity ($\sim$10 mrad), C_{Hb} as low as $\sim$0.5 g/dl may be detected (normal RBC C_{Hb}: 30–36 g/dl). Compared with attenuation-based spectral techniques [62], NLDS is two orders of magnitude more sensitive.

The full potential of NLDS is realized by analyzing the detailed dispersion spectra, allowing the accurate quantification of C_{Hb}. The procedure entails using the oxygenated-Hb dispersion coefficients in a matrix inversion with the computed dispersion [55]. Figure 14.12D shows a typical spectra along with the best fit corresponding to $C_{Hb} = 31.9$ g/dl. The average concentration across the entire cell was $C_{Hb} = 33.4 \pm 7.7$ g/dl, in good agreement with expected values of healthy RBCs [51,67]. Figure 14.13 shows a group of four RBCs with concentrations of (from left to right) 34.6 ± 9.1, 36.0 ± 11.9,

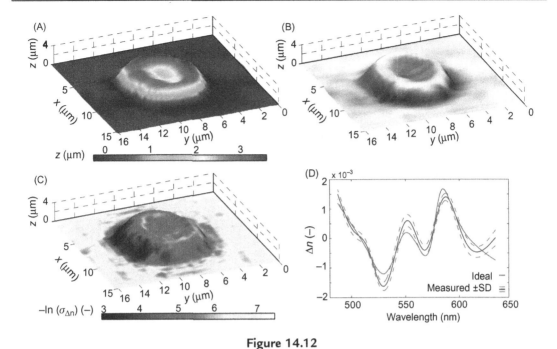

Figure 14.12
(A) Topological map of a healthy RBC. (B) True color representation superposed with the topological map. (C) Negative log of the standard deviation of the changes in the real part of the RI superposed with the topological map. (D) Representative spectral profile of the real part of the RI. (For interpretation of the references to color in this figure legend, the reader is referred to the web version of this book.) *Source: From Ref. [60].*

31.6 ± 13.0, and 36.7 ± 8.2 g/dl, respectively. These numbers are also in good agreement with the expected values.

As described above, NLDS is a novel technique which measures the dispersion profile of a sample across a wide spectral bandwidth and with high spectral resolution. Furthermore, the technique may be integrated with SDPM and SOCT to obtain topological maps of samples along with their absorptive properties, all from a single measurement [57]. We have also demonstrated NLDS's ability to detect and quantify different dispersion-inducing molecules, specifically a synthetic fluorophore and oxy-Hb. These results have important implications for the diagnosis of various RBC diseases, e.g., sickle-cell anemia, thalassemia, and malaria, using only intrinsic spectroscopic information from the cells. Additionally, NLDS may be also used for molecular imaging with exogenous agents to monitor cell receptor expression and disease states. Moreover, a recent theoretical treatment [68] showed that the mean free scattering path length and isotropic coefficient can be derived from the spatial variance and variance gradient of the phase, respectively. Looking forward, NLDS may be integrated with such analysis to provide a more complete picture of the

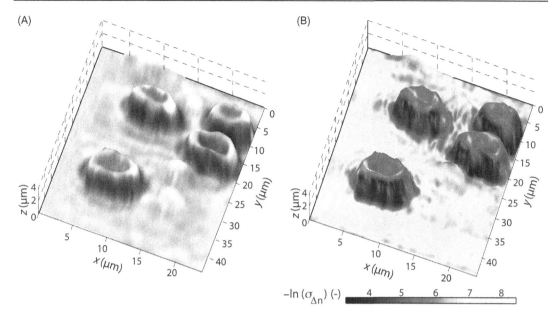

Figure 14.13
Four healthy RBCs. (A) True color representation superposed with the topological map. (B) Negative log of the standard deviation of the changes in the real part of the RI superposed with the topological map. (For interpretation of the references to color in this figure legend, the reader is referred to the web version of this book.)

biologically relevant properties of samples, which would include their scattering, anisotropy, absorption, and dispersion properties.

14.6 Conclusions

In contrast to the single-polarization, single-wavelength operation in conventional digital-holography-based QPM systems, probing a sample with additional polarizations and wavelengths generates a wealth of new information that can be exploited to markedly enhance system performance and functionality.

As we have demonstrated in this chapter, information from these extra dimensions can be acquired with a variety of configurations. Both the DQPM system and the SD-DIC system take advantage of a Wollaston prism to create an orthogonal polarization component, which shares similar noise characteristics with the original polarization, and hence greatly facilitates noise reduction and stability improvement. While the DQPM technique takes a full-field approach to capture high-speed dynamics across the entire object, SD-DIC is capable of monitoring selected points with extremely high resolution, with the help of a broadband light source.

Additional wavelengths also open the possibility for new system functions by exploiting spectral diversity. We have introduced a multichannel multiplexing scheme, which is conveniently based on commercial color cameras, for simultaneous recording of holographic images using multiple wavelengths to remove ambiguities in phase unwrapping and extend the range of phase measurements. We also used the technique to acquire coregistered label-free phase maps of a cell and its corresponding fluorescence signature, providing molecular information to complement the highly resolved structural information. A highly desired feature of this approach is the single-detector, common-path coregistration, which minimizes motion artifacts and maximizes imaging speed with reduced system cost.

While the first three techniques are focused on the measurement of optical/physical properties of a sample, specifically its OPL, the NLDS technique integrates optical spectroscopy into spectral-domain QPM, providing quantitative information regarding the sample's chemical composition and therefore introduces the potential for molecular specificity.

These are just a few examples that demonstrate the power of polarization and spectral techniques for extending the abilities of traditional QPM. Since both polarization and wavelength are among the most fundamental properties of light and can be manipulated, modified, and detected by numerous optical components and systems, we expect that efforts in this direction will continue to grow and that a multitude of novel techniques will further enhance performance and functionality for QPM.

Acknowledgments

Grant support was provided by the National Institutes of Health (National Cancer Institute R01CA138594) and the National Science Foundation (CBET-0651622 and MRI-1039562).

References

[1] M. Pluta, Advanced Light Microscopoy, vol. 2, Elsevier Science Publishing, New York, 1988.
[2] T. Zhang, I. Yamaguchi, Three-dimensional microscopy with phase-shifting digital holography, Opt. Lett. 23 (1998) 1221–1223.
[3] E. Cuche, F. Bevilacqua, C. Depeursinge, Digital holography for quantitative phase-contrast imaging, Opt. Lett. 24 (1999) 291–293.
[4] P. Marquet, B. Rappaz, P.J. Magistretti, E. Cuche, Y. Emery, T. Colomb, et al., Digital holographic microscopy: a noninvasive contrast imaging technique allowing quantitative visualization of living cells with subwavelength axial accuracy, Opt. Lett. 30 (2005) 468–470.
[5] G. Popescu, L.P. Deflores, J.C. Vaughan, K. Badizadegan, H. Iwai, R.R. Dasari, et al., Fourier phase microscopy for investigation of biological structures and dynamics, Opt. Lett. 29 (2004) 2503–2505.

[6] K.J. Chalut, W.J. Brown, A. Wax, Quantitative phase microscopy with asynchronous digital holography, Opt. Express 15 (2007) 3047−3052.
[7] T. Ikeda, G. Popescu, R.R. Dasari, M.S. Feld, Hilbert phase microscopy for investigating fast dynamics in transparent systems, Opt. Lett. 30 (2005) 1165−1167.
[8] J. Munoz, M. Strojnik, G. Paez, Phase recovery from a single undersampled interferogram, Appl. Opt. 42 (2003) 6846−6852.
[9] T. Kiire, S. Nakadate, M. Shibuya, Simultaneous formation of four fringes by using a polarization quadrature phase-shifting interferometer with wave plates and a diffraction grating, Appl. Opt. 47 (2008) 4787−4792.
[10] G. Rodriguez-Zurita, N.I. Toto-Arellano, C. Meneses-Fabian, J.F. Vazquez-Castillo, One-shot phase-shifting interferometry: five, seven, and nine interferograms, Opt. Lett. 33 (2008) 2788−2790.
[11] N. Brock, J. Hayes, B. Kimbrough, J. Millerd, M. North-Morris, M. Novk, et al., Dynamic interferometry, Proc. SPIE 5875 (2005) 58750F.
[12] Y. Zhu, L. Liu, Y. Zhi, Z. Luan, D. Liu, Phase detection from two phase-shifting interferograms, Proc. SPIE 6671 (2007) 66711D.
[13] N.T. Shaked, M.T. Rinehart, A. Wax, Dual-interference-channel quantitative-phase microscopy of live cell dynamics, Opt. Lett. 34 (2009) 767−769.
[14] M.G. Nomarski, Microinterferometre differentiel a ondes polarisees, J. Phys. Radium 16 (1955) S9−S13.
[15] D. Fu, S. Oh, W. Choi, T. Yamauchi, A. Dorn, Z. Yaqoob, et al., Quantitative DIC microscopy using an off-axis self-interference approach, Opt. Lett. 35 (2010) 2370−2372.
[16] B. Heise, D. Stifter, Quantitative phase reconstruction for orthogonal-scanning differential phase-contrast optical coherence tomography, Opt. Lett. 34 (2009) 1306−1308.
[17] S.V. King, A. Libertun, R. Piestun, C.J. Cogswell, C. Preza, Quantitative phase microscopy through differential interference imaging, J. Biomed. Optics 13 (2008) 024020.
[18] M.R. Arnison, K.G. Larkin, C.J.R. Sheppard, N.I. Smith, C.J. Cogswell, Linear phase imaging using differential interference contrast microscopy, J. Microsc. 214 (2004) 7−12.
[19] M.A. Choma, M.V. Sarunic, C.H. Yang, J.A. Izatt, Sensitivity advantage of swept source and Fourier domain optical coherence tomography, Opt. Express 11 (2003) 2183−2189.
[20] M.A. Choma, A.K. Ellerbee, C.H. Yang, T.L. Creazzo, J.A. Izatt, Spectral-domain phase microscopy, Opt. Lett. 30 (2005) 1162−1164.
[21] C.K. Hitzenberger, A.F. Fercher, Differential phase contrast in optical coherence tomography, Opt. Lett. 24 (1999) 622−624.
[22] J. Zhang, B. Rao, L.F. Yu, Z.P. Chen, High-dynamic-range quantitative phase imaging with spectral domain phase microscopy, Opt. Lett. 34 (2009) 3442−3444.
[23] Y. Zhu, N.T. Shaked, L.L. Satterwhite, A. Wax, Spectral-domain differential interference contrast microscopy, Opt. Lett. 36 (2011) 430−432.
[24] N.T. Shaked, L.L. Satterwhite, N. Bursac, A. Wax, Whole-cell-analysis of live cardiomyocytes using wide-field interferometric phase microscopy, Biomed. Opt. Express 1 (2010) 706−719.
[25] T. Korhonen, S.L. Hanninen, P. Tavi, Model of excitation-contraction coupling of rat neonatal ventricular myocytes, Biophys. J. 96 (2009) 1189−1209.
[26] R. Nielson, B. Koehr, J.B. Shear, Microreplication and design of biological architectures using dynamic-mask multiphoton lithography, Small 5 (2009) 120−125.
[27] S. Basu, P.J. Campagnola, Properties of crosslinked protein matrices for tissue engineering applications synthesized by multiphoton excitation, J. Biomed. Mater. Res. A 71A (2004) 359−368.
[28] F.L. Yap, Y. Zhang, Protein and cell micropatterning and its integration with micro/nanoparticles assembly, Biosens. Bioelectron. 22 (2007) 775−788.
[29] M.A. Schofield, Y.M. Zhu, Fast phase unwrapping algorithm for interferometric applications, Opt. Lett. 28 (2003) 1194−1196.
[30] T.J. Flynn, Two-dimensional phase unwrapping with minimum weighted discontinuity, J. Opt. Soc. Am. A Opt. Image Sci. Vis. 14 (1997) 2692−2701.

[31] J.M. Huntley, Noise-immune phase unwrapping algorithm, Appl. Opt. 28 (1989) 3268–3270.
[32] D. Ghiglia, M. Pritt, Two Dimensional Phase Unwrapping: Theory, Algorithms and Software, Wiley Interscience, New York, 1998.
[33] J.M. Huntley, H. Saldner, Temporal phase-unwrapping algorithm for automated interferogram analysis, Appl. Opt. 32 (1993) 3047–3052.
[34] J. Gass, A. Dakoff, M.K. Kim, Phase imaging without 2 pi ambiguity by multiwavelength digital holography, Opt. Lett. 28 (2003) 1141–1143.
[35] Y.Y. Cheng, J.C. Wyant, 2-Wavelength phase-shifting interferometry, Appl. Opt. 23 (1984) 4539–4543.
[36] D. Parshall, M.K. Kim, Digital holographic microscopy with dual-wavelength phase unwrapping, Appl. Opt. 45 (2006) 451–459.
[37] C.J. Mann, P.R. Bingham, V.C. Paquit, K.W. Tobin, Quantitative phase imaging by three-wavelength digital holography, Opt. Express 16 (2008) 9753–9764.
[38] J. Kuehn, T. Colomb, F. Montfort, F. Charriere, Y. Emery, E. Cuche, et al., Real-time dual-wavelength digital holographic microscopy with a single hologram acquisition, Opt. Express 15 (2007) 7231–7242.
[39] I. Yamaguchi, T. Matsumura, J. Kato, Phase-shifting color digital holography, Opt. Lett. 27 (2002) 1108–1110.
[40] P. Picart, D. Mounier, L.M. Desse, High-resolution digital two-color holographic metrology, Opt. Lett. 33 (2008) 276–278.
[41] P. Ferraro, L. Miccio, S. Grilli, M. Paturzo, S. De Nicola, A. Finizio, et al., Quantitative phase microscopy of microstructures with extended measurement range and correction of chromatic aberrations by multiwavelength digital holography, Opt. Express 15 (2007) 14591–14600.
[42] J. Muller, V. Kebbel, W. Juptner, Digital holography as a tool for testing high-aperture micro-optics, Opt. Lasers Eng. 43 (2005) 739–751.
[43] M.T. Rinehart, N.T. Shaked, N.J. Jenness, R.L. Clark, A. Wax, Simultaneous two-wavelength transmission quantitative phase microscopy with a color camera, Opt. Lett. 35 (2010) 2612–2614.
[44] E. Cuche, P. Marquet, C. Depeursinge, Spatial filtering for zero-order and twin-image elimination in digital off-axis holography, Appl. Opt. 39 (2000) 4070–4075.
[45] N.J. Jenness, R.T. Hill, A. Hucknall, A. Chilkoti, R.L. Clark, A versatile diffractive maskless lithography for single-shot and serial microfabrication, Opt. Express 18 (2010) 11754–11762.
[46] Y. Park, C.A. Best, K. Badizadegan, R.R. Dasari, M.S. Feld, T. Kuriabova, et al., Measurement of red blood cell mechanics during morphological changes, Proc. Natl. Acad. Sci. USA. 107 (2010) 6731–6736.
[47] Y. Park, G. Popescu, K. Badizadegan, R.R. Dasari, M.S. Feld, Diffraction phase and fluorescence microscopy, Opt. Express 14 (2006) 8263–8268.
[48] C.J. Mann, P.R. Bingham, H.K. Lin, V.C. Paquit, S.S. Gleason, Dual modality live cell imaging with multiple-wavelength digital holography and epi-fluorescence, 3D Res. 2 (2011) 1–6.
[49] S. Farsiu, M. Elad, P. Milanfar, Multiframe demosaicing and super-resolution of color images, IEEE. Trans. Image Proc. 15 (2006) 141–159.
[50] B.K. Gunturk, J. Glotzbach, Y. Altunbasak, R.W. Schafer, R.M. Mersereau, Demosaicking: color filter array interpolation, IEEE Signal Proc. Mag. 22 (2005) 44–54.
[51] Y. Park, T. Yamauchi, W. Choi, R. Dasari, M.S. Feld, Spectroscopic phase microscopy for quantifying hemoglobin concentrations in intact red blood cells, Opt. Lett. 34 (2009) 3668–3670.
[52] N.T. Shaked, L.L. Satterwhite, M.J. Telen, G.A. Truskey, A. Wax, Quantitative microscopy and nanoscopy of sickle red blood cells performed by wide field digital interferometry, J. Biomed. Opt. 16 (2011).
[53] U. Morgner, W. Drexler, F.X. Kartner, X.D. Li, C. Pitris, E.P. Ippen, et al., Spectroscopic optical coherence tomography, Opt. Lett. 25 (2000) 111–113.
[54] F.E. Robles, A. Wax, Measuring morphological features using light-scattering spectroscopy and Fourier-domain low-coherence interferometry, Opt. Lett. 35 (2010) 360–362.
[55] F.E. Robles, A. Wax, Separating the scattering and absorption coefficients using the real and imaginary parts of the refractive index with low-coherence interferometry, Opt. Lett. 35 (2010) 2843–2845.

[56] J. Yi, X. Li, Estimation of oxygen saturation from erythrocytes by high-resolution spectroscopic optical coherence tomography, Opt. Lett. 35 (2010) 2094–2096.

[57] F.E. Robles, C. Wilson, G. Grant, A. Wax, Molecular imaging true color spectroscopic optical coherence tomography, Nat. Photonics 5 (2011) 744–747.

[58] C.H. Yang, A. Wax, I. Georgakoudi, E.B. Hanlon, K. Badizadegan, R.R. Dasari, et al., Interferometric phase-dispersion microscopy, Opt. Lett. 25 (2000) 1526–1528.

[59] B. Rappaz, F. Charriere, C. Depeursinge, P.J. Magistretti, P. Marquet, Simultaneous cell morphometry and refractive index measurement with dual-wavelength digital holographic microscopy and dye-enhanced dispersion of perfusion medium, Opt. Lett. 33 (2008) 744–746.

[60] F.E. Robles, L.L. Satterwhite, A. Wax, Non-linear phase dispersion spectroscopy, Opt. Lett. 36 (2011) 4665–4667.

[61] J.A. Izatt, M.A. Choma, in: W. Drexler, J.G. Fujimoto (Eds.), Optical Coherence Tomography: Technology and Applications, Springer, New York, NY, 2008, pp. 47–72.

[62] F.E. Robles, S. Chowdhury, A. Wax, Assessing hemoglobin concentration using spectroscopic optical coherence tomography for feasibility of tissue diagnostics, Biomed. Opt. Express 1 (2010) 310–317.

[63] R.N. Graf, W.J. Brown, A. Wax, Parallel frequency-domain optical coherence tomography scatter-mode imaging of the hamster cheek pouch using a thermal light source, Opt. Lett. 33 (2008) 1285–1287.

[64] Y. Zhu, N.G. Terry, A. Wax, Scanning fiber angle-resolved low coherence interferometry, Opt. Lett. 34 (2009) 3196–3198.

[65] R.K. Ahrenkiel, Modified Kramers–Kronig analysis of optical spectra, J. Opt. Soc. Am. 61 (1971) 1651.

[66] D.J. Faber, T.G. van Leeuwen, Are quantitative attenuation measurements of blood by optical coherence tomography feasible? Opt. Lett. 34 (2009) 1435–1437.

[67] Y. Park, M. Diez-Silva, G. Popescu, G. Lykotrafitis, W. Choi, M.S. Feld, et al., Refractive index maps and membrane dynamics of human red blood cells parasitized by *Plasmodium falciparum*, Proc. Natl. Acad. Sci. USA. 105 (2008) 13730–13735.

[68] Z. Wang, H. Ding, G. Popescu, Scattering-phase theorem, Opt. Lett. 36 (2011) 1215–1217.

CHAPTER 15

Polarization Microscopy

Rudolf Oldenbourg
Marine Biological Laboratory, Woods Hole, MA and Physics Department, Brown University, Providence, RI

Editor: Lisa L. Satterwhite

15.1 Introduction

The polarized light microscope is used to analyze the anisotropy of a specimen's optical properties such as refraction and absorption. Optical anisotropy is a consequence of molecular order, which renders material properties such as absorption, refraction, and scattering dependent on the polarization of light. Polarized light microscopy exploits this dependency and provides a sensitive tool to analyze the alignment of molecular bonds or fine structural form in a specimen.

Most biological structures exhibit some degree of anisotropy that is characteristic of their molecular architecture, such as membranes and filament arrays. A membrane, for example, is composed of lipid molecules in which proteins are embedded, all of which maintain some degree of orientation with respect to the plane of the membrane. Hence, tissues, cells, and organelles that include extensive membranous structures, such as mitochondria, photoreceptors, and the retina, can exhibit birefringence (anisotropy of the refractive index) and dichroism (anisotropy of the absorption coefficient) that are characteristic of their molecular architecture.

In addition to membranes, biological structures commonly include filaments that are in themselves anisotropic, such as collagen fibrils, stress fibers made of filamentous actin and myosin, and microtubules (MTs) that form the mitotic spindle. Double-stranded DNA is a polymer with high intrinsic anisotropy due to the alignment of the aromatic rings that are part of the uniformly stacked base pairs. Figure 15.1 shows the birefringence observed in a sperm head in which the packing arrangement of DNA results in a series of birefringent domains inside each chromosome [1]. Figure 15.2 shows a series of birefringent figures that appear in the center of a freshly fertilized sea urchin egg observed with a polarizing

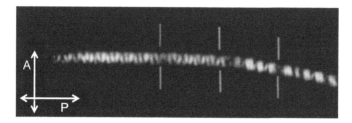

Figure 15.1
Live sperm head of cave cricket viewed between crossed polarizers. The helical regions of the DNA, wound in a coil of coil within the chromosomes, appear bright or dark depending on their orientation with respect to the crossed polarizer (P) and analyzer (A). Bars indicate junctions of chromosomes that are packed in tandem in the needle-shaped sperm head (width approximately 1 μm). This is the first (and virtually only) mode of microscopy by which the packing arrangement of DNA and the chromosomes have been clearly imaged in live sperm of any species. (The sperm head is immersed in DMSO (dimethyl sulfoxide) for index matching and imaged with a high-resolution polarizing microscope using rectified optics (97×/1.25NA).) Source: *From Ref. [1]*.

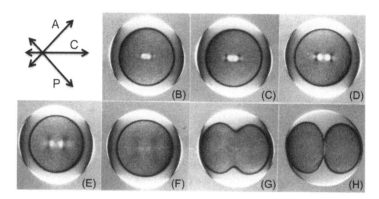

Figure 15.2
Birefringence of spindle, asters, and fertilization envelope in the early zygote of a sea urchin, *Lytechinus variegatus*, observed between crossed polarizers (A, P) and a compensator (C, the arrow indicates slow axis orientation), using a traditional polarizing microscope. In panels (B) through (F), near the center of the cell, birefringent figures appear that represent arrays of parallel MTs of varying density and orientation. Densely packed spindle MTs extend horizontally and appear bright, while the less dense arrays of astral MTs extend radially centered near the spindle poles. Because astral MT arrays are radially arranged, they appear either bright (horizontal orientation) or dark (vertical orientation) as their birefringence either adds to or subtracts from the birefringence of the compensator. The mitotic chromosomes exhibit negligible birefringence due to the tight winding of DNA around histones. The fertilization envelope that surrounds the cell is composed of tangentially aligned paracrystalline polymers inducing a strong birefringence that is visible in all panels and appears brighter or darker in different quadrants due to the action of the compensator. Source: *Reproduced from Ref. [2]*.

microscope [2]. These figures are generated by the birefringence of parallel arrays of MTs that form the mitotic spindle and astral rays in dividing eukaryotic cells [3]. In addition to the mitotic spindle, the images in Figure 15.2 also show the strong birefringence of the fertilization envelope that surrounds the cell and is formed by tangentially aligned filaments and paracrystalline polymer layers [4].

The birefringent structures seen in Figures 15.1 and 15.2 exhibit the orientation-dependent contrast that is a hallmark of the traditional polarizing microscope. Figure 15.3 shows an image recorded with a new type of polarized light microscope, the liquid-crystal polarization microscope (LC-PolScope), which avoids the confusion of orientation-dependent contrast. The LC-PolScope generates a map that represents the magnitude of birefringence and separately a map of its orientation. Figure 15.3 shows only the magnitude of birefringence, or retardance (see Section 15.7), independent of its orientation in the specimen. Black in a retardance image means that there is no anisotropy at that image point, and increasing brightness means increasing retardance, using a linear relationship between image brightness and the measured retardance values. This is in stark contrast to traditional polarized light micrographs, in which image brightness depends on both the magnitude of birefringence and its orientation with respect to the polarizers and compensator in the optical path. Using a traditional polarizing microscope one can interpret the specimen anisotropy only through a series of observations while stepwise rotating the specimen and/or the polarizers and compensator. The liquid crystal-based universal compensator in the LC-PolScope relieves the experimenter from manually rotating optical components and at the same time combines several polarization optical views of the specimen into one resultant image that represents the measured retardance and/or orientation of the specimen birefringence. Using this new technique, very small retardance values (down to 0.03 nm) can reliably be measured, in fast time intervals (less than 1 s), over the whole field of view at the highest resolution of the light microscope (200 nm). The retardance image of the dividing insect spermatocyte in Figure 15.3 illustrates the sensitivity and analytic power of the LC-PolScope that provides comprehensive measurements of intrinsic material properties of native anisotropic structures inside a living cell. In fact, the structural origins of the birefringence of many of the organelles visible in Figure 15.3, such as mitochondria and lipid droplets, remain mysterious.

The classic books on polarizing microscopy were published several decades ago, when optical phase microscopy was in its heyday [5,6]. Several decades before Hartshorne, Schmidt published two celebrated monographs that to this day report the most comprehensive surveys of biological specimens observed under polarized light [7,8]. Inspired by Schmidt's observations, Shinya Inoué significantly improved the sensitivity of the instrument and made seminal contributions to our understanding of the architectural dynamics inside living cells [9–11]. For many lucid discussions of polarized light microscopy and its application to biology, I refer the reader to the many articles and books by Inoué, including his recently published Collected Works [12–16]. Finally, I refer to my

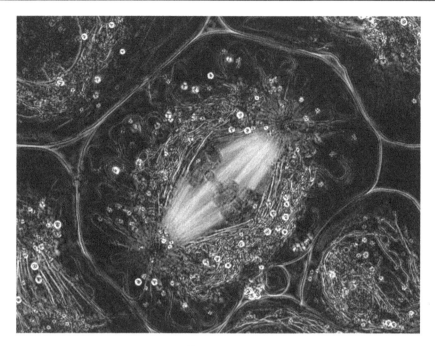

Figure 15.3
The living cell in the center is a primary spermatocyte from the crane fly, *Nephrotoma suturalis*, imaged with the LC-PolScope. Brightness is directly proportional to the birefringence of the specimen, independent of the orientation of the birefringence axis (slow axis). The flattened cell is suspended in the testicular fluid and its image reveals with great clarity the birefringence of spindle MTs extending from the chromosomes to the spindle poles and of astral MTs radiating from the centrosomes located near the spindle poles. The birefringence of other cell organelles, such as elongated mitochondria, surrounding the spindle like a mantle, and small spherical lipid droplets, is also evident against the dark background of the cytoplasm. The pole-to-pole distance is approximately 25 μm.

articles that discuss various aspects of polarizing microscopy, including practical guides to using the LC-PolScope [17–21].

15.2 Traditional Polarized Light Microscopy

Polarized light micrographs often render birefringent objects in beautiful, vivid colors, even though the objects are transparent and colorless when viewed without polarizers. The colors of those objects appear only when placing them between crossed (or parallel) polarizers, revealing the color to be as much a property of the object as it is a property of the method of observation. The colors originate from interference effects, similar to colors that appear upon reflection of white light off a thin film of oil floating on water. For a birefringent object, the hue of the interference color depends on the object's thickness and birefringence, more specifically on the product of thickness times birefringence, which is called retardance (see Section 15.7). If the retardance is less than 275 nm, the object appears gray between crossed polars, while thick or highly birefringent objects that induce a retardance of more than 275 nm appear colored. With increasing retardance, often associated with increasing thickness of an object, it appears yellow (retardance ~ 300 nm), orange (500 nm), red (530 nm), purple (560 nm), blue (650 nm), green (750 nm), and yellow again (900 nm). The color of the objects of even higher retardance roughly repeat in the same order, albeit the colors become less saturated. To identify the retardance of a birefringent object based on its interference color, one can use a so-called Michel-Lévy chart, which graphically represents the sequence of colors and their associated retardance values. The chart often includes additional lines that help relate the retardance to the birefringence of a specimen and its thickness measured in the direction of the light path.

Aside from the preceding remarks, this chapter discusses the instrumentation and methods that are optimized for analyzing the objects of low birefringence and retardance of less than 275 nm. When viewed in white light between crossed polars, objects of low retardance remain gray but appear increasingly brighter, the higher their retardance is. Biological specimens, such as the ones presented in Section 15.1, induce a retardance that is often only a few nanometers and therefore require sensitive methods to observe them and measure their retardance accurately. In this chapter, I discuss the methods for detecting, imaging, and measuring low retardance with high fidelity and sensitivity. I refer to the books by Hartshorne and Stuart [5,6] for a discussion of quantitative measurements of more highly birefringent objects and mention the chapter on microscopes in the *Handbook of Optics* [21].

15.2.1 Basic Setup

The polarized light microscope (also called polarizing microscope or polarization microscope, Figure 15.4) generally differs from a standard transilluminating microscope by the addition of a polarizer before the condenser; a compensator and analyzer behind the objective lens; strain-free optics; a graduated, revolving stage; centrable lens mounts; cross hairs in the ocular aligned parallel or at 45° to the polarizer axes; and a focusable Bertrand lens that can be inserted for conoscopic observation of interference patterns in the back aperture of the objective lens. (For a definition of polarization optical terms, such as polarizer, compensator, and birefringence, see Section 15.7.)

Polarizers

Most light sources (halogen bulb, arc burner, and light-emitting diode) generate unpolarized light, and hence the first polarizer located before the condenser optics polarizes the light that illuminates the specimen. The second polarizer serves to analyze the polarization of the light after it passed through the specimen; therefore, it is called the analyzer. In its most basic configuration, the polarizing microscope has no

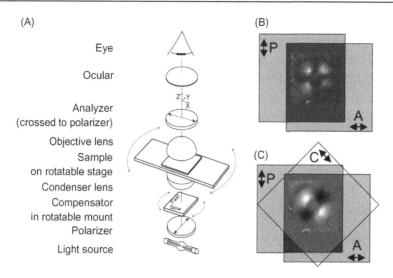

Figure 15.4
Traditional polarized light microscope and image cartoons. (A) Optical arrangement of a conventional polarizing microscope. (B) Cartoon depicting at its center the image of an aster as it appears when located between a crossed polarizer P and analyzer A. The arrows on the polarizer and analyzer sheet indicate their transmission directions. An aster is made of birefringent MT arrays radiating from a centrosome (aster diameter of 15 μm). MTs that run diagonal to the polarizer and analyzer appear bright, while MTs that run parallel to polarizer or analyzer appear dark. (C) Cartoon of aster as it appears when located between polarizer, analyzer, and a compensator C, which is made of a uniformly birefringent plate. The arrow in the compensator plate indicates its slow axis direction. MTs that are nearly parallel to the slow axis of the compensator appear bright, while those that are more perpendicular to the slow axis are dark. Therefore, the birefringence of MTs has a slow axis that is parallel to the polymer axis, as is the case for many biopolymers.

compensator and the polarizer and analyzer are in crossed position, so that the analyzer blocks (absorbs) nearly all the light that has passed through the specimen. In this configuration, the image of the specimen looks very dark, except for structures that are birefringent or otherwise optically anisotropic and appear bright against the dark background (Figures 15.1 and 15.4B). When the specimen is rotated on a revolving stage (around the axis of the microscope), the birefringent parts change brightness, changing from dark to bright and back to dark four times during a full 360° rotation. A uniformly birefringent specimen part appears darkest when its optical axes are parallel to polarizer and analyzer. This is called the extinction position. Rotating the specimen by 45° away from the extinction position makes the birefringent part appear brightest. When rotating the specimen, not all birefringent parts in the field of view will turn dark at the same position, because in general, each specimen part has different axis orientations.
In summary, by rotating the specimen between crossed polarizers, one can recognize birefringent components and determine their axis orientations.

Compensator

While not absolutely necessary for some basic observations, especially of highly birefringent objects, the compensator (a) can significantly improve the detection and visibility of weakly birefringent objects, (b) is required to determine the slow and fast axis of specimen birefringence, and (c) is an indispensable tool for the quantitative measurement of object birefringence. There are several types of compensators; most of them are named after their original inventors. For the observation of weakly birefringent objects, typically encountered in biological specimens, the Brace-Köhler compensator is most often used. It consists of a thin birefringent plate, often made from mica, with a retardance of a tenth to a thirtieth of a wavelength ($\lambda/10-\lambda/30$; see Section 15.7 for the definition of retardance). The birefringent plate is placed in a graduated rotatable mount and inserted either between the polarizer and condenser, as in Figure 15.4, or between the objective lens and the analyzer. The location varies between microscope manufacturers and specific microscope types. In either location, the effect of the Brace-Köhler compensator on the observed image is the same and its standard usage is independent of its location. In general, the birefringence of the compensator causes the image background to become brighter, while birefringent specimen parts can turn either brighter or darker than the background, depending on their orientation with respect to the compensator. If the birefringent structure turns brighter, its slow axis aligns more parallel to the compensator slow axis, while if the structure turns darker, its slow axis aligns more perpendicular to the compensator slow axis. Hence, the compensator with known slow axis orientation can be used to determine the slow axis of birefringent parts in the specimen. The compensator can also enhance specimen contrast, and it is used to quantify specimen birefringence by measuring its retardance, as discussed in Section 15.2.2.

15.2.2 Birefringence, Retardance, and Slow Axis

Birefringence occurs when there is a molecular order, when the average molecular orientation is nonrandom, as in crystals or in aligned polymeric materials. Molecular order usually gives the material two orthogonal optical axes, with the index of refraction along one axis being different from the other axis. The difference between the two indices of refraction is called birefringence and is an intrinsic property of the material:

$$\textbf{birefringence} = \Delta n = n_\| - n_\perp$$

where $n_\|$ and $n_\perp$ are the refractive indices for the light polarized parallel or perpendicular to the two optical axes.

The light polarized parallel to one axis travels at a different speed through the sample than the light polarized parallel to the orthogonal axis. As a result, these two light components, which were in phase when they entered the sample, are retarded at a different rate and exit the sample out of phase. Using the polarizing microscope, one can measure this differential retardation, also called retardance, and thereby quantify the magnitude and orientation of molecular order in the specimen. Retardance is an extrinsic property of the material and a product of the birefringence and the path length l through the material:

$$\textbf{retardance} = R = \Delta n \cdot l$$

Birefringence also has an orientation associated with it. The orientation refers to the specimen's optical axes of which one is called the *fast axis* and the other the *slow axis*. The light polarized parallel to the slow axis experiences a higher refractive index and travels slower than the light polarized parallel to the fast axis. In materials that are built from aligned filamentous molecules, the slow axis is typically parallel to the average orientation of the filaments. Birefringence orientation always correlates with molecular orientation, which usually changes from point to point in the specimen (see aster in Figure 15.4).

15.2.3 Quantitative Analysis of Specimen Retardance

Birefringence and retardance are optical properties that can be directly related to molecular order such as the alignment of polymeric material. By using the traditional polarizing microscope, Sato, Ellis, and Inoué measured the retardance of mitotic spindles in living cells and were able to conclude that the birefringence is caused by the array of aligned spindle microtubules [3]. Their measurements were made possible by the careful analysis of specimen birefringence using a traditional compensator in addition to a polarizer and analyzer in the microscope optical path. Using a Brace-Köhler compensator, the retardance of a resolved image point or uniformly birefringent area in the field of view can be measured by carrying out the following steps:

1. With the specimen in place and by viewing through the eye piece (or on the monitor attached to a video camera), make sure that polarizer and analyzer are crossed and the compensator is in the extinction position. Given those settings, background areas that have no birefringence appear dark, and birefringent structures change from dark to bright four times when rotating the specimen by 360°.
2. Again, using the rotatable stage, orient the birefringent structure of interest so that it appears darkest (extinction position), and note the orientation and then rotate the specimen 45° away from the extinction position. At that orientation, the structure appears brightest. (By eye, it is usually easier to determine the orientation that leads to the lowest intensity rather than the highest intensity. The two orientations are 45° apart.)
3. Now rotate the compensator either clockwise or counterclockwise, until the birefringent structure of interest appears dark. When rotating the compensator, any background area becomes brighter, while birefringent structures become either brighter or darker. For structures that turn darker than the background, the birefringence of the compensator is said to "compensate" the birefringence of the sample structure. Take the

aster image in Figure 15.4C as an example. The radial array of MTs shows two bright and dark quadrants. In the bright quadrants, the MTs' slow axis runs more parallel to the compensator slow axis, while in the dark quadrants, the slow axis of the MT birefringence is oriented nearly perpendicular to the slow axis of the compensator.

As the compensator is rotated from its extinction position, the specimen turns darkest at a characteristic angle θ_{min}. Given the known compensator retardance R_{cmp}, the angle θ_{min} is a direct measure of the specimen retardance R_{spc}:

$$R_{spc} = R_{cmp} \sin(2\theta_{min})$$

Figure 15.5A shows a graph of the intensity calculated for a uniformly birefringent specimen area (or single specimen point) as a function of the rotation angle θ of a Brace-Köhler compensator. At the rotation angle θ_{min}, the intensity is a minimum. For other rotation angles, the intensity varies according to a complex expression of trigonometric functions that can be derived using the Jones calculus. This expression can also be used to calculate the expected contrast of a birefringent object against its nonbirefringent background. The graph in Figure 15.5B shows the computed contrast of the birefringent

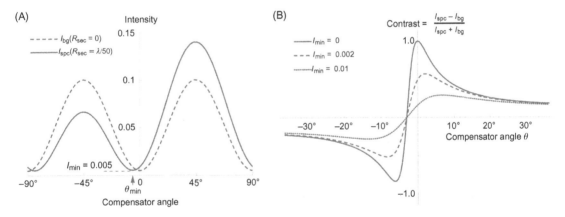

Figure 15.5
(A) The graph shows intensities detected in a single image point or uniform area of the specimen versus the rotation angle of a Brace-Köhler compensator. Dashed curve shows the background intensity I_{bg} of a specimen area that exhibits no birefringence. The solid curve shows the intensity I_{spc} of a specimen point that exhibits a small retardance ($\lambda/50$). The transmitted intensity is given as a fraction of the amount of light that has passed through the first polarizer. I_{min} represents the spurious intensity that is detected when the compensator is in the extinction position. (B) This graph shows the contrast of a birefringent specimen point ($R_{spc} = \lambda/50$) as a function of compensator angle, for three different values of I_{min}. I_{min} is affected by the quality of the polarizers used and the polarization distortions introduced by the intervening optics (condenser and objective lens). Both graphs were generated computationally using the Jones calculus and assuming a retardance of $\lambda/10$ for the birefringent crystal plate of the Brace-Köhler compensator and a specimen retardance of $\lambda/50$ with slow axis oriented at 45° to polarizer and analyzer.

specimen area versus θ. We define contrast as the ratio between the intensity difference between the specimen area and the background, divided by their sum:

$$\text{contrast} = \frac{I_{spc} - I_{bg}}{I_{spc} + I_{bg}}$$

As apparent from Figure 15.5B, the highest contrast and therefore the best visibility of the birefringent specimen are achieved when adjusting for a compensator angle of around θ_{min}.

Figure 15.5B further shows the importance of using so-called high-extinction optics when observing a specimen of low birefringence. Figure 15.5B introduces the parameter I_{min}, which represents the spurious intensity observed when polarizer and analyzer are crossed. The better the quality of the polarizers employed and of the intervening optical components (especially condenser and objective lenses), the lower I_{min} will be and the higher the contrast that can be achieved for a specimen of given retardance. To improve the extinction of microscope optics, it is possible to use the so-called polarization rectifiers, first introduced by Inoué and Hyde [9], that counteract the polarization distortions introduced by high numerical aperture (NA) objective and condenser lenses [22]. In the 1960s and 1970s, some microscope manufacturers (e.g., American Optical Spencer and Nikon) made rectified, high-extinction optics, with which high-contrast images of low-birefringent objects, like the sperm head shown in Figure 15.1, were recorded.

Since the 1980s, fluorescence microscopy has become the dominant technique for which the optics of biological microscopes are optimized, demanding high transmission and low image distortions over a wide range of wavelengths. These demanding specs are now met by complex lens designs incorporating many components, each of which has special antireflection coatings, in a single objective lens, compromising its polarization performance. As a result, many modern objectives lead to low extinction and inferior sensitivity when used in a polarizing microscope. While the LC-PolScope technique discussed in Section 15.3 alleviates some of the shortcomings in polarization performance of modern microscope objectives, it is still best to use microscope optics that is designated for differential interference contrast (DIC) or polarized light microscopy.

15.3 The Liquid-Crystal Polarization Microscope (LC-PolScope)

Over the years, several schemes have been proposed to automate the measurement process and exploit more fully the analytic power of the polarizing microscope. These schemes invariably involve the traditional compensator, which is either moved under computer control [23] or replaced by electro-optical modulators, such as Pockel cells [24], Faraday rotators [25], and liquid crystal variable retarders [26]. These schemes also involve quantitative intensity measurements using electronic light detectors, such as photomultipliers or charge-coupled device (CCD) cameras. For quantitative measurements, acquisition and processing algorithms relate measured image intensities and compensator settings to optical characteristics of the specimen (e.g., [18]). Here we discuss in more detail the LC-PolScope, a birefringence imaging system that was first developed at the Marine Biological Laboratory (MBL) and is commercially available from CRi Inc. (Now part of Caliper Life Sciences and PerkinElmer, http://www.caliperls.com/).

The optical design of the LC-PolScope builds on the traditional polarized light microscope, introducing two essential modifications: the specimen is illuminated with nearly circularly polarized light and the traditional compensator is replaced by a liquid crystal-based universal compensator. The LC-PolScope also requires the use of narrow bandwidth (≤ 40 nm) or monochromatic light. In Figure 15.6, the schematic of the optical train shows the universal

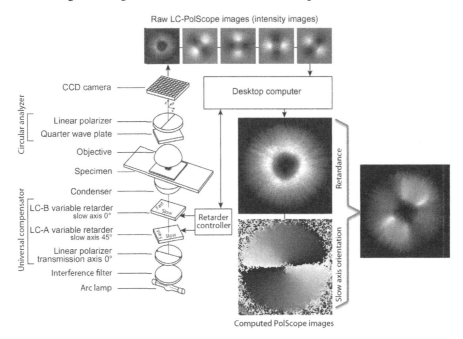

Figure 15.6

LC-PolScope. The optical design (left) builds on the traditional polarized light microscope with the conventional compensator replaced by two variable retarders LC-A and LC-B. The polarization analyzer passes circularly polarized light and is typically built from a linear polarizer and a quarter wave plate. Images of the specimen (top row, aster isolated from surf clam egg) are captured at five predetermined retarder settings, which cause the specimen to be illuminated with circularly polarized light (first, left-most image) and with elliptically polarized light of different axis orientations (second to fifth image). Based on the raw PolScope images, the computer calculates the retardance image and the slow axis orientation or azimuth image using specific algorithms [18]. The false-color image on the right was created by combining the retardance and slow axis orientation data, with the orientation encoded as hue and the retardance as brightness. (For interpretation of the references to color in this figure legend, the reader is referred to the web version of this book.)

compensator located between the light source (typically an arc lamp followed by an interference filter) and the condenser lens. The analyzer for circularly polarized light is placed after the objective lens. The universal compensator is built from two variable retarder plates and a linear polarizer. The variable retarder plates are implemented as electro-optical devices made from two liquid crystal plates. Each liquid crystal plate has a uniform retardance that depends on the voltage applied to the device. An electronic control box that in turn is connected to a computer supplies the voltage. The computer is also connected to the electronic camera, typically a CCD camera, for recording the specimen images projected onto the camera by the microscope optics. Specialized software synchronizes the image acquisition process with the liquid crystal settings and implements image-processing algorithms. These algorithms compute images that represent the retardance and slow axis orientation at each resolved image point. For a more in-depth discussion of the LC-PolScope technology, see Refs. [18,19,26]. Shribak also proposed a variance to the LC-PolScope technique that uses only one liquid crystal device instead of two [27].

The commercial LC-PolScope technique, developed and distributed by CRi Inc. (now part of Caliper Life Sciences and PerkinElmer), is available as an accessory to microscope stands of all the major microscope manufacturers (e.g., Leica, Nikon, Olympus, and Zeiss). It usually includes the universal compensator, circular polarizer, a camera with control electronics, and a computer with software for image acquisition and processing. Three slightly differing versions are available, each optimized for research in the life sciences (Abrio LS), industrial metrology (Abrio IM), and for in vitro fertilization and related laboratory techniques (Oosight).

Figure 15.7 shows retardance images recorded with the LC-PolScope, illustrating the clarity, resolution, and analytic potential for examining the birefringent fine structure in living cells. Figure 15.8 shows its use in procedures involving enucleation and cloning. All images were recorded without the need for exogenous stain or label in the specimen.

15.4 Practical Considerations

15.4.1 Choice of Optics

As indicated earlier, the polarization distortions introduced by the objective and condenser lenses limit the extinction that can be achieved in a polarizing microscope. Most microscope manufacturers offer lenses that are designated "Pol" to indicate low polarization distortions, which can arise from a number of factors, including stress or crystalline inclusions in the lens glass and the type of antireflection coatings used on lens surfaces. Some lens types are available only with "DIC" designation. "DIC" lenses do not meet the more stringent "Pol" requirements but pass for use in DIC and can also be used with the LC-PolScope (without the Wollaston or Nomarski prisms specific to DIC, of course).

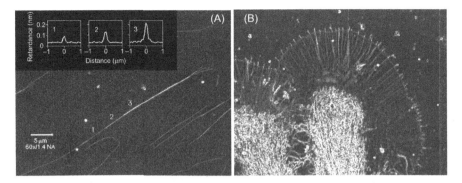

Figure 15.7
LC-PolScope images of single MTs and of actin bundles. (A) Spontaneously assembled MTs, stabilized with taxol, and adhered to the coverglass surface were imaged with the LC-PolScope (retardance image). Most filaments are single MTs, with the exception of one bundle containing one, two, and three MTs. The inset shows line scans across the filament axis at locations with one, two, and three MTs. Note that at the top right end, the bundle sprays into three individual MTs. (B) Living growth cone of an *Aplysia* bag cell neuron. The peripheral lamellar domain contains radially aligned fibers composed of 15–40 actin filaments. A detailed description of the birefringent fine structure including a time lapse movie of the growth cone dynamics is published in [41]. Image brightness represents measured retardance between 0 nm (black) and 0.5 nm or larger (white). Image width about 70 μm. *Source: Panel (A) Reproduced from Ref. [40].*

The polarization performance of the most highly corrected lenses, the so-called Plan Apochromat objectives, is often compromised by large number of lens elements, special antireflection coatings, and, in some cases, special types of glass used to construct the lenses [21]. For some applications, these high-quality lenses, which provide a large, highly corrected viewing field over a wide spectral range, might not be required. If the objective lens is to be used only with the PolScope that requires monochromatic light and typically acquires images from a region near the center of the viewing field, a less-stringent correction might well suffice. To find out what works best for a particular imaging situation, several lenses should be tested. It is also helpful to be able to select the best performing combination of condenser and objective lens from a batch of the same lens types.

Whenever possible or practical, oil-immersion lenses should be used. This is because the transition of a light ray between two media of different refractive indices ($n_{air} = 1.00$, $n_{glass} = 1.52$) introduces polarization distortions, especially for high NA lenses. The peripheral rays leaving the condenser front lens are highly tilted to the slide surface and their polarization is typically rotated when traversing the air–glass interface. Oil, and to a lesser degree water and other immersion liquids, greatly reduce polarization aberrations caused by air–glass interfaces between the specimen and the lenses. For a discussion of the origin of polarization distortions in lenses and optical systems, and of ways to reduce them, see Ref. [22], which also discusses the various conoscopic images that can be observed in a polarizing microscope equipped with crossed, or nearly crossed, polarizers and a Bertrand lens.

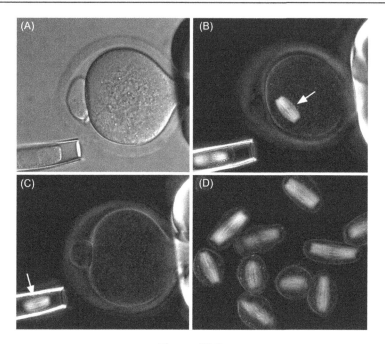

Figure 15.8
Mouse oocyte and enucleated spindles. (A) Mouse oocyte held in place by gentle suction of a holding pipette (oocyte diameter approximately 75 μm, DIC microscopy). (B) Retardance image of the same oocyte, with birefringent spindle of meiosis II (white arrow). (C) The spindle (arrow) is aspirated into an enucleation pipette. (D) A batch of enucleated spindles and chromosomal karyoplasts. Chromosomes are aligned in the middle of spindles. Source: *Courtesy of Dr. Lin Liu [42]*

Lenses and other optical elements that are part of the optical train but not located between the crossed polarizers do not affect the polarization performance of the microscope setup. Filter cubes for epi-illumination, for example, should be placed outside the polarization optical train, if possible. (The polarization optical train is defined as the stretch between the crossed polarizers.) In a dissecting microscope, the polarizing elements and compensator could be placed in front of the objective lens instead of behind it. The objective of a dissecting scope has very low NA and the image-forming rays have a small angle of divergence (in contrast to high NA lenses). Rays with a small tilt angle with respect to the normal of the polarizer and compensator plates do not appreciably affect the performance of these devices.

15.4.2 Specimen Preparation

The following are recommendations for preparing living cells for observation with a polarizing microscope. Other resources for polarized light microscopy of living cells include Refs. [14] and [17], which have sections on suitable cell types and their preparation.

- If cells have to be kept in plastic Petri dishes during observations, only use dishes with glass coverslip bottoms. The strong and random birefringence of a plastic bottom ruins the polarization of the light.
- Prepare specimens such that cells or structures of interest are as close to the coverslip as possible. When using high NA (>0.5) oil immersion or dry optics, a layer of more than 10 µm of aqueous medium introduces enough spherical aberration to noticeably reduce the resolution of cell images. The use of water-immersion optics alleviates this problem.
- For observations of structures near the cell surface, it might be necessary to increase the medium's refractive index to match the refractive index of the cytoplasm and reduce the effect of edge birefringence [28]. Adding polyethyleneglycol, polyvinylpyrrolidone, or some other harmless polymer or protein substitute to the medium will increase its refractive index to match that of the cytoplasm inside the cell. The concentration of substitute required might vary from a few percent to over 10%, depending on the average refractive index of the cytoplasm of the cells of interest. It is best to test various media containing different polymer concentrations by suspending cells in them and examining the refractive index mismatch in a slide and coverslip preparation with DIC or polarized light. At the optimal polymer concentration, no distinct cell boundary is visible, giving the impression that the organelles and cytoskeleton are freely suspended. After determining the optimal concentration for imaging, the medium should be tested for compatibility with growing cells. Many cells grow and develop normally in media with inert additives. The molecular weight of the polymer needs to be around 40 kD or more to prevent its uptake into cells.
- For observations with the LC-PolScope, prepare the specimen so that when mounted in the microscope, at least one clear area without cells or birefringent material can be identified and moved into the viewing field when required. A clear area is needed for calibrating the PolScope and for recording background images. The background images are used to remove spurious background retardance from specimen images, allowing precise measurement and imaging of the cells and structures of interest. Many cell cultures and cell-free systems can be prepared without special attention to this requirement. For example, a free area can often be found around sparsely plated cells. Cultures of free-swimming cells might have to be diluted before mounting a small drop of the suspension between slide and coverslip to observe free areas. Sometimes it is helpful to add a tiny drop of oil or other nontoxic, immiscible liquid to the preparation. This clear drop can provide an area for calibrating the instrument and taking background images in a preparation that is otherwise dense with birefringent structures.

15.4.3 Combining Polarized Light with DIC and Fluorescence Imaging

Polarization and DIC microscopy both rely on the use of polarized light for creating contrast in images of phase objects. While polarized light imaging highlights changes of refractive index as a function of the polarization state of the illuminating light, DIC imaging highlights refractive index changes as a function of position in the specimen. Since DIC employs polarized light, its image contrast tends to combine both effects, i.e., image features in DIC reflect both types of changes in the specimen refractive index. Shribak has developed powerful new DIC methods, described in Chapter 2, which includes a variance that combines DIC with polarized light imaging [29]. Nevertheless, many microscopes are equipped with traditional polarization and DIC optics, or LC-PolScope and DIC optics, and the operator might wonder how to combine both types of observations and get optimal results. Therefore, I briefly comment on their combination.

For DIC imaging, two matching prisms are added to the polarization optical train. The two prisms, called Wollaston or Nomarski prisms, are specially designed to fit in specific positions, one before the condenser and the other after the objective lens, and to match specific lens combinations. The prisms in their regular positions are usually combined with the linear polarizers of a traditional polarizing microscope. The prisms can also be combined with the LC-PolScope setup, in which case, the universal compensator and circular analyzer take the places of the linear polarizer and analyzer in a standard DIC arrangement. In fact, the same DIC prisms function equally well using either linearly or circularly polarized light. Note, however, that Carl Zeiss has recently combined linear polarizers with their condenser DIC prisms preventing their use with circularly polarized light.

In combining DIC with the LC-PolScope, the universal compensator is first calibrated without prisms to find the extinction setting. With the compensator set to extinction, the prisms are entered into the optical path. Typically, one of the two prisms has a mechanism for finely adjusting its position, which in turn affects the brightness and contrast of the DIC image. In a PolScope setup, the mechanical fine adjustment or the universal compensator can be used to add a bias retardance to change the brightness and contrast of the DIC image. When using the universal compensator, retardance must be added or subtracted from the extinction setting of either the LC-A or LC-B retarder (Figure 15.6), depending on the orientation of the shear direction of the prisms.

For live cell imaging, it is very useful to be able to switch easily between polarized light and DIC imaging. DIC provides good contrast of many morphological features in cells using direct viewing through the eyepiece or by video imaging. When viewing specimens that have low polarization contrast, it is more effective to align the optics, including the visualization of the specimen, using first DIC. After the optics has been aligned and the specimen is in focus, the DIC prisms are removed from the optical path and the PolScope specific adjustments are completed.

Fluorescence imaging can be combined with the PolScope in several ways. Fluorescence is commonly observed using epi-illumination, which requires a filter cube in the imaging path. The filter cube includes a dichromatic mirror and interference filters for separating the excitation and emission wavelengths. For best results, the filter cube should be removed for PolScope observations to avoid the polarization distortions caused by the dichromatic mirror. For observing fluorescence, on the other hand, the PolScope analyzer should be removed because it attenuates the fluorescence emission by at least 50%. To meet both requirements, the PolScope analyzer can be mounted in an otherwise empty filter cube holder, which is moved into the optical path as the fluorescence cube is moved out and vice versa.

If, however, the option of removing the fluorescence cube is unavailable, the cube can remain in the optical path for observations with polarized light and the LC-PolScope. In this case, the following points should be considered:

- For the light source of the PolScope, choose a wavelength that is compatible with the emission wavelength filter in the fluorescence cube. For example, fluorescein isothiocyanate (FITC) requires excitation with blue light (485 nm) and fluoresces in green. The FITC dichromatic beam splitter and a barrier filter passing green fluorescence light with wavelength longer than 510 nm would be compatible with 546 nm light for polarized light observations using a mercury arc burner for the transmission light path.
- The light source for one imaging mode must be blocked while observing with the other imaging mode.
- Removing the polarization analyzer, while observing fluorescence, more than doubles the fluorescence intensity.
- The LC-PolScope must be calibrated with the fluorescence filter cube in place. The polarization distortions caused by the dichromatic mirror are partially counteracted by the calibrated settings of the universal compensator.

15.5 Polarized Light Imaging in Three Dimensions

In this last section, we briefly address the question of how the three-dimensional (3D) nature of a specimen affects the images and measurements recorded with a polarizing microscope. I consider two issues here: (a) the effect of optical anisotropy located in sections of the specimen that are not in focus, and (b) the fact that anisotropy itself is a 3D property.

A polarizing microscope generates 2D images of 3D objects. Each image contains information from all specimen planes simultaneously, from the in-focus plane and the planes that are out of focus. The out-of-focus blur superimposes on the sharp details of structures that are located in the in-focus plane. This characteristic also applies to the measurement of birefringence of a particular structure in the image. In Figure 15.9, the retardance image of a meiotic spindle clearly delineates the birefringence of the kinetochore fibers (k-fibers) that extend between the chromosomes and the spindle poles. However, the retardance of the k-fibers is superimposed on the retardance of the whole spindle, which more than doubles the retardance value that is attributable to a k-fiber alone. Nevertheless,

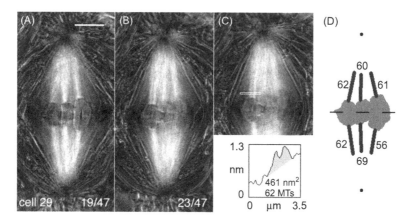

Figure 15.9

Meiosis I spindle in a spermatocyte of the crane fly *Nephrotoma suturalis* imaged with the LC-PolScope. (A) and (B) are two sections (19 and 23 out of 43) from a series of optical sections made through the cell at focus steps of 0.26 μm. Each section shows chromosomes as areas of reduced birefringence and the high birefringence of the dense arrays of MTs forming the kinetochore fibers. In addition, the birefringence of all the parallel spindle MTs superimposes on the birefringence of the k-fibers, more than doubling the measured k-fiber retardance. In viewing these LC-PolScope images, brightness represents the retardance (black = 0 nm and white = 2.5 nm retardance), irrespective of the orientation of the slow axis. The bar in panel A is 5 μm. (C) is a duplicate image of (B) and includes a rectangle that indicates the location of the image data that were used to generate the retardance plot in the lower half of the panel. The shaded area in the plot is the retardance associated with the k-fiber, which was further evaluated for the number of kinetochore MTs. (D) The metaphase positions of the three bivalent chromosomes in this cell upon projection of all images within the Z-focus series to make a 2D profile. Two dots indicate the positions of the flagellar basal bodies within the centrosomes at the two spindle poles. The numbers on each kinetochore fiber indicate the number of kinetochore MTs in that fiber, based on the retardance analysis. *Source: Reproduced from Ref. [30].*

as explained in the caption to Figure 15.9, the retardance measured in areas adjacent to the k-fiber can be used to estimate the spindle retardance at the location of the k-fiber and subtract its contribution from that of the k-fiber [30].

In the previous example, simple subtraction of the spindle retardance leads to a correct estimate of the k-fiber retardance because the slow axes of the k-fiber and spindle birefringence are co-aligned. The superposition of birefringent structures whose slow axes are not parallel to each other is complex and their individual retardance values do not simply add. The effect of one birefringent structure, whose image overlaps with that of another birefringent structure, depends on their mutual orientation. So far, we lack a reliable method to unambiguously identify the individual contributions of birefringent structures that are separated along the optical axis of the microscope.

The situation becomes even more complex when considering that optical anisotropy is really a 3D property. Using a conventional polarized light microscope, one measures only a projection of the 3D property onto the plane that is perpendicular to the optical axis of the microscope. We have recently devised two different approaches that attempt to analyze the 3D aspect of birefringence in a biological sample. The first method relies on oblique illumination, essentially tilting the beam path through the specimen. By combining measurements made with four different tilt angles, it is possible to reconstruct the 3D birefringence properties [31,32].

The other method involves a micro lens array that is placed in the image plane of the objective lens. The camera is placed behind the micro lenses and captures a hybrid image of the specimen (Figure 15.10). Since the camera is located in a plane that is conjugate to the back focal plane of the micro lenses, it records a large array of small conoscopic images, each specific to a small sample area. While this imaging method reduces the lateral resolution, it systematically records the specimen birefringence as a function of tilt angle of the beam path [33]. Both methods are related to each other but require vastly different implementation. The first method is implemented through a variable aperture mask in the condenser aperture. It only slightly decreases the resolution of the microscope, but raw images are difficult to interpret directly. The second method reduces the resolution by about a factor 10, producing an array of conoscopic images that are readily interpreted based on the classic approach of conoscopic imaging.

15.6 Conclusion

Polarized light microscopy allows one to nondestructively follow the dynamic organization of living cells and tissues at the microscopic as well as submicroscopic levels. Imaging with polarized light reveals information about the organization of the endogenous molecules that built the complex and highly dynamic architecture of cells and tissues. While polarized light microscopy is not sensitive to the chemical nature of the constituent molecules, it is sensitive to the structural, anisotropic nature of macromolecular assemblies such as the submicroscopic alignment of molecular bonds and filaments.

Polarization analysis can also be applied to fluorescence imaging, combining the molecular specificity of fluorescence labeling with the structural specificity of polarization. Polarized fluorescence has its origin in the dipole radiation patterns of most fluorophores, including fluorescent proteins. Recently, we have extended the use of the liquid crystal universal compensator to analyzing the assembly dynamics of higher-order septin structures in yeast, fungi, and mammalian cells [34,35]. The study took advantage of the polarized fluorescence emitted by every single green fluorescent protein (GFP) fluorophore. GFP was fused to septin with a rigid linker that limits the GFP's ability to

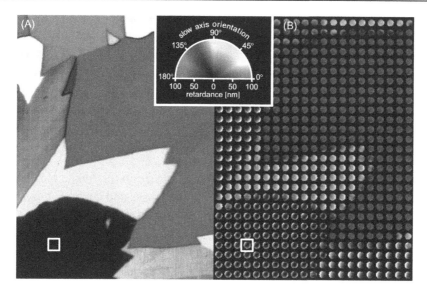

Figure 15.10
(A) LC-PolScope image of a thin, polycrystalline calcite film of uniform thickness (300 nm). In this false-color image, hue represents the orientation of the slow axis and brightness is proportional to the retardance (see inset). Each crystalline domain has nearly uniform hue and brightness, due to the uniform birefringence of each domain. (B) The same sample area imaged with the LC-PolScope with a micro lens array in front of the camera. Each small disk represents a conoscopic image that is projected by a single micro lens. Each micro lens captures rays that traversed a small sample region. The white frame in (A and B) identifies a disk in (B) and the corresponding sample region in (A). The framed disk image has a black center that is surrounded by a bright ring of changing color. Such a conoscopic interference figure is compatible with the fact that the thin film is made of calcite that has a uniaxial negatively birefringent crystal structure. In the black crystal domain, the optic axis of calcite is oriented parallel to the microscope axis, while in other domains, the optic axis is tilted away from the microscopic axis. The colors of the different domains indicate the azimuth orientation of the tilt angle. Both images in (A and B) were recorded using a 63×/1.4NA objective lens. The micro lens array has a pitch of 2 μm in object space. For more details see Ref. [33]. (For interpretation of the references to color in this figure legend, the reader is referred to the web version of this book.)

rotate relative to the septin molecule. When septin was assembled into an ordered structure, a consistent orientation of the GFP dipoles was detected throughout the structure.

With this chapter, I intended to give a short survey of traditional and modern approaches to polarized light microscopy and its biomedical applications. I believe, with current instrumentation, we have only scratched the surface of the potential of polarized light microscopy, and I look forward to many more advances that will come with combining hardware components and software tools to decipher ever finer detail that will help reveal the structural basis of cell function.

15.7 Glossary of Polarization Optical Terms

The following is a brief introduction to terms that are relevant for observations with a polarized light microscope. A more detailed explanation of these and other terms that describe physical phenomena or optical devices can be found in Refs. [36–39].

15.7.1 Analyzer

An analyzer is a polarizer that is used to analyze the polarization state of light (see Section 15.7.13).

15.7.2 Azimuth

The azimuth is an angle that refers to the orientation of the slow axis of a uniformly birefringent region. The azimuth image refers to the array of azimuth values of a birefringent specimen imaged with the LC-PolScope. The azimuth is typically measured from the horizontal orientation with values increasing for counterclockwise rotation. Angles range between 0° and 180°, with both endpoints indicating horizontal orientation.

15.7.3 Birefringence

Birefringence is a material property that can occur when there is molecular order, that is, when the average molecular orientation is nonrandom, as in crystals or in aligned polymeric materials. Molecular order usually renders the material optically anisotropic, leading to a refractive index that changes with the polarization of the light. Birefringence is defined as the difference in refractive index for light of different polarization.

Many types of molecular order lead to what is called uniaxial anisotropy, i.e., light propagating through a uniformly aligned material suffers phase retardation that depends on the polarization of the light, except when the light propagates along one unique direction, which is called the optic axis. Light propagating through the anisotropic material along the optic axis experiences the so-called ordinary refractive index (n_o), which is the same for all polarization directions. However, light that propagates in any other direction through the material experiences differences in the refractive index, depending on the light's polarization. Light that propagates perpendicular to the optic axis experiences the largest difference in refractive index when polarized parallel versus perpendicular to the optic axis. When polarized perpendicular to the optic axis, it is the ordinary refractive index n_o, and when polarized parallel to the optic axis, the light experiences the extraordinary refractive index n_e. Birefringence is the difference $n_e - n_o = \Delta n$ and characterizes the anisotropy of refraction of the uniaxial material.

There are also materials whose anisotropy is called biaxial. For their explanation, the reader is referred to Refs. [36–39].

15.7.4 Compensator

A compensator is an optical device that includes one or more retarder plates and is commonly used to analyze the birefringence of a specimen. For a traditional polarizing microscope, several types of compensators exist that typically use a single fixed retarder plate mounted in a mechanical rotation stage. With the help of a compensator, it is possible to distinguish between the slow and fast axis direction and to measure the retardance of a birefringent object after orienting it at 45° with respect to the polarizers of the microscope.

The LC-PolScope employs a universal compensator that includes two electro-optically controlled, variable retarder plates. Using the universal compensator, it is possible to measure the retardance and slow axis orientation of birefringent objects that have any orientation in the plane of focus.

15.7.5 Dichroism

Dichroism is a material property that can occur in absorbing materials in which the light-absorbing molecules are arranged in a nonrandom orientation. Dichroism refers to the difference in the absorption coefficients for light polarized parallel and perpendicular to the principal axis of alignment.

The measurement of optical anisotropy by the LC-PolScope is affected by the dichroism of absorbing materials. In nonabsorbing, clear specimens, however, dichroism vanishes and birefringence is the dominant optical anisotropy measured by the LC-PolScope. Like absorption, dichroism is strongly wavelength dependent, while birefringence only weakly depends on wavelength.

15.7.6 Extinction

Extinction (or extinction coefficient) is defined as the ratio of maximum to minimum transmission of a beam of light that passes through a polarization optical train. Given a pair of linear polarizers, for example, the extinction is the ratio of intensities measured for parallel versus perpendicular orientation of the transmission axes of the polarizers (extinction = $I_{\parallel}/I_{\perp}$). In addition to the polarizers, the polarization optical train can also include other optical components, which usually affect the extinction of the complete optical train. In a polarizing microscope, the objective and condenser lenses are located between the polarizers and significantly reduce the extinction of the whole setup.

15.7.7 Fast Axis

The fast axis describes an orientation in a birefringent material. For a given propagation direction, light that is polarized parallel to the fast axis experiences the lowest refractive index, and hence travels the fastest in the material (for the given propagation direction). See also Section 15.7.16.

15.7.8 Optic Axis

The optic axis refers to a direction in a birefringent material. Light propagating along the optic axis does not change its polarization, hence for light propagating along the optic axis, the birefringent material behaves as if it were optically isotropic.

15.7.9 Polarized Light

A beam of light is said to be polarized when its electric field is distributed nonrandomly in the plane perpendicular to the beam axis. In unpolarized light, the orientation of the electric field is random and unpredictable. In partially polarized light, some fraction of the light is polarized, while the remaining fraction is unpolarized. Most natural light is unpolarized (sun, incandescent light) but can become partially or fully polarized by scattering, reflection, or interaction with optically anisotropic materials. These phenomena are used to build devices to produce polarized light (see Section 15.7.13).

15.7.10 Linearly Polarized Light

In a linearly polarized light beam, the electric field remains oriented along a single direction that is perpendicular to the propagation direction.

15.7.11 Circularly Polarized Light

In circularly polarized light, the electric field direction rotates either clockwise (right circularly) or counterclockwise (left circularly) when looking toward the source. While the field direction rotates, the field strength remains constant. Hence, the endpoint of the field vector describes a circle.

15.7.12 Elliptically Polarized Light

In elliptically polarized light, as in circularly polarized light, the electric field direction rotates either clockwise or counterclockwise when looking toward the source. However, while the field direction rotates, the field strength varies in such a way that the end point of the field vector describes an ellipse. The ellipse has long and short principal axes that are orthogonal to each other and have fixed orientation.

Any type of polarization (linear, circular, or elliptical) can be transformed into any other type of polarization by means of polarizers and retarders.

15.7.13 Polarizer

A polarizer, sometimes called a polar, is a device that produces polarized light of a certain kind. The most common polar is a linear polarizer made from dichroic material (e.g., a plastic film with small, embedded iodine crystals that have been aligned by stretching the plastic), which transmits light of one electric field direction while absorbing the orthogonal field direction. Crystal polarizers are made of birefringent crystals that split the light beam into orthogonal linear polarization components. A polarizer that produces circularly polarized light, a circular polarizer, is typically built from a linear polarizer followed by a quarter wave plate.

The LC-PolScope employs a universal compensator that also serves as a universal polarizer in that it converts linear polarization into any other type of polarization by means of two variable retarders.

15.7.14 Retardance

Retardance is a measure of the relative optical path difference, or phase change, suffered by two orthogonal polarization components of light that has passed through an optically anisotropic material. Retardance is also called differential or relative retardation. Retardance is the primary quantity measured by a polarizing microscope. Assume a nearly collimated beam of light traversing a birefringent material. The light component that is polarized parallel to the high refractive index axis travels at a slower speed through the birefringent material than the component polarized perpendicular to that axis. As a result, the two components, which were in phase when they entered the material, exit the material out of phase. The relative phase difference, expressed as the distance between the respective wave fronts, is called the retardance:

$$\text{retardance } R = (n_e - n_0) \cdot l = \Delta n \cdot l,$$

where l is the physical path length or thickness of the birefringent material. Hence, retardance has the dimension of a distance and is often expressed in nanometers. Sometimes, it is convenient to express that distance as a fraction of the wavelength λ, such as $\lambda/4$ or $\lambda/2$. Retardance can also be defined as a differential phase angle, in which case $\lambda/4$ corresponds to $90°$ and $\lambda/2$ to $180°$ phase difference.

As a practical example, consider a mitotic spindle observed in a microscope that is equipped with low NA lenses (NA ≤ 0.5). When the spindle axis is contained in the focal plane, the illuminating and imaging beams run nearly perpendicular to the spindle axis. Under those conditions, the retardance measured in the center of the spindle is proportional to the average birefringence induced by the dense array of aligned spindle MTs. To determine Δn, it is possible to estimate the thickness, l, either by focusing on spindle fibers located on top and bottom of the spindle and noting the distance between the two focus positions, or by measuring the lateral extent of the spindle when focusing through its center. The latter approach assumes a rotationally symmetric shape of the spindle. Typical values for the spindle retardance of crane fly spermatocytes (Figure 15.4B) and of other cells is 3–5 nm and the spindle diameter is about 30–40 μm, leading to an average birefringence of around 10^{-4}. It has been found that the retardance value of the spindle is largely independent of the NA for imaging systems using NA ≤ 0.5 [3].

On the other hand, when using an imaging setup that employs high NA optics (NA > 0.5) for illuminating and imaging the sample, the measured retardance takes on a somewhat different context. For example, the retardance measured in the center of an MT image recorded with an LC-PolScope equipped with a high NA objective and condenser lens is 0.07 nm. A detailed study showed that the peak retardance decreased inversely with the NA of the lenses. However, the retardance integrated over the cross-section of the MT image was independent of the NA [40]. While a conceptual understanding of the measured retardance of submicroscopic filaments has been worked out in the aforementioned publication, a detailed theory of these and other findings about the retardance measured with high NA optics has yet to materialize.

15.7.15 Retarder

A retarder, or waveplate, is an optical device that is typically made of a birefringent plate. The retardance of the plate is the product of the birefringence of the material and the thickness of the plate. Fixed retarder plates are either cut from crystalline materials such as quartz, calcite, or mica, or they are made of aligned polymeric material. If the retardance of the plate is $\lambda/4$, for example, the retarder is called a quarter waveplate.

A variable retarder can be made from a liquid crystal device. A thin layer of highly birefringent liquid crystal material is sandwiched between two glass windows, each bearing a transparent electrode. A voltage applied between the electrodes produces an electric field across the liquid crystal layer that reorients the liquid crystal molecules. This reorientation changes the birefringence of the layer without affecting its thickness or the direction of its slow axis.

15.7.16 Slow Axis

The slow axis describes an orientation in a birefringent material. For a given propagation direction, light polarized parallel to the slow axis experiences the highest refractive index and hence travels the slowest in the material. Also see Section 15.7.7.

15.7.17 Waveplate

See Section 15.7.15.

Acknowledgments

I gratefully acknowledge many years of illuminating discussions on polarized light microscopy with Shinya Inoué and Michael Shribak of the MBL.

This work was supported by funds from the National Institute of Biomedical Imaging and Bioengineering (grant EB002045).

References

[1] S. Inoué, H. Sato, Deoxyribonucleic acid arrangement in living sperm, in: T. Hayashi, A.G. Szent-Gyorgyi (Eds.), Molecular Architecture in Cell Physiology, Prentice Hall, Englewood Cliffs, NJ, 1966, pp. 209–248.
[2] E.D. Salmon, S.M. Wolniak, Role of microtubules in stimulating cytokinesis in animal cells, Ann. N. Y. Acad. Sci. 582 (1990) 88–98.
[3] H. Sato, G.W. Ellis, S. Inoue, Microtubular origin of mitotic spindle form birefringence. Demonstration of the applicability of Wiener's equation, J. Cell Biol. 67(3) (1975) 501–517.
[4] D.E. Chandler, J. Heuser, The vitelline layer of the sea urchin egg and its modification during fertilization. A freeze-fracture study using quick-freezing and deep-etching, J. Cell. Biol. 84(3) (1980) 618–632.
[5] N.H. Hartshorne, A. Stuart, Crystals and the Polarising Microscope: A Handbook for Chemists and Others, third ed., Arnold, London, 1960.
[6] N.H. Hartshorne, A. Stuart, Practical Optical Crystallography, American Elsevier Publishing Co., Inc., New York, NY, 1964.
[7] W.J. Schmidt, Die Bausteine des Tierkörpers in polarisiertem Lichte, Cohen, Bonn, 1924.
[8] W.J. Schmidt, Die Doppelbrechung von Karyoplasma, Zytoplasma und Metaplasma. Protoplasma Monographien, vol. 11, Bornträger, Berlin, 1937.
[9] S. Inoué, W.L. Hyde, Studies on depolarization of light at microscope lens surfaces II. The simultaneous realization of high resolution and high sensitivity with the polarizing microscope, J. Biophys. Biochem. Cytol. 3(6) (1957) 831–838.
[10] S. Inoue, H. Sato, Cell motility by labile association of molecules. The nature of mitotic spindle fibers and their role in chromosome movement, J. Gen. Physiol. 50(Suppl. 6) (1967) 259–292.
[11] S. Inoue, J. Fuseler, E.D. Salmon, G.W. Ellis, Functional organization of mitotic microtubules. Physical chemistry of the in vivo equilibrium system, Biophys. J. 15(7) (1975) 725–744.
[12] S. Inoué, Video Microscopy, Plenum Press, New York, NY, 1986.
[13] S. Inoue, R. Oldenbourg, Microtubule dynamics in mitotic spindle displayed by polarized light microscopy, Mol. Biol. Cell. 9(7) (1998) 1603–1607.
[14] S. Inoue, Polarization microscopy, Curr. Protoc. Cell Biol. (2002) (Chapter 4: Unit 4.9).

[15] S. Inoué, Exploring living cells and molecular dynamics with polarized light microscopy, in: P. Török, F. J. Kao (Eds.), Optical Imaging and Microscopy, Springer-Verlag, Berlin, Heidelberg, New York, NY, 2003, pp. 3–20.

[16] S. Inoué, Collected Works of Shinya Inoué: Microscopes, Living Cells, and Dynamic Molecules, World Scientific, Singapore, 2008.

[17] R. Oldenbourg, Polarized light microscopy of spindles, Methods Cell Biol. 61 (1999) 175–208.

[18] M. Shribak, R. Oldenbourg, Techniques for fast and sensitive measurements of two-dimensional birefringence distributions, Appl. Opt. 42(16) (2003) 3009–3017.

[19] R. Oldenbourg, Polarization microscopy with the LC-PolScope, in: R.D. Goldman, D.L. Spector (Eds.), Live Cell Imaging: A Laboratory Manual, Cold Spring Harbor Laboratory Press, Cold Spring Harbor, NY, 2005, pp. 205–237.

[20] R. Oldenbourg, Analysis of microtubule dynamics by polarized light, Methods Mol. Med. 137 (2007) 111–123.

[21] R. Oldenbourg, M. Shribak, Microscopes, in: C.D. Michael Bass, J.M. Enoch, V. Lakshminarayanan, G. Li, C. Macdonald, V.N. Mahajan, et al. (Eds.), Handbook of Optics, McGraw-Hill, Inc., New York, NY, 2010, pp. 28.1–28.62.

[22] M. Shribak, S. Inoué, R. Oldenbourg, Polarization aberrations caused by differential transmission and phase shift in high NA lenses: theory, measurement and rectification, Opt. Eng. 41(5) (2002) 943–954.

[23] A.M. Glazer, J.G. Lewis, W. Kaminsky, An automatic optical imaging system for birefringent media, Proc. R. Soc. London A Math. Phys. Sci. 452 (1996) 2751–2765.

[24] R.D. Allen, J. Brault, R.D. Moore, A new method of polarization microscopic analysis I. Scanning with a birefringence detection system, J. Cell Biol. 18 (1963) 223–235.

[25] J.R. Kuhn, Z. Wu, M. Poenie, Modulated polarization microscopy: a promising new approach to visualizing cytoskeletal dynamics in living cells, Biophys. J. 80(2) (2001) 972–985.

[26] R. Oldenbourg, G. Mei, New polarized light microscope with precision universal compensator, J. Microsc. 180 (1995) 140–147.

[27] M. Shribak, Complete polarization state generator with one variable retarder and its application for fast and sensitive measuring of two-dimensional birefringence distribution, J. Opt. Soc. Am. A Opt. Image Sci. Vis. 28(3) (2011) 410–419.

[28] R. Oldenbourg, Analysis of edge birefringence, Biophys. J. 60(3) (1991) 629–641.

[29] M. Shribak, J. LaFountain, D. Biggs, S. Inouè, Orientation-independent differential interference contrast microscopy and its combination with an orientation-independent polarization system, J. Biomed. Opt. 13 (2008) 014011.

[30] J.R. LaFountain Jr., R. Oldenbourg, Maloriented bivalents have metaphase positions at the spindle equator with more kinetochore microtubules to one pole than to the other, Mol. Biol. Cell. 15(12) (2004) 5346–5355.

[31] M. Shribak, R. Oldenbourg, Three-dimensional birefringence distribution in reconstituted asters of Spisula oocytes revealed by scanned aperture polarized light microscopy, Biol. Bull. 205(2) (2003) 194–195.

[32] M. Shribak, R. Oldenbourg, Mapping polymer birefringence in three dimensions using a polarizing microscope with oblique illumination, Biophotonics Micro- and Nano-Imaging, SPIE Proceedings, Strasbourg, France, 2004.

[33] R. Oldenbourg, Polarized light field microscopy: an analytical method using a microlens array to simultaneously capture both conoscopic and orthoscopic views of birefringent objects, J. Microsc. 231(3) (2008) 419–432.

[34] B.S. DeMay, N. Noda, A.S. Gladfelter, R. Oldenbourg, Rapid and quantitative imaging of excitation polarized fluorescence reveals ordered septin dynamics in live yeast, Biophys. J. 101(4) (2011) 985–994.

[35] B.S. DeMay, X. Bai, L. Howard, P. Occhipinti, R.A. Meseroll, E.T. Spiliotis, et al., Septin filaments exhibit a dynamic, paired organization that is conserved from yeast to mammals, J. Cell. Biol. 193(6) (2011) 1065–1081.

[36] E. Hecht, Optics, fourth ed., Pearson/Addison-Wesley, San Francisco, CA, 2002.

[37] M. Born, E. Wolf, Principles of Optics: Electromagnetic Theory of Propagation, Interference and Diffraction of Light, sixth ed., Pergamon Press, Elmsford, NY, 1980.
[38] W.A. Shurcliff, Polarized Light, Production and Use, Harvard University Press, Cambridge, MA, 1962.
[39] R.A. Chipman, Polarimetry, in: M. Bass (Ed.), Handbook of Optics, McGraw-Hill, Inc., New York, NY, 1995, pp. 22.1–22.37.
[40] R. Oldenbourg, E.D. Salmon, P.T. Tran, Birefringence of single and bundled microtubules, Biophys. J. 74(1) (1998) 645–654.
[41] K. Katoh, K. Hammar, P.J. Smith, R. Oldenbourg, Birefringence imaging directly reveals architectural dynamics of filamentous actin in living growth cones, Mol. Biol. Cell. 10(1) (1999) 197–210.
[42] L. Liu, R. Oldenbourg, J.R. Trimarchi, D.L. Keefe, A reliable, noninvasive technique for spindle imaging and enucleation of mammalian oocytes, Nat. Biotechnol. 18(2) (2000) 223–225.

PART 4
Phase Nanoscopy

CHAPTER 16

Is There a Fundamental Limit to Spatial Resolution in Phase Microscopy?

Stephen G. Lipson

Physics Department, Technion-Israel Institute of Technology, Haifa and Department of Physics and Optical Engineering, Ort-Braude College, Karmiel, Israel

Editor: Zeev Zalevsky

16.1 Introduction

Abbe's well-known criterion for spatial resolution in incoherent imaging, $\delta x = \lambda/2\text{NA}$, has been the benchmark for all discussions of resolution since its publication in 1873. The term "super-resolution" was coined in 1952 by Toraldo di Francia [1] who showed, theoretically, that there is actually no resolution limit to imaging because a point spread function with arbitrarily small diameter can be created by means of an appropriately designed phase mask. This idea has recently been rediscovered and given the name "super-oscillation" by Berry and Popescu [2] because the phase of the light in the central spot changes spatially at a rate larger than $k_0 = 2\pi/\lambda$. What is clear from Toraldo di Francia's papers is that the light utilization is very poor, an overwhelming part of the light being diffracted into bright outer rings of the point spread function; however, within a certain dark field of view, bounded by the bright rings, super-resolution can be achieved. This method was first used to improve resolution in a confocal microscope by Hegedus and Sarafis [3], where the outer bright rings of the point spread function could be obstructed by the light collection aperture. Improvement of resolution of a two-point image using this technique was demonstrated by Leiserson and Lipson [4], and in principle their method shows that phase images could also be created the same way. More recent work on resolution enhancement using coded apertures is reviewed by Willett et al. [5].

The relationship between field of view and resolution had already been emphasized in 1966 by Lukosz [6], who showed for a general imaging system that the real limitation is the number of degrees of freedom in the image or the *space-bandwidth product*. Thus, when an object is known to occupy a limited region of the field of view, an optical system can be built to "beat" the Abbe limit. His "gedankenexperiment" to prove this is not altogether convincing and cannot exceed the resolution obtained with unit NA (in air); however, the result is correct and can be reached by other means. The interplay between field of view restriction and resolution has more recently been employed to improve resolution in linear systems by Shapiro et al. [7] and Gazit et al. [8]. The new term for these approaches is "sparsity," implying that there is known to be relatively little information in the image (e.g., it fills a small known region of the field of view), which harks back to Lukosz's approach.

In recent decades, the idea that resolution is unlimited for incoherent self-emitting, particularly fluorescent objects has been underlined by developing the powerful techniques of STED [9], PALM, and STORM [10], all of which employ inherent features of the fluorescence itself. They can be considered as coherent-to-incoherent conversion, and, therefore, are of no relevance to phase imaging. In this chapter, I will have little to say about these techniques (see the review by Lipson et al. [11]) because I want to concentrate on linear, and in particular phase microscopy, where the images produced are properties of the object itself and not of stains.

16.2 Where Was Abbe's Theory Incomplete?

It is interesting to ask how Abbe's theory should be modified in the light of subsequent developments. One view of Abbe's theory is obtained by looking at it as an application of Heisenberg's uncertainty principle. Actually, the argument below was first used by Heisenberg as an explanation of the uncertainty principle; he called it the "gamma-ray microscope" gedankenexperiment. Consider a one-dimensional system. If a photon with momentum $p_0 = h/\lambda$ from a point object at position x in the microscope field of view is received by the imaging detector, it must have passed through the microscope lens. If this lens subtends semi-angle α at the object, the photon can have transverse momentum p_x in the range between $\pm p_0 \sin \alpha$. Thus, the uncertainty is $\delta p_x = 2 p_0 \sin \alpha$. It follows from Heisenberg that

$$\delta x \geq h/p_x = \lambda/2 \sin \alpha = \lambda/2\text{NA} \qquad (16.1)$$

where NA $= n \sin \alpha$ is the NA and $n = 1$ for imaging in air. This is Abbe's limit. But clearly, it refers to a single photon, and we can consider the use of N independent photons, to which Poisson statistics apply, so that they impinge on the lens aperture at random positions. Then, the combined uncertainty in p_x is $\delta p_x = 2\sqrt{N} p_0 \sin \alpha$, from which it follows that

$$\delta x = \lambda/(2\text{NA}\sqrt{N}) \qquad (16.2)$$

This is the principle behind the stochastic super-resolution techniques PALM and STORM.

Another point of view claims that a basic flaw was that Abbe's theory considered the image of an *infinite* periodic grating. Abbe's basic idea was that the resulting diffraction orders are delta-functions in Fourier space (which is represented more or less by the lens aperture) and therefore either two orders were within the aperture, in which case the object period could be resolved, or only one, in which case it could not be. However, if the object is a finite grating, the orders are not delta-functions, and even if the center of an order is outside the aperture, its sidebands may be within the aperture and can allow resolution, although with less accuracy. If the shape of the object envelope is known, mathematical techniques can be used to improve this accuracy. Such techniques can be described as deconvolution, which will be discussed later. We can take this to the extreme in near-field scanning optical microscopy (NSOM) [12], where the instantaneous field of view (the illuminated spot) is much smaller than λ so that the diffraction order picture is not relevant. Some radiation is always received by the detector, and the image is built up sequentially. However, knowledge of the position of the imaged point is lost in the diffraction but is known from the scanning mechanism. As a result, scanning microscopy is not diffraction limited.

16.3 Parallel Full-Field Linear Imaging

From the point of view of spatial resolution, phase imaging is not different from intensity imaging. This has been demonstrated by an interferometric NSOM, in which the detector receives light simultaneously from the near-field probe and reference beam, and a phase-dependent interference signal is recorded [13]. However, the question to be addressed here refers to resolution limitations in linear imaging when the whole field is recorded by an imaging device. This is important generally for observing dynamic processes or in industry where point scanning methods are too slow for high throughput. Two main directions have been explored for linear super-resolution: image reconstruction from coherent diffraction patterns and structured illumination microscopy (SIM) using incoherent illumination.

16.4 Reconstruction from Diffraction Patterns

The basis of this method is like X-ray diffraction: the diffraction pattern of an object is recorded, using a coherent incident beam, and the structure is determined by inverse Fourier transformation. Its use for X-ray imaging by nonperiodic objects was first suggested in 1998 by Sayre et al. [14]. It is important to recall that if the diffraction pattern is not centro-symmetric, the object must include nontrivial phase information. The phase problem—that only the diffracted amplitudes are observed and not the phases—which seemed to be an obstacle to such microscopy is solved algorithmically by applying constraints to the object and finding by iterative techniques an object which fits both the constraints and the observed diffraction pattern [11,15]. In crystallography, the constraints of the unit cell properties and non-negativity of electron density are used [16], while in optics, the bounding shape of the object may be defined [7,8]. In these experiments, the object fills a small region of the coherently illuminated field of view so that the diffraction pattern is over-sampled [17], and both amplitude and phase information about the object can be found algorithmically. The basis of super-resolution in this approach is that because the object is bounded, its diffraction pattern can be written as a convolution between a diffraction function and the Fourier transform of the support (a function which is unity within the object's boundary and zero outside); this extra knowledge allows the diffraction function to be determined with super-resolution. Figure 16.1 shows an example of a phase image obtained this way using X-rays from a diffraction pattern which is not centro-symmetrical, although in this case the resolution was limited by the detector array and does not exceed the diffraction limit.

This problem has been studied in depth from the mathematical point of view by Donoho et al. [18], and I will try to summarize their main conclusions regarding resolution in one dimension. Suppose we have an object which is described in full by a vector with n components. We observe only $m < n$ components of its Fourier transform, which in full would also have n complex components. The ratio $\varepsilon \equiv m/n$ is called the "incompleteness ratio." Then, Donoho et al. [18] showed that if the number of nonzero elements in the object is less than $\tfrac{1}{2}m$ (i.e., $\tfrac{1}{2}\varepsilon n$), the full n components of the object can be reconstructed from that part of the Fourier transform, even in the presence of noise. This is often described as super-resolving a "sparse" object. If the number of nonzero elements is between $\tfrac{1}{2}\varepsilon n$ and εn, reconstruction may be possible if the noise is weak and the m elements can be chosen optimally. In other words, if the object is known to have only $\tfrac{1}{2}m$ nonzero elements, but their positions and complex values are unknown, there are m unknowns which can be determined from measurements of m complex Fourier components. Applying this to the imaging problem, the m components of the Fourier transform may be limited to the region sampled by the aperture of the optics, but the n components of the field of view may represent information on a much finer scale than the resolution limit. This is essentially implied in Lukosz's paper [6], where the object fills only a fraction $\tfrac{1}{2}\varepsilon$ of the field of view.

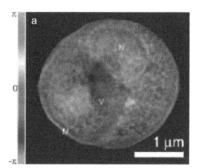

Figure 16.1
Image of a yeast cell derived from its diffraction pattern using 750 eV X-rays. The color image shows both phase (hue) and intensity (saturation) [7]. (For interpretation of the references to color in this figure legend, the reader is referred to the web version of this book.) Source: *Courtesy of C. Jacobsen, SUNY; copyright (2005) National Academy of Sciences, USA.*

The concept of image retrieval from minimum data, in the presence of noise, is extremely important in telescope imaging. Because astronomical fields are mainly black anyhow, deconvolution with the known point spread function is very effective [19,20]. In terms of a microscope image, biological objects satisfying the conditions are rare; however, it is possible to isolate an object by using a field stop provided that the object is illuminated by a much broader coherent beam so that the coherence within the field stop is complete and therefore its diffraction pattern is well determined. In the work of Shapiro et al. [7], the support was derived iteratively rather than being known exactly a priori. Moreover, the authors emphasize that since the object is small compared with the size D of the coherent illuminating X-ray beam, the Fourier transform can be sampled on a fine lattice (at angular intervals of λ/D), which makes the reconstruction possible.

16.5 Structured Illumination Microscopy

The method of structured illumination, which can be applied to transmitted or reflected light imaging, requires illuminating the object not with uniform light but with an illumination pattern containing high spatial frequencies. The resulting image has a Fourier transform, which is basically the transform of the required image convolved with that of the illumination pattern. Because of the high spatial frequencies in the latter, similar high frequencies in the object transform are transferred by the convolution operation to low frequencies in the image transform and are therefore clearly represented. The image is then reconstructed algorithmically from this transform. The effect is well known to us as the moiré effect, where two superimposed high-frequency gratings give rise to low-frequency "beats" in the image (Figure 16.2). This method is hinted at in Lukosz's [6] paper but was developed by Ben-Levy and Pelac [21] for semiconductor wafer inspection and by Heintzmann and Cremer [22] and Gustafsson [23] for microscopy. The usual implementation is to use an illumination grating with spatial frequency k_g close to the resolution limit of the microscope, which we will call k_m, approximately $\lambda/2NA$ (Figure 16.3). Now you can see that information in the image with frequency $2k_m$, normally well outside the imaging capabilities of the microscope, provides a component at $(2k_m - k_g) = k_m$ when it is modulated by the grating frequency, and this is just at the resolution limit of the microscope and can therefore contribute to the image. *The resolution limit has thus been increased by a factor of two.* To carry this out isotropically for a general two-dimensional image containing information at many spatial frequencies, it is necessary to record a series of modulated images with the illumination grating at several angles and each at least in three different phases. The process is shown in Figure 16.3.

SIM has been developed for both bright-field and fluorescence imaging, although mainly for the latter. This is because the modulation transfer function (MTF) for spatially incoherent illumination, which is usually used for nonfluorescent microscopy to avoid speckle, approaches zero at the resolution limit (Figure 16.4). This can be overcome by

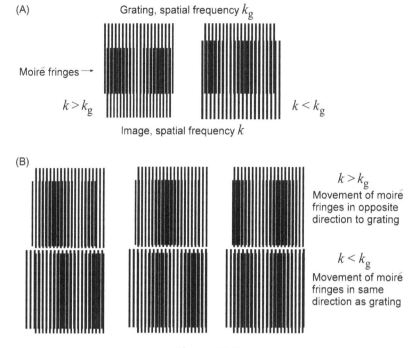

Figure 16.2
Moiré fringes are produced by the superposition of two gratings with different spatial frequencies. (A) Similar fringes are produced when the image frequency is larger or smaller than the grating by the same amount but (B) these can be distinguished by shifting the grating (changing its phase) with respect to the image. Source: *After Lipson et al. [11].*

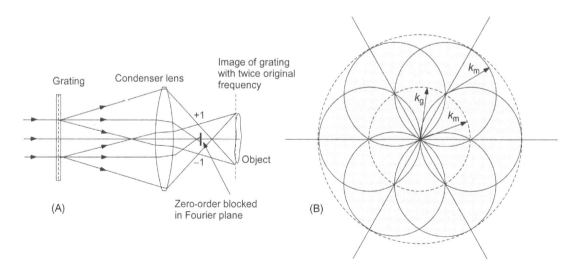

Figure 16.3
Optics of SIM illumination [23]. (A) Laser illumination and (B) frequency plane coverage using three different grating orientations with $k_g = k_m$. Each image is recorded with three different phases of the illumination pattern created by linear translation of the grating.

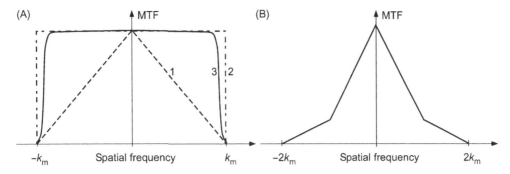

Figure 16.4
MTFs for SIM. (A) Illumination MTF with (1) incoherent illumination, (2) coherent laser excitation, and (3) partially-coherent illumination; (B) one-dimensional MTF for SIM imaging with coherent illumination.

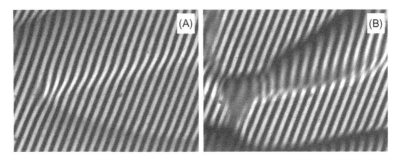

Figure 16.5
Phase objects observed with SIM, where there is a small focal offset between the object and the superimposed grating pattern. (A) Small phase variation ($<\pi$) and (B) substantial phase variation ($>\pi$).

using spatially partially-coherent illumination. On the other hand, fluorescence imaging uses coherent excitation, and the coherent MTF remains unity out to the resolution limit. The MTF for the complete imaging process has been studied by Somekh et al. [24].

The question remains—can SIM be modified to record phase information at super-resolved frequencies? To our knowledge, this has not been discussed specifically in the literature. First, we note that the product of the illumination pattern with an amplitude image retains the phase of the illumination pattern. Now suppose we have a thick object with phase gradients. Then, when the illumination grating is projected onto the object, the phase information is recorded as phase modulation in the image (Figure 16.5), which can be extracted algorithmically in the following manner.

Although strictly only three images with different phases are necessary, it is easier to appreciate the phase-imaging algorithm when four SIM images are acquired, with phases 0, $\pi/2$, π, and $3\pi/2$, and with k_g parallel to the x axis. We will call these images $s_1,\ldots,s_4$, and their Fourier transforms $S_1,\ldots,S_4$. These are described by:

$$s_1 = I(x,y) \cdot [1 + \cos(k_g x + \phi(x,y))] \tag{16.3}$$

$$s_2 = I(x,y) \cdot [1 + \sin(k_g x + \phi(x,y))] \tag{16.4}$$

$$s_3 = I(x,y) \cdot [1 - \cos(k_g x + \phi(x,y))] \tag{16.5}$$

$$s_4 = I(x,y) \cdot [1 - \sin(k_g x + \phi(x,y))] \tag{16.6}$$

$$S_1 = \tilde{I}(\vec{k})^* \left\{ \delta(\vec{k}) + \frac{1}{2}\tilde{\Phi}(\vec{k}-\vec{k}_g) + \frac{1}{2}\tilde{\Phi}^*(-\vec{k}+\vec{k}_g) \right\},\ldots \tag{16.7}$$

where $I(x,y)$ expresses the image intensity and $\phi(x,y)$ is its phase; * represents convolution and the vector $\vec{k}_g$ is parallel to the x axis. $\tilde{I}(\vec{k})$ and $\tilde{\Phi}(\vec{k})$ are the transforms of $I(x,y)$ and $\exp[i\phi(x,y)]$, respectively. It then follows that the transform of the complex image $I(x,y)\exp[i\phi(x,y)]$ is given by

$$\tilde{I}(\vec{k}) = [S_1 - S_3 + i(S_2 - S_4)]^* \delta(\vec{k}-\vec{k}_g) \tag{16.8}$$

This way, the complete phase structure of the image can be constructed with super-resolution approaching 2.

We now ask whether $2k_m$ is the limiting spatial frequency of SIM. If the modulation grating is projected through the microscope condenser onto the object, then $k_g = k_m$ is indeed the largest shift which can be projected since it uses the whole NA of the optical system (Figure 16.4A). But if the grating can be put in contact with the object, then there is the possibility of using a subwavelength grating with spatial frequency $k_g > k_m$, in which case the illumination wave propagates evanescently into the object. Then, the interaction of the evanescent wave with object structure having frequency k_{obj} close enough to k_g will give rise to a propagating wave of frequency $|k_{obj} - k_g| < k_m$. This frequency can be imaged by the microscope aperture. Clearly, such an imaging process would not be easy to carry out, particularly because to cover a wide range of object frequencies, it would be necessary to use many subwavelength gratings in contact with the object, but the process is not forbidden for fundamental reasons. Figure 16.6 shows a possible way in which this might be implemented to gain a modest degree of super-resolution of a thin air-immersed object; use of an actual subwavelength grating in contact with the object would be better. Obviously, this implementation requires that the object be thinner than the evanescent penetration depth, $2\pi/\sqrt{k_{obj}^2 - k_0^2}$. Work using this approach has been done by Gur et al. [25], who used a flowing suspension of nanoparticles in contact with an object to superimpose on it a series of independent subwavelength illumination patterns with varying

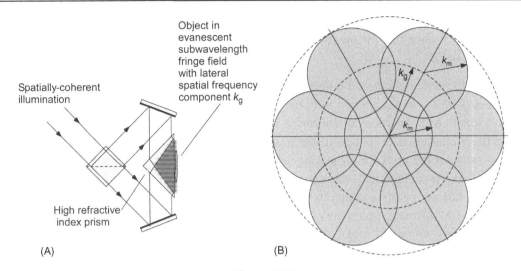

Figure 16.6
A possible way in which evanescent subwavelength periodic illumination can give super-resolution. (A) Illumination optics and (B) frequency plane coverage using three different illumination orientations.

spatial frequencies. They showed that an improvement in image resolution could be obtained by using the analysis of many different images of this type.

16.6 Imaging Three-Dimensional Phase Objects

Now, the question is what techniques might allow super-resolution of phase objects considerably thicker than the wavelength. This is essentially a question of three-dimensional imaging of the optical density. Since there is clearly no way of creating extended volume fringes with subwavelength spacing, structured illumination cannot be extended beyond the $2k_m$ limit. On the other hand, in parallel with the determination of three-dimensional crystal structures from diffraction data, we would expect diffraction synthesis to be possible and the limitations to be similar to those in one and two dimensions. Now, in a single diffraction pattern, the three-dimensional Fourier transform of the object is sampled only on the Ewald sphere, which has radius $2k_0$ and passes through the origin of k space. In Fourier space, the Ewald sphere represents geometrically the requirement for conservation of photon energy during elastic scattering (see, e.g., Ref. [11]). As the incident beam (direction of the vector k_0) is rotated with respect to the sample, the total accessible region of Fourier space is thus a sphere with radius $2k_0$ and volume $(32/3)\pi k_0^3$. A coherent incident beam size of D allows this in principle to be sampled in k space on a scale of $(2\pi/D)^3$, which means that the maximum number of data points in the three-dimensional transform is about $m = (2D/\lambda)^3$. Applying the results of Donoho et al. [18] to three dimensions, this would suggest that for a compact sample with volume d^3, the resolution limit attainable by deconvolution is such that the number of resolvable elements in the sample is less than half m, i.e., $\delta x \approx \lambda d/D$.

The above discussion assumes the ideal situation in which diffraction patterns of the object can be accessed in every direction with the required dense angular sampling at intervals λ/D in azimuth and elevation; an enormous task. The inevitable practicalities essentially caused the limitation of resolution of the image shown in Figure 16.1. However, the analysis provides the tools for estimating achievable resolution under practical circumstances. As a numerical example, we might take a coherent field of view of diameter $D = 1000\lambda$, diffracting off a sample whose extent is known to be $d = 10\lambda$ (i.e., $d/D = 0.01$). Then, using 100 diffraction patterns of the object in distinctly different orientations, each one recorded in a unit solid angle, one could theoretically obtain resolution of about 0.03λ, which exceeds the Abbe's diffraction limit by an order of magnitude. Of course, here, we have ignored the considerable computational effort required for each image, and the fact that recording numerous diffraction patterns with high signal to noise might make the process as slow as scanning, which we were trying to avoid!

16.7 Conclusions

Without the use of fluorescent markers, microscopic resolution is limited by the physics of the diffraction process. However, this can be extended beyond the classical Abbe's limit. Two proven methods of increasing resolution are SIM and reconstruction from sampling the coherent diffraction pattern of an object. Both can create phase contrast. The former is restricted to two-dimensional objects, while the latter requires the object to fill a known small region of the field of view in order to make deconvolution possible in the presence of noise.

Acknowledgments

I wish to thank Dr. E. Ribak for numerous comments on this manuscript.

References

[1] G. Toraldo di Francia, Super-gain Antennae and optical resolving power, Suppl. Nuovo Cimento 9 (1952) 426–438.
[2] M.V. Berry, S. Popescu, Evolution of quantum superoscillations and optical superresolution without evanescent waves, J. Phys. A: Math. Gen. 39 (2006) 6965–6977.
[3] Z.S. Hegedus, V. Sarafis, Superresolving filters in confocally scanned imaging systems, J. Opt. Soc. Am. A3 (1986) 1892–1896.
[4] I. Leiserson, S.G. Lipson, Superresolution in far-field imaging, Opt. Lett. 25 (2000) 209–211.

[5] R.M. Willett, R.F. Marcia, J.M. Nichols, Compressed sensing for practical optical imaging systems: a tutorial, Opt. Eng. 50 (2011) 072601-1–072601-13.

[6] W.J. Lukosz, Optical systems with resolving powers exceeding the classical limit, J. Opt. Soc. Am. 56 (1966) 1463–1472.

[7] D. Shapiro, P. Thibault, T. Beetz, V. Elser, M. Howells, C. Jacobsen, et al., Biological imaging by soft x-ray diffraction microscopy, Proc. Natl. Acad. Sci. USA 102 (2005) 15343–15346.

[8] S. Gazit, A. Szameit, Y.C. Eldar, M. Segev, Super-resolution and reconstruction of sparse sub-wavelength images, Opt. Express 17 (2009) 23920–23946.

[9] S.W. Hell, J. Wichmann, Breaking the diffraction resolution limit by stimulated emission: stimulated-emission-depletion fluorescence microscopy, Opt. Lett. 19 (1994) 780–782.

[10] X. Zuang, Nano-maging with STORM, Nat. Photonics 3 (2009) 365–367.

[11] A. Lipson, S.G. Lipson, H. Lipson, Optical Physics, fourth ed., Cambridge University Press, Cambridge, 2011.

[12] A. Lewis, M. Isaacson, A. Harootunian, A. Murray, Development of a 500Å resolution light microscope, Ultramicroscopy 13 (1984) 227–230.

[13] M. Vaez-Iravani, F.L Toledo-Crow, Phase contrast and amplitude pseudoheterodyne interference near field scanning optical microscopy, Appl. Phys. Lett. 62 (1992) 1044–1046.

[14] D. Sayre, H.N. Chapman, J. Miao., On the extendibility of X-ray crystallography to non-crystals, Acta Cryst. A54 (1998) 232–239.

[15] R.W. Gershberg, W.O. Saxton, A practical algorithm for the determination of phase from image and diffraction plane pictures, Optik 35 (1972) 237–246.

[16] H.A. Hauptmann, The phase problem of X-ray crystallography, Rep. Prog. Phys. 54 (1991) 1427–1454.

[17] R.H.T. Bates, Fourier phase problems are uniquely solvable in more than one dimension, I: underlying theory, Optik 61 (1982) 247–262.

[18] D.L. Donoho, I.M. Johnstone, J.C. Hoch, A.S. Stern, Maximum entropy and the nearly black object, J. R. Stat. Soc. B54 (1992) 41–81.

[19] J. Högbom, Aperture synthesis with a non-regular distribution of interferometer baselines, Astron. Astrophys. Suppl. Ser. 15 (1974) 417–426.

[20] J.L. Starck, E. Pantin, F. Murtagh, Deconvolution in astronomy: a review, Publ. Astr. Soc. Pac. 800 (2002) 1051–1069.

[21] M. Ben-Levy, E. Pelac, E. Imaging measurement system. 1997, patent WO/1997/006509

[22] R. Heintzmann, C. Cremer, Laterally modulated excitation microscopy: improvement of resolution by using a diffraction grating, Proc. SPIE 3568 (1998) 185–195.

[23] M.G.L. Gustafsson, Surpassing the lateral resolution limit by a factor of two using structured illumination microscopy, J. Microsc. 198 (2000) 82–87.

[24] M.G. Somekh, K. Hsu, M.C. Pitter, Effect of processing strategies on the stochastic transfer function in structured illumination Microscopy, J.Opt. Soc. Am. A28 (2011) 1925–1934.

[25] A. Gur, D. Fixler, V. Mic, J. Garcia, Z. Zalevsky, Linear optics based nanoscopy, Opt. Express 18 (2010) 22223–22231.

CHAPTER 17

Nano-Holographic Interferometry for In-Vivo Observations

Federico M. Sciammarella[1], Cesar A. Sciammarella[2], Luciano Lamberti[2]
[1]*College of Engineering and Engineering Technology, Northern Illinois University, DeKalb, IL*
[2]*Dipartimento di Ingegneria Meccanica e Gestionale, Politecnico di Bari, Bari, Italy*

Editor: Natan T. Shaked

17.1 Introduction: Theoretical Basis of Nano Fourier Transform Holography

With the introduction of holographic recording in the Fourier transform (FT) space, applications for reconstructing objects have gained considerably in spatial resolution [1–4]. The utilization of Fourier transform holography (FTH) is well documented both in the visible light realm [5,6] and in other frequencies of the electromagnetic spectrum [7,8].

The work presented in this chapter is a variant of FTH that utilizes evanescent illumination to increase the spatial bandwidth of the recorded holograms and includes Fourier transform holographic interferometry (FTHI). Following the principle of reverse interference postulated by Toraldo di Francia [9] and using numerical procedures based on Fourier image analysis, image components can be analyzed in different regions of the spatial frequency spectrum.

In the scheme of plane-wave solutions of the Maxwell equations available in classical optics, monochromatic waves with definite frequencies and wave numbers are usually considered. However, this idealized condition does not apply in the present case. One of the foundations of this new method is the observation of objects that are self-luminous. The objects themselves will generate electromagnetic waves that are related to the object's material properties and depend on the geometry of the object: hence, these waves carry information on the object configuration. The object is excited by an evanescent electromagnetic field and under conditions of resonance will emit light. A superposition of waves will be generated and it will be distributed over the frequency bandwidth Δ thus producing a sequence of wave packets of length $2\pi/\Delta$.

A simple one-dimensional model can be developed to illustrate the abovementioned point. By utilizing the scalar theory of light wave propagation, one can start from the Fourier solution of the Maxwell equations:

$$E(x,\tau) = \frac{1}{\sqrt{2\pi}} \int_{-\infty}^{+\infty} A(k) e^{ikx - i\omega(k)\tau} dk \qquad (17.1)$$

where $E(x,\tau)$ is the scalar representation of the propagating electromagnetic field, x is the direction of propagation of the field, τ is the time, $A(k)$ is the amplitude of the field, k is the wave number $2\pi/\lambda$, and $\omega(k)$ is the angular frequency. $A(k)$ provides the linear superposition of the different waves that propagate and can be expressed as:

$$A(k) = \sqrt{2\pi} \delta(k - k_o) \qquad (17.2)$$

where $\delta(k - k_o)$ is the Dirac's delta function. This amplitude corresponds to a monochromatic wave, i.e., $E(x,\tau) = e^{ikx - i\omega(k)\tau}$. At time $\tau = 0$, $E(x,0)$ represents (Figure 17.1C) a finite wave-train of length L_{wt} where $A(k)$ is not a delta function but a function that spreads a certain length Δk. The dimension of L_{wt} depends on the object size which, in the present case, is smaller than the wavelength of the light.

In Ref. [11], it is stated that if L_{wt} and Δk are defined as the RMS deviations from the average values of L_{wt} and Δk defined in terms of the intensities $|E(x,0)|^2$ and $|A(k)|^2$, then:

$$L_{wt} \Delta k \geq \frac{1}{2} \qquad (17.3)$$

Since L_{wt} is very small, the spread of wave numbers of monochromatic waves must be large. There is a quite different scenario from the classical context in which the length L_{wt} is large compared to the wavelength of light. This simple model indicates that the observed objects can be considered as electromagnetic oscillators emitting a spectrum of frequencies that will play an important role in the developments that follow.

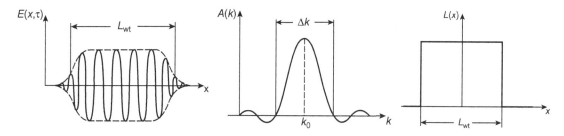

Figure 17.1
(A) Harmonic wave-train of finite extent L_{wt}; (B) corresponding Fourier spectrum in wave numbers k; (C) representation of a spatial pulse of light whose amplitude is described by the rect(x) function [10].

To simplify the notation, one can reason in one dimension without loss of generality. The spatial pulse of light represented in Figure 17.1C is defined as follows:

$$A(x) = A_o \, \text{rect}(x) \tag{17.4}$$

where:

$$\text{rect}(x) = \begin{cases} 1 & |x| < \dfrac{1}{2} \\ \dfrac{1}{2} & |x| = \dfrac{1}{2} \\ 0 & \text{elsewhere} \end{cases} \tag{17.5}$$

The FT of $A(x)$ is equal to $\text{sinc}(x)$. Therefore, the FT of the light intensity $[A(x)]^2$ is $[\text{sinc}(x)]^2$. To the order 0, it is necessary to add the shifted orders ± 1. The function $A(x \pm \Delta x)$ can be represented through the convolution relationship:

$$A(x \pm \Delta x) = \int_{-\infty}^{+\infty} A(x') \cdot \delta(x \pm \Delta x) \cdot dx' \tag{17.6}$$

where $x' = x \pm \Delta x$. The FT of the function $A(x \pm \Delta x)$ will be:

$$\text{FT}[A(x \pm \Delta x)] = A(f_x) \cdot e^{i[2\pi f_x(\mp \Delta x/2)]x} \tag{17.7}$$

where f_x is the spatial frequency. The real part of Eq. (17.7) is:

$$\text{Re}\{\text{FT}[A(x \pm \Delta x]\} = A(f_x) \cdot \cos\left\{\left[2\pi f_x\left(\frac{\pm \Delta x}{2}\right)\right]x\right\} \tag{17.8}$$

By taking the FT of Eq. (17.8), one can return to Eq. (17.6). Then the object will emit wave trains with a bandwidth that is influenced by the size of the emitting object. Since L_{wt} is very small, the spread of wave numbers of monochromatic waves must be large. Hence, there is a quite different scenario with respect to the classical optics context, in which the length L_{wt} is large when compared to the wavelength of light.

17.2 Gratings Illuminated by Evanescent Waves

Gratings play a fundamental role in the proposed FTH technique. In view of this, the present section will introduce arguments originally developed by Toraldo di Francia concerning evanescent solutions of the grating diffraction equations [9].

The equations of the plane-wave solution for the scalar form of the Maxwell equations must be considered. The general solution derived by Toraldo di Francia [9] is also utilized in this study. For a three-dimensional problem, it is assumed that the direction cosines of the wavefronts are complex quantities of the form:

$$\begin{cases} \cos\alpha = a_{1e} + ia_{2e} \\ \cos\beta = b_{1e} + ib_{2e} \\ \cos\gamma = c_{1e} + ic_{2e} \end{cases} \quad (17.9)$$

where all coefficients a_{1e}, b_{1e}, c_{1e}, a_{2e}, b_{2e}, and c_{2e} are real numbers corresponding to evanescent waves. For a plane wavefront such that results are independent of the Y coordinate (i.e., $\cos\beta = 0$), by replacing the direction cosines defined above in the plane wavefront equation one obtains:

$$E_e(x, z) = A e^{-k(a_{2e}x + c_{2e}z)} \, e^{ik(a_{1e}x + c_{1e}z)} \quad (17.10)$$

Recalling the property of the direction cosines, $\cos^2\alpha + \cos^2\gamma = 1$, one gets:

$$a_{1e}^2 + c_{1e}^2 = 1 + a_{2e}^2 + c_{2e}^2 \quad (17.11)$$

and

$$a_{1e}a_{2e} + c_{1e}c_{2e} = 0 \quad (17.12)$$

The evanescent field solution (real part) shows that the amplitude of electromagnetic field decreases with depth; this means that the field cannot propagate, as ordinary waves do, in the direction perpendicular to the surface. However, experiments show that evanescent waves can propagate in their own plane, even if the medium of propagation changes. Furthermore, the planes of constant phase corresponding to the imaginary component of the electromagnetic field are orthogonal to the planes of constant amplitude as shown by Eq. (17.12). This is a very important result. Since the energy goes with the amplitude of the electromagnetic field (Poynting vector), it is possible to show that the Poynting vector of the field that goes beyond the boundary of the two media is equal to zero, and no energy is transmitted to the second medium.

In the analysis of diffraction-order formation including both the transmission of wavefronts and diffraction orders generated by evanescent illumination, Toraldo di Francia [9] showed that while there is a limited number of ordinary diffraction orders as the maximum value of the spatial angular spectrum is $\pi/2$, the diffraction orders corresponding to evanescent illumination are infinite (imaginary solutions of the diffraction equation).

17.3 Basic Setup Utilized for the Observation of Nano-Objects

The basic setup shown in Figure 17.2 consists of a cell formed by two pieces of microscopic slides that are kept together by a metallic frame. Following the arrangement of total internal reflection (TIR), a helium–neon (He–Ne) laser beam with nominal

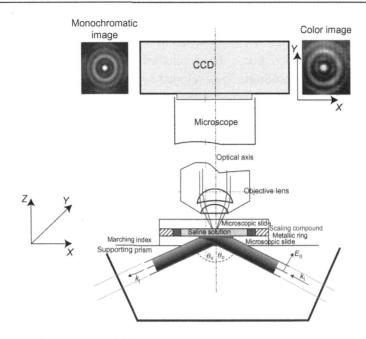

Figure 17.2
Optical setup utilized in the measurements on nano-sized objects and details on polystyrene microsphere and saline solution [10].

Parameter	Value	Note
Polystyrene microsphere diameter D_{sph}	6 ± 0.042 μm	Tolerance specified by the manufacturer
Refraction index of polystyrene sphere n_p	1.57 ± 0.01	Value specified by the manufacturer
Refraction index of saline solution n_s	1.36	Computed from NaCl concentration for the nominal wavelength of $\lambda = 590$ nm

wavelength 632.8 nm impinges normally to the face of a prism designed to produce limit-angle illumination at the interface between the bottom microscopic slide (supported by the prism itself) and a saline solution of sodium chloride contained in a small cell supported by this slide. Consequently, evanescent light is generated inside the saline solution. The objects observed with the microscope are supported by the upper face of the microscopic slide. Inside the cell filled with the NaCl solution, there is a polystyrene microsphere that is 6 μm in diameter. The microsphere is fixed to the face of the slide through chemical treatment of the contact surface in order to avoid Brownian motions. The polystyrene sphere acts as a relay lens which collects the light wavefronts generated by the nano-sized objects under observation. For more details on the polystyrene sphere and the saline solution properties see the table given in Figure 17.2.

The image of the diffraction pattern is focused by a microscope with NA = 0.95 and recorded with a charge-coupled device (CCD) attached to the microscope. The CCD is a square pixel camera with 1600 × 1152 pixels. Image analysis is performed with the Holo-Moiré Strain Analyzer software (HoloStrain™) developed by Sciammarella and collaborators [12].

The first step to understanding the process of recording the FTs of the observed objects is to consider the microsphere as a relay lens (Figure 17.3). The focal distance f of this lens can be computed using the equations of geometric optics. The microsphere generates a diffraction pattern. The source of the illumination generating the diffraction pattern of the microsphere is the resonant electromagnetic oscillations of interaction between the microsphere and the supporting microscope slide. The whole region of contact between the particle and the microscopic slide becomes the equivalent of a microlaser. The light is generated in the sphere around the first 100 nm in depth and focused in the focal plane of the sphere. The microscope is focused to the plane of best contrast of the diffraction pattern, which must be very close to the microsphere focal plane. The diffraction pattern is not formed by a plane wavefront illuminating the microsphere, as in the case of the classical Mie's solution for a diffracting sphere. In the present case, a large number of evanescent wavefronts are generated in a very small region (radius 400 nm) that is in contact with the microsphere and then generate propagating wavefronts that are seen as interference fringes.

The formation of the interference fringes has been explained by utilizing an approximate solution of the diffraction equation [10,13] reproducing the observed experimental pattern. These results were confirmed following a formal approach to the diffraction problem of a dielectric spherical particle [14].

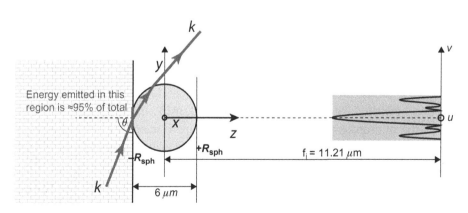

Figure 17.3
Polystyrene microsphere modeled as a small spherical lens [10].

Figure 17.4 shows the diffraction pattern recorded by the sensor. Besides the diffraction pattern of the polystyrene microsphere, the image recorded by the optical microscope also contains a system of rectilinear fringes. All the FT observations were made in the central region (i.e., limited by the first dark ring) of the pattern that has a radius of 1.05 μm. Figure 17.4 also shows the FT of the recorded image. By filtering this FT, different families of fringes were extracted. Recalling the complex argument of the diffraction evanescent orders presented in Section 17.2, Figure 17.4 also shows the values of complex sine for the different evanescent diffraction orders which are up to 120. It is also possible to see that the diffraction rectilinear fringes are modulated by the diffraction pattern of the particle; an inset in Figure 17.4 shows the cross-section of these fringes in the region corresponding to the central crown of the diffraction pattern of the sphere. In this cross-section, it is possible to see the modulation of a system of parallel fringes produced by the presence of the polystyrene sphere.

17.4 System of Fringes Contained in the Recorded Image: Multi-k Vector Fields

Aside from the diffraction pattern of the polystyrene microsphere, the image recorded by the optical microscope also contains a system of rectilinear fringes. Understanding the

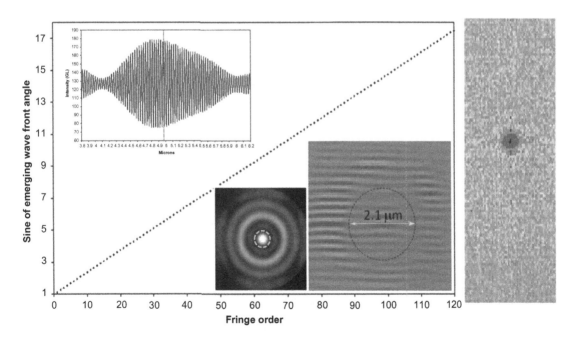

Figure 17.4
System of fringes observed in the image of the 6 μm diameter polystyrene microsphere. The dotted circle represents the first dark ring in the particle diffraction pattern [10].

formation of these fringes is fundamental because they play an important role in the process of retrieving information from the recorded images. Figure 17.5 presents a model for the process of generation of the evanescent electromagnetic fields that provide the energy required for the formation of the images. In this model, a grating simulates the diffraction effect arising from the presence of residual stresses in the upper layer of the prism. This effect is discussed in detail in Ref. [15].

Guillemet [16] analyzed the process of interference fringe formation originated by evanescent illumination in the presence of residual stresses on glass surfaces. The glass in the neighborhood of its surface can be treated as a layered medium and the fringe orders depend on the refraction index gradients.

Figure 17.6A represents the outer layer of the prism that is subject to a quasi isotropic state of compressive stresses [15]. This outer layer undergoes a biaxial stress state. The ellipsoids of the index of refraction of the equivalent birefringent crystal are surfaces of revolution with the optical axis directed horizontally as is shown in Figure 17.6A. In this figure, the intersection of the ellipsoids with the element of the glass prism surface is represented. There are two indices of refraction: the ordinary beam refraction index n_o and the extraordinary beam refraction index n_e. Since glass is a negative crystal, the ordinary wavefront has the largest index of refraction. The axes of the ellipsoids are along the lines where $n_e = n_o$.

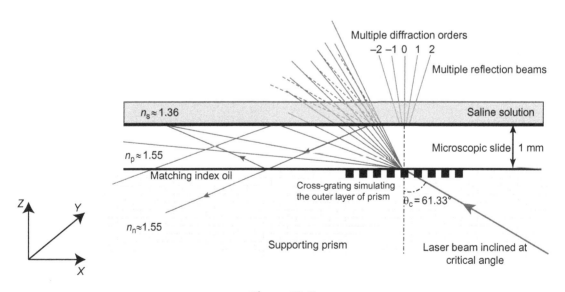

Figure 17.5
Model of the interface between the supporting prism and the microscope slide as a diffraction grating causing the impinging laser beam to split into different diffraction orders [10].

Figure 17.6A shows the coordinate system used in the current analysis; principal stresses act as indicated in the schematic. The polarization vector of the incident beam is contained in the plane of incidence (p-polarization), which at the same time is the plane of the algebraically smallest stress (i.e., the highest negative). The highest stress (i.e., the smallest negative) is in the perpendicular direction.

Figure 17.6B shows the optical path of the two beams generated by the glass birefringence: the ordinary beam (p-polarized) and the extraordinary beam (s-polarized), respectively. The optical paths of the plane wavefronts from the source on the prism surface to the arrival at the interface between the microscope slide and saline solutions as well as their propagation in the saline solutions are presented. The light beams that originated at the outer layer of the prism entering the glass slide have a different index of refraction than the prism bulk

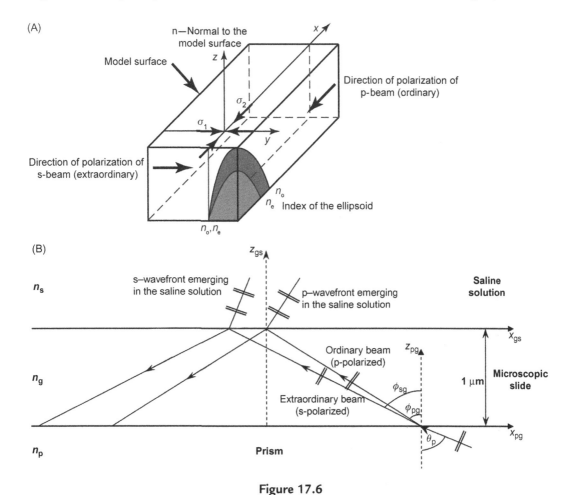

Figure 17.6
(A) Schematic representation of the prism element with indication of polarized beams and corresponding indices of refraction; (B) optical path of evanescent wavefronts [13].

glass and hence experience changes of direction. The interface between the glass slide and prism is filled with an index-matching fluid.

At the interface between the prism and the glass slide, from Snell's law the p-beam is:

$$\sin \phi_{pg} = \frac{n_g \sin \phi_p}{n_o} \tag{17.13}$$

And the s-beam is:

$$\sin \phi_{sg} = \frac{n_g \sin \phi_p}{n_e} \tag{17.14}$$

Since the two beams enter the glass-saline solution interface with incident angles corresponding to total reflection, they originate evanescent waves. These evanescent waves produce propagating waves in the saline solution. Figure 17.6B shows the two wavefronts corresponding to the ordinary and extraordinary beams. From Eq. (17.12) it follows that the emerging beams are orthogonal to the direction of the incoming beams' photons that, according to the preservation of momentum, continue their trajectory along the incident directions. The electromagnetic field that penetrates the saline medium generates scattered light in the dielectric medium hence a change of direction of $\pi/2$ takes place for the emerging wavefronts. It should be noted that these wavefronts are orthogonally polarized and hence do not interfere. However, they produce interference fringes whose spacing can be measured experimentally; hence, inferences on wavefronts can be made. The emerging wavefronts have the same wavelength of the incident beam and then, assuming that the fringe spacing is determined by the incidence angles, one can write:

$$\begin{cases} \sin \phi_{pm}^{exp} = \dfrac{\lambda}{p_{pm}} \\ \sin \phi_{pm}^{exp} = \dfrac{\lambda}{p_{sm}} \end{cases} \tag{17.15}$$

where the values p_{pm} and p_{sm} are the measured pitch values for the observed fringes.

After a long derivation [17], the following relationships are obtained:

$$\begin{cases} \sin \phi_{pm} = \dfrac{\sqrt{n_o^2 - n_g^2}}{n_s} \\ \sin \phi_{sm} = \dfrac{\sqrt{n_e^2 - n_g^2}}{n_s} \end{cases} \tag{17.16}$$

A propagation model of the beams in the interfaces was developed to also account for the effect of birefringence. Approximate values are available for the following quantities: thickness of the microscope slide (t_s) and the measured values of p_{pm} and p_{sm}. From the latter values and from Eq. (17.16), it is possible to compute ϕ_{pm} and ϕ_{sm}. The optical paths of the beams from the prism to the emergence in the saline solution can hence be determined. At this point, one must add the Maxwell–Neumann equations that relate stresses to the observed retardations:

$$\begin{cases} n_o - n_g = A\sigma_1 + B\sigma_2 \\ n_e - n_g = B\sigma_1 + A\sigma_2 \end{cases} \quad (17.17)$$

The above equations also contain the constants A and B and the residual compressive stresses σ_1 and σ_2 present in the prism. Approximate values for A and B are also known. The utilization of the above equations to formulate an optimization problem leads to the best values that define the posed problem. The set of values listed in Table 17.1 were reported in Refs. [13,17]. The optimization process was started from three different initial points in order to converge to the global optimum: (i) lower bounds of design variables (Run A); (ii) mean values of design variables (Run B); (iii) upper bounds of design variables (Run C).

Besides the residual error between the model and experimental values, Table 17.1 also shows the photoelastic constant $C = A - B$ and the shear stress τ in the prism determined as $|(\sigma_1 - \sigma_2)/2|$. Remarkably, the residual error on emerging wave angles is always smaller than 0.4%.

The above developments provide a proof of the consistency of the proposed model. A formal derivation of the fringe formation process is omitted for the sake of brevity.

Table 17.1: Results of Optimization Runs Carried Out to Analyze the Effect of Prism's Artificial Birefringence

Parameters	Run A	Run B	Run C
n_o	1.5716	1.5709	1.5699
n_e	1.5447	1.5447	1.5468
n_g	1.5399	1.5399	1.5405
A (m^2/N)	-0.5167×10^{-11}	-0.3017×10^{-11}	-0.6994×10^{-11}
B (m^2/N)	-2.8096×10^{-11}	-2.9645×10^{-11}	-3.1390×10^{-11}
$C = A - B$ (m^2/N)	2.2929×10^{-11}	2.6628×10^{-11}	2.4396×10^{-11}
σ_1 (MPa)	-135.30	-142.50	-161.85
σ_2 (MPa)	-145.01	-145.83	-164.90
τ (MPa)	4.855	1.666	1.528
Residual error Ψ (%)	0.387	0.273	0.339

The microscopic slide acts as a Fabry–Pérot interferometer and generates the multiple diffraction orders that are shown in Figure 17.5. Hence, the residual stresses and the microscopic slide produce the effect equivalent of the grating shown in Figure 17.5.

17.5 Determination of the Pitch of the Gratings

This section describes the fringe pitch measurement process formed by the wavefronts corresponding to the ordinary and extraordinary beams. Whilst in the preceding analysis only the real orders were considered, in this section, the orders corresponding to the imaginary solutions of the diffraction equation derived by Toraldo di Francia [9] are included. This information is retrieved from the FT of the image captured by the CCD camera (see the inset of Figure 17.4). Sine values of the emerging wavefronts resulting from the evanescent waves are plotted versus the fringe orders extracted from the FT shown in Figure 17.4.

As mentioned before, orders were measured in the FT starting from the zero order which is taken as the origin of coordinates. Sine values are computed utilizing the equation developed by Toraldo di Francia ([9], chapter III, section 47):

$$\sin \phi = \frac{n_g}{n_s} \sin \theta_c + \frac{\lambda}{n_s(p_o/N_o)} \quad (17.18)$$

where ϕ is the complex angle corresponding to the evanescent orders: the angle of diffraction is evaluated with respect to the direction of the incoming laser beam, p_o is the pitch of the fringes, N_o is the fringe order obtained from the FT, n_s is the index of refraction of the saline solution, and n_g is the index of refraction of the microscope slide.

All the emerging orders except the first one are in the range of the complex sine function. Since the sine values are greater than 1, the corresponding angles are complex numbers with a real part and an imaginary part. Utilizing the plane-wave complex solutions of the Maxwell equations for both ordinary and extraordinary wavefronts, one arrives to a system of fringes whose variable intensity, which is finally recorded by the sensor, can be expressed as follows:

$$I(x) = I_o + I_1 \cos\left[\frac{2\pi}{p_o} N_o \frac{n_s}{n_g}\right] x \quad (17.19)$$

Figure 17.7 shows the fringe pitch variation for the different orders extracted from the FT. The spatial frequency (i.e., pitch) was determined by dividing the field of view of the recorded image by the fringe order. By fitting the experimental data, one obtains hyperbolic-type trend functions that define the fundamental frequency of the fringe pattern. A detailed frequency analysis of the observed images revealed that there are two spatial

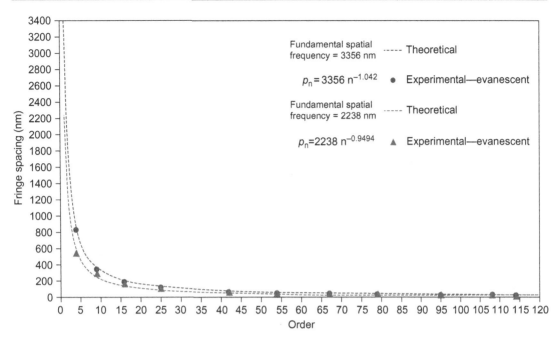

Figure 17.7
Variation of spatial frequency of rectilinear fringes with evanescent orders [10].

frequencies that can be separated everywhere in the image and have average values 2.238 and 3.356 μm, respectively.

The presented derivations define a fundamental family of interference fringes referred to the zero-order wavefronts corresponding to the ordinary and extraordinary beams. The different orders define other families of fringes.

Another very important aspect of these results shows that despite being in the far optical field, it was possible to observe optical phenomena that do not fit within the classical diffraction-limited optical system's restrictions in resolution. As such, we were successful in making metrological measurements well beyond the classical resolution limits. A first of its kind!

17.6 Formation of the Holograms at the Nanoscale

The optical system utilized to observe nano-objects was represented in Figure 17.2 and described in Section 17.3. In Sections 17.2 and 17.4, aspects concerning the illumination process of the observation region with wavefronts possessing a large angular spectrum of **k** vectors were discussed and the formation of families of rectilinear fringes present in the formed images was analyzed. The formation of equivalent FT holograms at a nanometric scale and with properties that are unlike classical FT holograms must be discussed. However, the power of the original idea of Gabor is still present under a different form.

The first step is to analyze the optical path of the wavefronts that lead to the recording of FTH images. The basic process will be illustrated in this section with a simple example of a prismatic object that later on will be one of the nano-objects actually observed in the measurements. In Section 17.1, the idea of the observed objects as optical resonators was introduced and a simple one-dimensional model was developed. The optical resonator is a nano-object smaller in size than the wavelength of light that arrives to it. The object itself becomes a light source emitting different frequencies that are determined by the geometry and by the properties of the material of the object.

What is received by the sensor of the optical setup shown in Figure 17.2 is an inline FT lens hologram where the source of illumination is the observed object. This is a fundamental aspect to understand the process of information recording in the CCD sensor. The source of illumination and the observed object coincides. This is a variant of the original idea but still preserves the basic feature: information of an object is retrieved in the FT field providing an increase in the spatial resolution.

Figure 17.8 shows the optical circuit bringing the images to the CCD detector. The object, in this example a prismatic dielectric material resting on the upper surface of the microscopic slide, is excited by the electromagnetic field of the evanescent waves and at the interface with the microscope slide emits wavefronts. These wavefronts are diffraction orders of the object. Assuming that the prism (Figure 17.8, Part 1) is approximately parallel to the image plane of the CCD, the successive diffraction orders emerge at different angles with respect to the normal of the slide surface that is assumed to be approximately parallel to the normal of the sensor. The largest fraction of energy is concentrated in the zero order and the first order [10]. Part 2 of Figure 17.8 shows the trajectories of the zero and first diffraction orders entering and emerging the relay lens. Part 3 of Figure 17.8 shows the image formation at the focal plane of the relay lens. As stated in [18] (section 5.2, page 103), the formed image is the FT of the object placed against the lens, with the presence of an additional quadratic phase factor.

Part 4 of Figure 17.8 shows the image that the microscope projects into the CCD sensor. Since the size of the object is small compared to the size of the relay lens, it is possible to make the assumption of plane wavefronts. Consequently, the quadratic phase factor becomes negligible in the current analysis of the image formation. The sensor hence displays the FT of the observed object. Since the hologram is an inline hologram, the zero order and the first order overlap—this is a characteristic of the inline holograms. However, in the present case, the zero-order diffraction pattern is simultaneously the source and the object itself, and hence the zero order contains information relevant to the object.

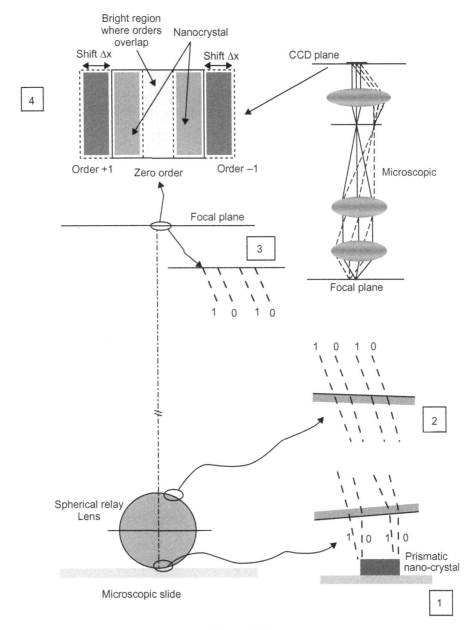

Figure 17.8
Schematic representation of the optical system leading to the formation of lens hologram: (1) Prismatic nano-object of dielectric material; (2) wavefronts entering and emerging from the polystyrene microsphere acting as a relay lens; (3) wavefronts arriving at the focal plane of the spherical lens; (4) wavefronts arriving at the image plane of the CCD. The simulation of the overlapping of orders 0, +1, and −1 in the image plane of the CCD is also shown [10].

The orders 0 and 1 overlap in an area that depends on the process of formation of the image (see Figure 17.8). The order 0 produces an image on the sensor, which is centered at a value x of the horizontal coordinate; let us call this image $S(x)$. The order $+1$ will create a shifted image $S(x - \Delta x)$. The shift implies a change in the optical path between corresponding points of the surface. In the present case, the trajectories of the beams inside the prismatic crystals are straight lines and the resulting phase changes are proportional to the observed image shifts.

The formation of the diffraction pattern was described for the X direction: a similar argument can be made for the Y direction. The diffraction pattern of a prismatic object was obtained. The goal is to extract metrological information from this pattern.

17.7 Procedures to Extract Metrological Information from the Recorded Images

A digital form of the observed objects FT was obtained and it is necessary to develop digital procedures to obtain metrological information on the object. There are a number of ways to process the recorded images. Two direct procedures to obtain metrological information are outlined in this section.

17.7.1 Image Shift

The zero-order image $S(x)$ was defined in Section 17.6 and it was remarked that the order $+1$ creates a shifted image of the particle, $S(x - \Delta x)$. The shift implies a change in optical path between corresponding points of the surface. Since beams inside the prismatic nano-sized object are straight lines, the resulting phase change $\Delta \phi$ is proportional to the observed image shift:

$$\Delta \phi(x, \Delta x) - K_p[S(x) - S(x - \Delta x)] \qquad (17.20)$$

where K_p is a coefficient of proportionality. Equation (17.20) corresponds to a shift of the image of the amount Δx. If the FT of the image is computed numerically, one can apply the shift theorem of the FT. For a function $f(x)$ shifted by the amount Δx, the Fourier spectrum remains the same but the linear term $\omega_{sp}\Delta x$ is added to the phase—ω_{sp} is the angular frequency of the FT. It is necessary to evaluate this phase change. The shift can be measured on the image by determining the number of pixels representing the displacement between corresponding points of the image (see Figure 17.8, Part 4). Through this analysis and by using FT, it is possible to compute the thickness t of the prismatic nano-object.

17.7.2 Change of Optical Path Through the Observed Object

Let us now consider the quasi-monochromatic coherent wave emitted by a nano-sized prismatic object. The actual formation of the image is similar to a typical lens hologram of a phase object illuminated by a phase grating [19]. The FT of the image of the nano-object extended to the complex plane is an analytical function. If the FT is known in a region, then, by analytic continuation, $F(\omega_{sp})$ can be extended to the entire domain. The resolution obtained in this process is determined by the frequency ω_{sp} captured in the image. The image can be reconstructed by a combination of phase retrieval and suitable algorithms. The image can be reconstructed from an $F(\omega_{sp})$ such that $\omega_{sp} < \omega_{sp,max}$, where $\omega_{sp,max}$ is determined by the wavefronts captured by the sensor.

The fringes generated by different diffraction orders that were analyzed in Sections 17.4 and 17.5 experience phase changes through the passage of the nano-object that provide depth information. This type of optical setup to observe phase objects was used in phase hologram interferometry as a variant of the original setup proposed by Burch and Gates [20] and Spencer and Anthony [21]. When the index of refraction in the medium is constant, the rays going through the object are straight lines. If a prismatic object is illuminated with a beam normal to its surface, the optical path s_{op} through the object is given by the integral:

$$s_{op}(x, y) = \int n_i(x, y, z) dz \qquad (17.21)$$

where the direction of propagation of the illuminating beam is the z coordinate and the analyzed plane wavefront is the plane $X-Y$; $n_i(x, y, z)$ is the index of refraction of the object through which light propagates.

The change experienced by the optical path is:

$$\Delta s_{op}(x, y) = \int_0^t [n_i(x, y, z) - n_o] dz \qquad (17.22)$$

where t is the thickness of the medium. By assuming that $n_i(x,y,z) = n_c$ where n_c is the index of refraction of the observed objects, Eq. (17.22) then becomes:

$$\Delta s_{op}(x, y) = (n_c - n_o)t \qquad (17.23)$$

By transforming Eq. (17.23) into phase differences and making $n_o = n_s$, where n_s is the index of refraction of the saline solution containing the observed objects, one can write:

$$\Delta \phi = \frac{2\pi}{p}(n_c - n_s)t \qquad (17.24)$$

where p is the pitch of the fringes present in the image and modulated by the thickness t of the specimen. In general, the change in the optical path is small and no fringes can be observed. In order to solve this problem, carrier fringes can be added. An alternative procedure is to introduce a grating in the illumination path [19]. In the present case, carrier fringes can be obtained from the FT of the lens hologram of analyzed nano-objects. In holography of transparent objects, one can start with recording an image without the transparent object of interest. In a second stage, one can add the object and then superimpose both holograms in order to detect the phase changes introduced by the object of interest. In the present study, reference fringes can be obtained from the background field away from the observed objects. This procedure presupposes that the systems of carrier fringes are present in the field independently of the self-luminous objects. This assumption is verified in the present case since one can observe fringes that are in the background and enter the object experiencing a shift.

17.8 Analysis of the Scales

In order to follow the process of information retrieval, it is necessary to understand the scales involved in the relationship between the object space and the FT space. As stated in Section 17.3, the focused plane is the focal plane of the relay lens. Hence, the images that are within the relay lens are scaled according to FT transform properties of lenses. At the same time, there is a system of fringes produced by the interference of plane wavefronts that occupies a volume and are focused by the microscope. The scale of these fringes is independent of the scale of the FT of the lens. This scale can be found by observing a grating at the same focal plane of the FT. Using this procedure, it was found that the scale corresponding to a 1024 × 1024 pixels image utilized for the FT analysis is $S_c = 29.18$ μm/1024 = 0.0285 μm where 29.18 μm is the size of field of view. The image was then re-pixelated to 2048 × 2048 by utilizing a bicubic spline—the corresponding pixel size is 14.25 nm.

Figure 17.9 shows a portion of the interference fringe pattern corresponding to the maximum order of rectilinear fringes that can be resolved by the sensor according to the Nyquist condition.

Following a classical approach of Fourier optics, it was concluded that the back focal plane of the spherical lens displays the Fraunhofer diffraction pattern of the field that exists in the first principal plane of the relay lens ($z = -R_{\text{sph}}$, Figure 17.3). In the present case, since the lens is in a saline solution, the Fourier components correspond to frequencies given by:

$$f_x = \frac{u}{n_s \lambda f}, \quad f_y = \frac{v}{n_s \lambda f} \tag{17.25}$$

where u and v are the coordinates in the Fourier plane (see Figure 17.3).

Figure 17.9
View of an enlarged portion of a fringe pattern extracted from the FT shown in Figure 17.4 and corresponding to the highest evanescent order.

To get a scale parameter for the FT plane, it is possible to use dimensionless variables and a screen with a known aperture, e.g., a calibrated pinhole. The reduced coordinate system can be used:

$$\begin{cases} \tilde{\xi} = \dfrac{u}{M} \\ \tilde{\eta} = \dfrac{v}{M} \end{cases} \quad (17.26)$$

where M is the magnification of the image.

The reduced coordinate system connects the object space (in this case, the first principal plane of the relay lens, $z = -R_{sph}$, Figure 17.3) with the focal plane of the spherical lens ($z = R_{sph} + f$, Figure 17.3).

The values of parameters involved in the definition of the scale are $n_s = 1.36$, $\lambda = 0.6328$ μm, and $f = 11.214$ μm. Between the object plane and the Fourier plane, there is an inverse relationship; hence, the diffraction pattern of a known pinhole consisting of circular fringes is connected to the diameter of the aperture through the dimensionless variables such that by measuring the minima or maxima of the diffraction pattern it is possible to determine the actual radius of the aperture. This procedure was utilized in the present study to determine the scale to connect the diffraction fringes of the polystyrene microsphere with the actual radius of the sphere. The product $n_s \lambda f$ is equal to $C = 1.36 \times 0.6328 \times 11.214 = 9.651$ μm^2.

If one is dealing with objects that have central symmetry with respect to the center of coordinates, the C factor must be utilized to reduce coordinates in the FT space to coordinates in the object space. In this way, it is possible to convert frequency space information into metrological data. If the object does not have central symmetry, the inverse relationship between the object space and the frequency space must be applied.

17.9 Observation of Nano-Sized Objects

The next step is the verification of the proposed model by carrying out observation of prismatic nano-sized objects and spherical particles. The prismatic objects present in the field of view are different isomers of sodium chloride (NaCl) that precipitate on the surface of the microscopic slide. Isomer dimensions were theoretically computed and experimentally verified in literature [22]. The other observed nano-objects are small polystyrene spheres that were injected into the NaCl solution. Among the large amount of objects present in the region under observation, only a number of them are analyzed.

17.9.1 Prismatic NaCl Nanocrystals

The NaCl nanocrystals are located in the recorded images by observing the image of the lens hologram with increasing numerical zooming. A square region around a selected particle is cropped and the image is digitally re-pixelated to either 512×512 pixels or 1024×1024 pixels to increase the numerical accuracy of the FT of the crystals' images. Image intensities are also normalized from 0 to 255. Figure 17.10 illustrates the case of a square cross-section crystal of size 86 nm. Figure 17.10A is the recorded image; Figure 17.10B and C shows the FT of the image and components that are filtered. Figure 17.10D shows the filtered zero order while Figure 17.10E shows the result of the filter shown in Figure 17.10C. Finally, Figure 17.10F is the addition of Figure 17.10D and E.

In order to further understand the structure of the image, the pattern shown in Figure 17.10F was numerically generated. As mentioned above, this pattern results from the sum of different filtered components (Figure 17.10D and E). Utilizing edge-detection digital techniques, the outlines of the zero order and the first order of diffraction are indicated in Figure 17.10A. In that figure, the scale transformation between the actual image and the FT was introduced, thus directly providing the space information measures. Since the pattern presents asymmetry in the right side, only the shift in the left side was evaluated. Later study of the geometry of the nanocrystal shows that there is a step on the right side of the particle.

A numerical simulation of the image pattern is shown in Figure 17.11B. This pattern was created by summing over the following: (i) a square area of uniform intensity corresponding

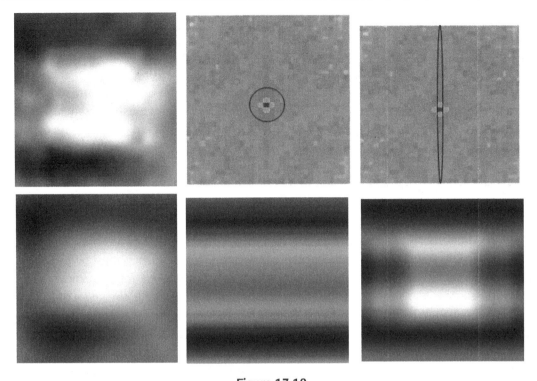

Figure 17.10
(A) Gray-level image recorded for the square cross-section nanocrystal; (B) FT pattern of the nanocrystal and filtering of the 0 order; (C) FT showing filtering of vertical orders; (D) zero-order image; (E) vertical-order image; (F) addition of images (D) and (E).

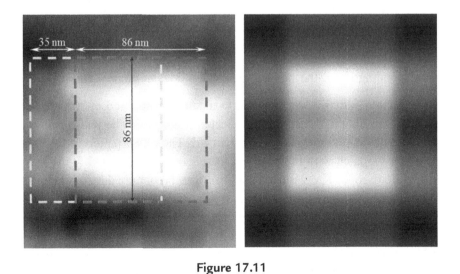

Figure 17.11
(A) Recorded image showing orders 0 and −1 separated by edge detection algorithms; (B) reconstruction of the 0 order, ±1 orders, and sinusoidal fringe pattern.

to the brightest region of the image in Figure 17.10A which is the zero order; (ii) the bright areas shifted in the $\pm X$ direction corresponding to orders ± 1; (iii) the bright areas shifted in the $\pm Y$ direction corresponding to orders ± 1; and (iv) a sinusoidal fringe. The actual image (Figure 17.10A), the sum of the filtered orders (Figure 17.10F), and the numerical simulation (Figure 17.11B) show a close resemblance. The model of image formation outlined in Section 17.6 (Figure 17.8) is fully supported by the previously presented developments. The details of the procedure to relate observed shifts to the thickness of the crystal are provided in Ref. [23].

17.9.2 Determination of Thickness of Nanocrystals Through Evaluation of the Optical Path Change

The procedure to obtain depth information from the recorded images is now described. For each analyzed crystal, a particular frequency is selected. This frequency must be present in the FT of the image and should be such that the necessary operations for frequency separation are feasible. This means that frequencies that depend on the thickness of nanocrystal must not be near the selected frequency. The frequency of interest is individualized in the background of the observed object. Then, the selected frequency is located in the FT of the object. A proper filter size is then selected to pass a number of harmonics around the chosen frequency. Those additional frequencies carry the information on the change in phase produced by the change in the optical path. In the next step, the phases of the modulated and unmodulated carriers are computed. The change of phase is introduced in Eq. (17.24) and the value of t is determined. The phase difference is not constant throughout the prismatic crystal face since it is unlikely that the crystal face is parallel to the image plane of the camera. Therefore, an average thickness is computed. This process was repeated for all prismatic nanocrystals analyzed in the study.

Figure 17.12A shows the recorded image for a prismatic nanocrystal, while Figure 17.12B shows the corresponding FT of the image. The crystal does not have central symmetry.

The "theoretical" phase pattern corresponding to a carrier composed of straight fringes is obtained by taking the filter 1×1 about the selected frequency. The phase pattern is modulated because of the presence of the nanocrystal. The phase pattern thus obtained must be masked in order to match with the edges of the nanocrystal detected in the image recorded by the microscope. The value of thickness can be obtained by averaging the thickness distributions over the whole surface. By filtering the FT pattern of nanocrystal shown in Figure 17.12B, one can obtain fringe patterns corresponding to different spatial frequencies, e.g., 8.3 nm (Figure 17.13A), 5.6 nm (Figure 17.13B), and 5.3 nm (Figure 17.13C), respectively. These pitches are fractions of the pixels of the original figure and hence are fractional orders resulting from the bicubic spline interpolation of the recorded image. To determine the thickness of the nanocrystal, harmonics were selected

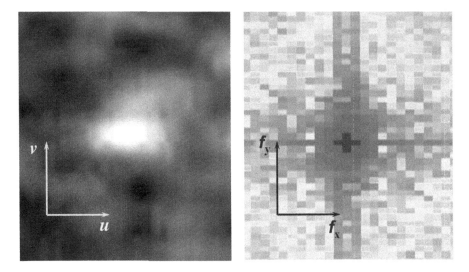

Figure 17.12
(A) Gray-level image recorded for a prismatic nanocrystal without central symmetry;
(B) FT pattern of the image.

along the vertical axis f_y of the FT space (see Figure 17.12B). This is done because the pattern shown in Figure 17.12A actually is the FT of the nanocrystal.

The results of thickness measurements carried out for four different nanocrystals are listed in Table 17.2. The table also shows the average of the thicknesses obtained by the two procedures outlined in Section 17.7 (see Ref. [23] for more details).

Table 17.3 shows the experimentally measured aspect ratios of nanocrystal dimensions compared with the corresponding theoretical values [23]—the average error on aspect ratios is 4.59% while the corresponding standard deviation is ±6.57%.

Table 17.4, on the basis of aspect ratios, shows the theoretical dimensions of the sides of nanocrystals and compares them with the measured values. Thicknesses reported in Table 17.3 are the average values shown in Table 17.2. The average absolute error on dimensions is 3.06 nm, the mean error is −1.39 nm, and the standard deviation of absolute errors is ±3.69 nm. A conservative assumption to estimate the accuracy of measurements is to adopt the smallest dimensions of the crystals as given quantities from which the other dimensions are then estimated. The smallest dimensions are the ones that will have the larger absolute errors.

Another example of the determination of the thickness of a nanocrystal through the change of optical path is the determination of the step present in the cross-square crystal. The theoretical structure $5 \times 4 \times 4$ corresponding to this crystal (side length $L = 86$ nm) has one step in the depth dimension.

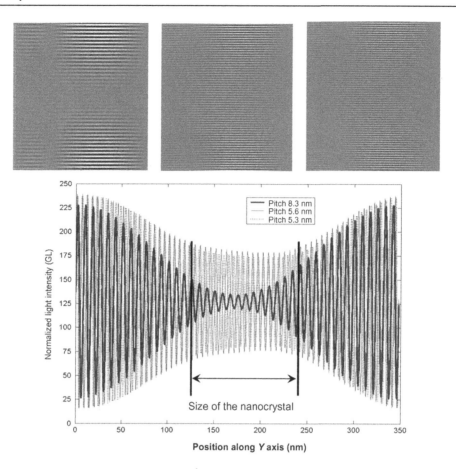

Figure 17.13

Fringe patterns corresponding to higher order frequencies of the FT of the nanocrystal without central symmetry: the spatial frequency of line patterns are 8.3 nm (A), 5.6 nm (B), and 5.3 nm (C); (D) frequency modulation of interference fringes due to the distortion of plane wavefronts produced by the crystal deposition on the upper surface of the microscopic slide [23].

Table 17.2: Determination of Thickness of Analyzed Nanocrystals

Nanocrystal Length (nm)	t_{fs} (nm) Optical Path Change	t_s (nm) Image Shift	$t_{av} = (t_{fs} + t_s)/2$ (nm)	Standard Deviation (nm)
54	58	48	53	±5
55	30	37	33.5	±3.5
86[a]	101	107	104	±3
120	46	46	46	±0

[a]Image size was 1024 × 1024 pixels; size of all other images was 512 × 512 pixels. Since the nanocrystal L = 86 nm has a stepped face, t_{av} corresponds to the average of the face profile shown in Figure 17.15D that gives an average value of 105 nm.

Table 17.3: Aspect Ratio of the Observed Nanocrystals: Experiments versus Theory

Nanocrystal Length (nm)	Experimental Dimensions (nm)	Experimental Aspect Ratio	Theoretical Aspect Ratio
54	72 × 54 × 53	5.43 × 4.08 × 4	5 × 4 × 4
55	55 × 45 × 33.5	4.93 × 4.03 × 3	5 × 5 × 3
86	104 × 86 × 86	4.84 × 4 × 4	5 × 4 × 4
120	120 × 46 × 46	7.83 × 3 × 3	8 × 3 × 3

Table 17.4: Main Dimensions of the Observed Nanocrystals: Experiments versus Theory

Nanocrystal Length (nm)	Dimensions Measured	Dimensions Theoretical	Difference (nm)
54	72	66.3	+5.7
	54	53	+1
	53	53	—
55	55	55.8	−0.8
	45	55.8	−10.8
	33.5	33.5	—
86	104	107.5	−3.5
	86	86	0
	86	86	—
120	120	122.7	−2.7
	46	46	0
	46	46	—

The obtained experimental results are consistent with the theoretical structure (Figure 17.14A). Figure 17.14C shows the level lines of the top face. Figure 17.14D shows a cross-section where each horizontal line corresponds to five elementary cells of NaCl. The crystal is inclined with respect to the camera plane that was corrected by means of an infinitesimal rotation. This allowed the actual thickness jump in the upper face of the crystal (see the theoretical structure in Figure 17.14B) to be obtained. The jump in thickness is 26 nm out of a side length of 86 nm, this corresponds to a ratio of 0.313 which is very close to theory. In fact, the theoretical structure predicted a vertical jump of one atomic distance versus three atomic distances in the transverse direction, i.e., a ratio of 0.333.

17.9.3 Observation of the Polystyrene Nanospheres

Microspheres and nanospheres made up of transparent dielectric media are excellent optical resonators. Unlike the NaCl nanocrystals, whose resonant modes have not been previously studied in the literature, both theoretical and experimental studies on the resonant modes of microspheres and nanospheres can be found in the literature. Of particular interest are the modes localized at the surface, along a thin equatorial ring. These modes are called

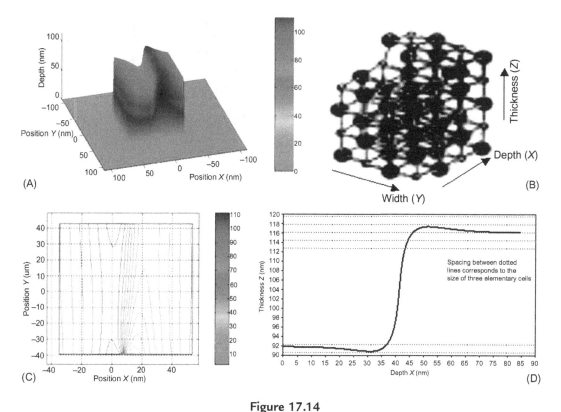

Figure 17.14
(A) Reconstruction of the NaCl nanocrystal of length 86 nm; (B) theoretical structure; (C) level lines; (D) rotated cross-section of the upper face of the nanocrystal: the spacing between dotted lines corresponds to size of three elementary cells [10].

whispering-gallery modes (WGM). WGM result from light confinement due to TIR inside a high-index spherical surface within a lower index medium and from resonance as the light travels a round trip within the cavity with phase matching [24]. The WG modes are within the Mie's family of solutions for resonant modes in light scattering by dielectric spheres. The WG modes can be also derived from Maxwell's equations by imposing adequate boundary conditions [25]. They can also be obtained as solutions of the Schrodinger-like equation in quantum mechanics describing the evolution of a complex angular momentum of a particle in a potential well. The WG modes form standing waves that can be recorded.

In the present case, since the FT of the spheres are available but the direct image is not, one has to remember that, because of spherical symmetry, the FT will also be circular and the WG modes will appear as bright spots since they are standing waves.

Figure 17.15A shows the diffraction pattern of a spherical nanoparticle of diameter 150 nm. This image presents the typical whispering-gallery intensity distribution. Waves are

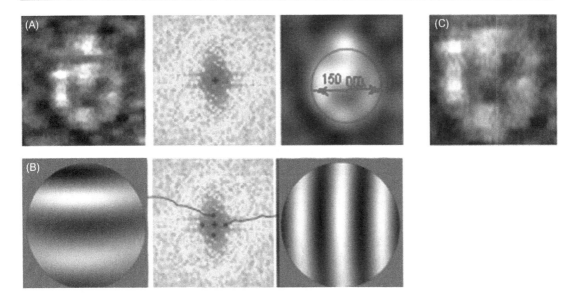

Figure 17.15
Spherical nanoparticle of estimated diameter 150 nm. (A) FT and zero-order filtered pattern; (B) systems of fringes modulated by the particle; (C) color image of the particle [13].

propagating around the diameter in opposite directions thus producing a standing wave with seven nodes and six maxima. The light is trapped inside the particle and there is basically a surface wave that only penetrates a small amount into the radial direction. The signal recorded for this particle is noisier compared with the signal recorded for the prismatic crystals. The noise increase is probably due to the Brownian motion of the spherical particles. While the NaCl nanocrystals seem to grow attached to the supporting surface, the nanospheres are not in the same condition. Of all resonant geometries, a sphere has the capability of storing and confining energy in a small volume.

The method of depth determination utilized for the nanocrystals can also be applied to the nanospheres. While in prismatic bodies made out of plane surfaces the pattern interpretation is straightforward, in the case of curved surfaces, the analysis of the patterns is more complex since light beams experience changes in trajectories determined by the laws of refraction. In the case of a sphere, the analysis of the patterns can be performed in a way similar to what is done in the analysis of the Ronchi test for lens aberrations. Figure 17.15 shows the distortion of a grating of pitch $p = 83.4$ nm as it goes through the nanosphere. The appearance of the observed fringes is similar to that observed in a Ronchi test. The detailed application of this process is not included in this chapter for the sake of brevity.

Figure 17.16 shows a spherical particle of estimated diameter 187 nm, while Figure 17.16B shows the average intensity. Figure 17.16C is taken from Pack [26] and shows the

 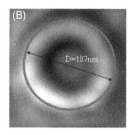 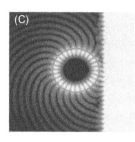 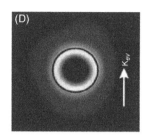

Figure 17.16

Spherical nanoparticle of estimated diameter 187 nm and numerical simulations. (A) Original image; (B) average intensity image; (C) numerical simulation of whispering modes; (D) average intensity of numerical simulation [13].

numerical solution for the WGM of a polystyrene sphere of diameter 1.4 μm, while Figure 17.16D shows the average intensity. There is good agreement between the experimental results and numerical simulation. The electromagnetic resonance occurs at the wavelength $\lambda = 386$ nm which corresponds to UV radiation. The color camera is sensitive to this frequency and Figure 17.15C shows the color picture of the $D = 150$ nm nanosphere; the observed color corresponds approximately to the resonance wavelength mentioned above.

Four spherical particles with radius ranging between 150 and 228 nm were analyzed. For example, Figure 17.17 shows the different stages involved by the determination of the $D = 150$ nm nanoparticle. The FT of the image is shown in Figure 17.17A while Figure 17.17B shows the FT of the real part of the image (see Section 17.1, Eq. (17.8)). Figure 17.17C shows the intensity distribution of the order 0. Different peaks corresponding to fringe systems present in the image can be observed. Figure 17.17D and E shows the cross-sections of the zero order with the corresponding peaks observed in Figure 17.17C. These patterns are similar to those observed in the case of the prismatic nanocrystals [23]. The gray-level intensity decays from 255 to 20 within 75 nm; this quantity corresponds to the radius of the nanosphere.

An alternative way to determine diameter from experimental data is based on the WGM properties that relate the diameter or radius of the nanosphere to the standing waves which in turn are characterized by the number of zero nodes or the number of maxima. These numbers depend on the index of refraction and on the radius of the nanosphere. The numerical solution of the WG mode of polystyrene sphere developed by Pack [26] was utilized in this study as the resonance modes occur approximately at the same wavelength, $\lambda = 386$ nm. Figure 17.18A shows the equatorial wavelength of the WG plotted versus the particle radius. The experimentally measured wavelengths are plotted in a graph that includes the numerically computed value (radius of 700 nm). Very good correlation was obtained showing that the edge detection gradient utilized yields values that are consistent

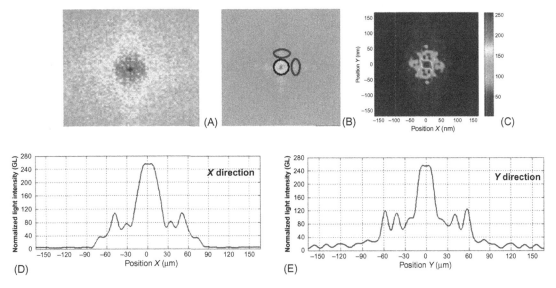

Figure 17.17
(A) FT pattern of the image of the 150 nm polystyrene nanosphere; (B) FT of the real part of the nanosphere; (C) two-dimensional view of the zero-order extracted from the real part of the FT pattern of the nanosphere; (D) cross-section of the zero order in the X direction; (E) cross-section of the zero order in the Y direction [13].

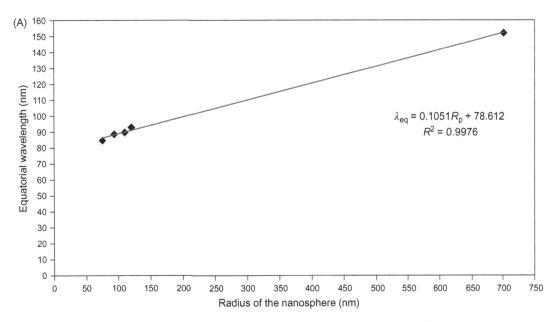

$\lambda_{eq} = 0.1051 R_p + 78.612$
$R^2 = 0.9976$

Figure 17.18
Relationships between nanosphere radius and (A) equatorial wavelength of the WG mode or (B) normalized equatorial wavelength of the WG mode [23].

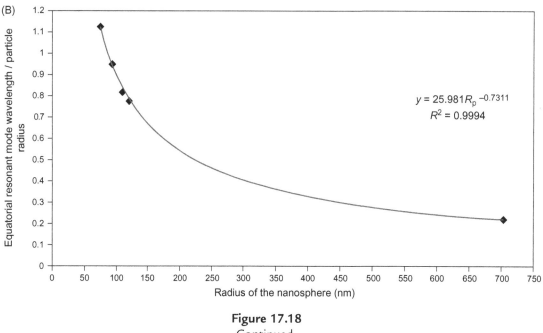

Figure 17.18
Continued.

with the numerically computed WG wavelengths. The graph shown in Figure 17.18B which is the ratio of equatorial wavelength to radius is plotted versus the radius; this seems to provide a correlation with a better sensitivity. Once again the correlation was very good.

17.10 Conclusions

New versions of FTH and FTHI were introduced. The basic idea behind these new versions is the self-illumination process of the observed objects that are at the same time light sources and imaged objects. The self-luminosity of the observed objects is caused by electromagnetic resonance. This effect is similar to the self-luminosity of quantum dot particles, and like in the case of quantum dots, the emitted light frequencies depend on the size of the nano-object (in this particular case, NaCl nanocrystals) as well as on the electronic structure of the object material. The observed frequencies correspond to the band of peak frequencies of NaCl [27]. The presence of a diffracting element in the optical circuit generates a large amount of **k** vectors that provide a wide spectrum of exciting frequencies. This aspect is discussed to some length in Ref. [23] including supporting evidence.

Another basis of the developed methodologies, supported by the experimental evidence of the obtained results, is the fact that the self-luminous wavefronts generate propagating waves that go through the whole process of image formation beyond the restrictions imposed by diffraction-limited optical instruments. This possibility was analyzed in many publications [28–31] where examples are included. The full discussion of this problem is beyond the scope of this chapter. This chapter also included FTHI holographic interferograms of phase objects that were formed and analyzed along the lines utilized in holographic interferometry.

In this chapter, procedures that are connected to numerical super-resolution methods are utilized [32]. These procedures include basic theorems on analytic functions and the numerical generation of additional pixels. The latter, called re-pixelation of the images, is a standard procedure also utilized in numerical super-resolution. Furthermore, the fact that the process of image formation includes different replications of the image due to the different diffraction orders is an additional factor that results in an increase in the information contained in an image. The effect of the successive images depends on the structure of the image plane array of the CCD, i.e., on the so-called "factor of fullness" of the detector array. However, it is equivalent to microscanning, which is to obtain successive frames of displaced images. Finer sampling procedures produce higher accuracies and better modulation transfer functions together with higher Nyquist frequencies [33].

The degree of accuracy in measurements was ascertained on the basis of nanocrystals of sodium chloride. The actual sizes of crystals utilized in the theoretical computations presented in Ref. [22] are of the same order of magnitude of the crystals observed in this chapter. The geometric proportions of the crystals can be obtained from theoretical considerations. The mean error of the measured aspects ratios is 4.6% with a standard deviation of ±6.6%. From the actual sizes of the measured sides, the mean absolute error in the length measurements is on the order of 3 nm and the standard deviation is ±3.7 nm. The measured lengths of the crystals agree very well with the lengths computed from the sodium chloride elementary cell size ($d = 0.573$ nm at room temperature). The overall measurements performed show an overall standard deviation that is of the order of ±5 elementary unit cells.

In this chapter, the problem of super-resolution was approached based on the theory of information and it is possible to further appreciate the power of the Gabor's original idea when he invented holography. The fact that with the holographic method one can measure the amplitude and the phase of the signal leads to results showing that the obtainable resolution depends basically on the atomic structure of the observed objects. In the case of simple objects as the prismatic isomeric nanocrystals of sodium chloride, the average standard deviation (±5) of all of the performed measurements is within the abovementioned quantity.

In a final summary, the possibility of observing near-field events in the far field was experimentally verified for the first time ever. Mathematical models were developed and adopted to explain within the proposed framework the feasibility of such an endeavor. Numerical results support the adopted models, even if it is true that statistical verifications require the laws of large numbers, and in the present case only a limited number of measurements were carried out. However, the results of a limited number of measurements provide a fairly good idea of the observed phenomena as it is well established by the theory of statistical sampling. The present work is a first step in a direction of research that can be very fruitful in the study of phenomena in the nano-size range since it illustrates the possibility of performing nanoscale measurements with the usual procedures of optical microscopy. Until now performing these measurements with such tremendous accuracy was considered impossible.

Many of the derivations included in the chapter are similar to those included in classical holography and holographic interferometry. The only difference with classic holography is that the observations are made beyond the limits of diffraction-limited optical systems, thus opening a whole new world of applications. This will prove very useful in the field of biological sciences where super-resolution is of considerable value. At the present time, there are problems in viewing biological materials with the utilization of electromagnetic radiations of shorter wavelength particularly if in vivo is desired. This restriction has now been removed due to the discoveries made and presented in this chapter.

References

[1] G.W. Stroke, D.G. Falconer, Attainment of high resolutions in wavefront-reconstruction imaging, Phys. Lett. 13 (1964) 306–309.
[2] G.W. Stroke, Lensless Fourier-transform method for optical holography, Appl. Phys. Lett. 6 (1965) 201–203.
[3] D. Gabor, G.W. Stroke, R. Restrick, A. Funkhouser, D. Brumm, Optical image synthesis (complex amplitude addition and subtraction) by holographic Fourier transformation, Phys. Lett. 18 (1965) 116–118.
[4] G.W. Stroke, An Introduction to Coherent Optics and Holography, second ed., Academic Press, New York, 1969. 978-0126739565.
[5] M. Gustafsson, M. Sebesta, B. Bengtsson, S.G. Pettersson, P. Egelberg, T. Lenart, High-resolution digital transmission microscopy—a Fourier holography approach, Opt. Lasers Eng. 41 (2004) 553–563.
[6] L. Granero, V. Micó, Z. Zalevsky, J. García, Superresolution imaging method using phase shifting digital lensless Fourier holography, Opt. Express 17 (2009) 15008–15022.
[7] I. McNulty, J. Kirz, C. Jacobsen, E.H. Anderson, M.R. Howells, D.P. Kern, High-resolution imaging by Fourier transform x-ray holography, Science 256 (1992) 1009–1012.
[8] P. Thibault, M. Dierolf, A. Menzel, O. Bunk, C. David, F. Pfeiffer, High-resolution scanning X-ray diffraction microscopy, Science 321 (2008) 379–382.
[9] G. Toraldo di Francia, La Diffrazione della Luce, Edizioni Scientifiche Einaudi, Torino, Italy, 1958 (In Italian).
[10] C.A. Sciammarella, Experimental mechanics at the nanometric level, Strain 44 (2008) 3–19.
[11] J.D. Jackson, Classical Electrodynamics, third ed., Wiley, New York, 2001. 978-0471309321.
[12] General Stress Optics Inc., Holo-Moiré Strain Analyzer Version 2.0, Chicago. <http://www.stressoptics.com>, 2007.

[13] C.A. Sciammarella, L. Lamberti, F.M. Sciammarella, Optical holography reconstruction of nano-objects, in: J. Rosen (Ed.), Holography, Research and Technologies, INTECH, Vienna, Austria, 2011. 978-9533072272, pp. 191–216 (Chapter 9, Invited book chapter).

[14] R.S. Ayyagari, S. Nair, Scattering of P-polarized evanescent waves by a spherical dielectric particle, J. Opt. Soc. Am. B Opt. Phys. 26 (2009) 2054–2058.

[15] C.A. Sciammarella, L. Lamberti, Observation of fundamental variables of optical techniques in the nanometric range, in: E.E. Gdoutos (Ed.), Experimental Analysis of Nano and Engineering Materials and Structures, Springer, The Netherlands, 2007. 978-1402062384.

[16] C. Guillemet, L'interférométrie à Ondes Multiples Appliquée à Détermination de la Répartition de L'indice de Réfraction dans un Milieu Stratifié, Ph.D. Dissertation, University of Paris, Paris, France, 1970 (In French).

[17] C.A. Sciammarella, L. Lamberti, F.M. Sciammarella, Light generation at the nano-scale, key to interferometry at the nano-scale, in: T. Proulx (Ed.), Conference Proceedings of the Society for Experimental Mechanics Series; Experimental and Applied Mechanics, vol. 6, Springer, New York, 2010, pp. 103–115. ISSN: 2191-5644

[18] J.W. Goodman, Introduction to Fourier Optics, third ed., Roberts & Co. Publishers, Englewood, NJ, 2004, ISBN: 0-974707724.

[19] L.H. Tanner, The scope and limitations of three-dimensional holography of phase objects, J. Sci. Instrum. 7 (1974) 774–776.

[20] J.W. Burch, C. Gates, R.G.N. Hall, L.H. Tanner, Holography with a scatter-plate as a beam splitter and a pulsed ruby laser as light source, Nature 212 (1966) 1347–1348.

[21] R.C. Spencer, S.A. Anthony, Real time holographic moiré patterns for flow visualization, Appl. Opt. 7 (1968) 561.

[22] R.R. Hudgins, P. Dugourd, J.N. Tenenbaum, M.F. Jarrold, Structural transitions of sodium nanocrystals, Phys. Rev. Lett. 78 (1997) 4213–4216.

[23] C.A. Sciammarella, L. Lamberti, F.M. Sciammarella, The equivalent of Fourier holography at the nanoscale, Exp. Mech. 49 (2009) 747–773.

[24] B.R. Johnson, Theory of morphology-dependent resonances—shape resonances and width formulas, J. Opt. Soc. Am. A Opt. Image Sci. Vis. 10 (1993) 343–352.

[25] C.F. Bohren, D.R. Huffman, Absorption and Scattering of Light by Small Particles, Wiley, New York, 1998. 978-0471293408.

[26] A. Pack, Current Topics in Nano-Optics, Ph.D. Dissertation, Chemnitz Technical University, Chemnitz, Germany, 2001.

[27] D.M. Roessler, W.C. Walker, Electronic Spectra of Crystalline NaCl and KCl, *NASA-R* 88200, 1988.

[28] J. Durnin, J.J. Miceley, J.H. Eberli, Diffraction free beams, Phys. Rev. Lett. 58 (1987) 1499–1501.

[29] Z. Bouchal, Non diffracting optical beams: physical properties, experiments, and applications, Czechoslovak J. Phys. 53 (2003) 537–578.

[30] R.I. Hernandez-Aranda, M. Guizar-Sicairos, M.A Bandres, Propagation of generalized vector Helmholtz-Gauss beams through paraxial optical systems, Opt. Express 14 (2006) 8974–8988.

[31] J.C. Gutiérrez-Vega, M.D. Iturbe-Castillo, G.A. Ramirez, E. Tepichin, R.M. Rodriguez-Dagnino, S. Chávez-Cerda, et al., Experimental demonstration of optical Mathieu beams, Opt. Commun. 195 (2001) 35–40.

[32] S. Borman, R. Stevenson, Spatial Resolution Enhancement of Low Resolution Image Sequences, A Comprehensive Review with Directions for Future Research, Laboratory for Image and Signal Analysis, IM 46556, University of Notre Dame, IN, 1998.

[33] G.D. Boreman, Modulation Transfer Function in Optical and Electro-Optical Systems, SPIE Press, Bellingham, 2001.

CHAPTER 18

Fluorescence Phase Microscopy (FPM) and Nanoscopy

Alberto Bilenca[1], Brett Bouma[2,3], Guillermo Tearney[2,4], Iwan Märki[5], Noelia Bocchio[5], Stefan Geissbuehler[5], Theo Lasser[5]

[1]*Department of Biomedical Engineering, Ben Gurion University of the Negev, Be'er-Sheva, Israel*
[2]*Harvard Medical School and Wellman Center for Photomedicine, Massachusetts General Hospital, Boston, MA* [3]*The Harvard-MIT Division of Health Sciences and Technology, Cambridge, MA*
[4]*Department of Pathology, Massachusetts General Hospital, Boston, MA*
[5]*Laboratoire d'Optique Biomédicale, École Polytechnique Fédérale de Lausanne, Lausanne, Switzerland*

Editor: Natan T. Shaked

18.1 Introduction

Fluorescence imaging at the nanoscale and mesoscale level is a rapidly evolving research field that provides a unique set of technological tools for tackling quantitatively measurement problems in the natural and the life sciences with high specificity and at various spatial resolution scales ranging from $1-10$ μm (mesoscopic resolution) to smaller than ~ 100 nm (nanoscopic resolution) in all three dimensions (3D). In general, state-of-the-art 3D optical nanoscopy offers sub-100 nm spatial resolution in all 3D and penetration depth of several micrometers [1–11], whereas state-of-the-art 3D fluorescence microscopy provides a 3D optical resolution of several micrometers along an extended penetration depth of up to a few millimeters [12–17]. Typically, 3D fluorescence imaging techniques aim at discriminating fluorescence photons emerging from one given point within the specimen against most of other fluorescent light. To accomplish this task, various fluorescence-discrimination mechanisms have been employed. In 3D nanoscopy, these mechanisms include (i) one or two photon activatable fluorescence that enables lateral centroid-based localization of individual fluorophores with nanometer precision [1–5,7–10]. Nanometer-scale localization of the fluorophores in the axial (depth) dimension can be achieved by interferometry [2,4] or by detecting an aberrated point-spread function (PSF) of single fluorophores [1,3,5,9,10], (ii) stimulated emission depletion of fluorescence for engineering of 3D PSFs that are narrowed down to subdiffraction dimensions [6], and (iii) structured illumination to obtain a 3D optical resolution of ~100 nm [11]. In 3D microscopy with mesoscopic resolution, fluorescence-discrimination mechanisms involve selective axial (depth) illumination and off-axis collection of light to improve the background suppression of fluorescence in 3D extended specimens [12,13,15,16] and multiple-view imaging that allows for 3D image reconstruction of the sample by using backprojection processing techniques [12,17] or mathematical models of photon propagation in turbid media together with application of inversion theory [12,14]. Common to most existing and proposed fluorescence imaging methods is their reliance on the detection of the fluorescence intensity (rather than the fluorescent field) to provide axial (depth) information about the specimen. The reason for these intensity-based approaches is simple: photodetectors can extract only the magnitude of the incident optical field. As we discuss later, however, the phase of the fluorescent wave can also encode information, which would be useful for a broad spectrum of application in the natural and life sciences.

In this chapter, we introduce the concept of fluorescence interferometry and present its usefulness in converting phase variations of fluorescent light into amplitude variations. As a result, this amplitude-to-phase conversion offers a new form of fluorescence imaging that encodes axial (depth) information about the sample in the phase of the fluorescent waves and yields a new paradigm of 3D imaging at the nanoscale and mesoscale levels. Here, we demonstrate that the treatment of fluorescent waves as low-temporal coherence optical fields and their manipulation by self-referencing interferometry is useful for metrology and imaging at the mesoscopic to nanoscopic resolution scales. Interestingly, manipulation of low-temporal coherence optical fields is largely employed in imaging through tissue and cells and has had a profound impact on the field of biomedicine. Examples include optical coherence tomography (OCT; [18–23]) that uses coherence gating to simultaneously provide micron-scale optical sectioning along penetration depths of a few millimeters and quantitative phase microscopy [24–28] that enables measurements with nanometer-level accuracy. We point out that the phenomenon of fluorescence interference has been investigated in fundamental studies of molecules in front of reflecting surfaces [29,30], applied to the assessment of nanometer-level displacements of a fluorescent molecule above a reflector [31–33], employed in 4Pi and I5M microscopy for achieving subdiffraction-limited resolution along the axial (depth) dimension [34–37], and also used in spectroscopy [38].

This chapter is organized as follows. We first describe the process of fluorescence self-interference and its potential use for providing new measurement capabilities, such as fluorescence tomography with mesoscopic resolution and nanometer-level localization of individual fluorescent markers in all 3D. Next, we present time-domain (TD) and spectral-domain (SD) experimental realizations of fluorescence phase microscopy (FPM) and demonstrate their application to optical tomography with mesoscopic resolution along an extended penetration depth (hundreds of micrometers), as well as their use in 3D localization of single fluorescent quantum dots (QDs) with nanometer-scale precision. Finally, we provide a summary and outlook.

18.2 The Fluorescence Self-Interference Process

Single fluorescent emitters such as fluorescent proteins and QDs are intrinsic quantum sources. As a result, a single fluorescent photon can provide both a phase-modulated signal and a local reference beam that are combined interferometrically to convert phase information into intensity information. Specifically, an emitted fluorescent photon can simultaneously travel two different optical paths, thereby generating two beams that are subsequently combined to produce an interference pattern. The relative phase (or path-length difference) between the two beams encodes the axial (depth) position of the emitter and can be directly retrieved from the detected self-interference pattern (fringe).

Figure 18.1 shows a single-photon fluorescence self-referencing interferometer. Consider a fluorescent point source placed between two opposing lenses such that the emitted fluorescent light is collected from both sides of the source and then directed to a beam splitter using mirrors where the clockwise and counterclockwise fluorescence fields are

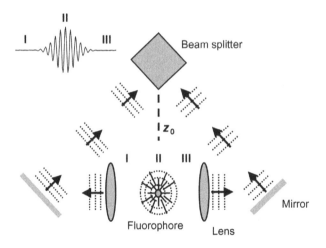

Figure 18.1

The fluorescence self-interference process. Fluorescent light waves emitted from an excited fluorophore located between two matched, opposing lenses are directed using mirrors to a beam splitter where they recombine. An interference pattern appears when the fluorophore is close to point II—near the zero differential path-length point (z_0) of the self-referencing interferometer. For positions far from z_0 (points I and III) only the constant fluorescence intensity is recorded. Source: *This figure is reproduced from figure 2 of Ref. [40] with permission of John Wiley & Sons Ltd.*

combined. A fluorescence interference pattern is detected only when the optical path-length difference of the clockwise and counterclockwise fluorescent beams is equal to or less than the coherence length of the fluorophore, which is typically on the order of a few micrometers. For a fluorescent emitter located far from the zero differential optical path-length point (z_0) of the self-referencing interferometer (positions I and III), no interference pattern is detected and the received signal is proportional to the emission intensity of the fluorophore. As the fluorescent source comes close to z_0 (position II) to within its coherence length, a clear interference pattern appears. The z position of a point source z (relative to z_0) can be extracted from the self-interference pattern with two distinct precision scales by using white-light interferometry. For example, a mesoscopic precision scale (1–10 µm) can be achieved by spectrally resolved interferometry [39–41] or by axially scanning the fluorescent point source through the depth of field (which is kept greater than the coherence length of the source or the z dimension of the object under observation) followed by the identification of the central fringe position [39–41], whereas a nanoscopic precision scale (<100 nm) can be obtained by phase-shifting [2,4] or spectral interferometric techniques [32,33,40,42].

The localization of fluorophores with mesoscopic or nanoscopic precision scales has important implications for fluorescence phase imaging. First, small fluorescent markers (i.e., those having width smaller than one half of the fluorophore's center emission wavelength) which are located at different axial positions and separated by greater than one half of the fluorophore's coherence length along the same axial line (depth) can be distinguished unmistakably because an interference pattern is detected only for markers residing at $|z - z_0| < l_c/2$, where z is the axial (depth) position of the fluorophore and l_c is its coherence length. As a result, this capability opens up the new possibility for depth-resolved optical imaging of thick specimens with a large confocal length at the mesoscopic scale. Secondly, phase-shifting analysis or spectral interferometry can yield the capability for localizing individual fluorescent probes along the axial (depth) dimension with nanometer-level accuracy—an important task for accomplishing fluorescence imaging with nanoscopic resolution in all 3D.

Section 18.3 deals with various experimental designs of FPM, such as setups employing time-resolved interferometry and spectrally resolved interferometry, and describes in detail their capabilities and limitations.

18.3 Experimental Setups of FPM

The design and experimental realization of a fluorescence phase microscope require the combination of a self-referencing interferometer (such as the two-opposing-lenses interferometer employed in 4Pi and I5M microscopy systems [34–37]) with appropriate excitation optics and a TD or SD detection module. In this section, we will describe in detail the experimental implementations of FPM in both TD and SDs.

18.3.1 Time-Domain Fluorescence Phase Microscopy

A TD-FPM system is shown in Figure 18.2. The detector output in the case of a single fluorescent emitter is related to:

$$I(x, y, z) = I_0(x, y)\{1 + |\gamma(2z)|\cos(2k_0 z)\} \tag{18.1}$$

Here, k_0 is the central emission wavenumber of the source, $I_0(x,y)$ is the transversal image of the source, $\gamma(\cdot)$ is the absolute value of the normalized source autocorrelation function, and z is the position of the source (relative to the zero differential optical path-length point z_0 of the self-referencing interferometer). The axial (depth) position of the source (z) can be retrieved by axially scanning the source through the depth of field (which is kept greater than the coherence length of the source or the z dimension of the object under observation) followed by the identification of the central fringe position. As a result, the axial (depth) localization precision is determined by the coherence length of the source (which is on the order of a few micrometers), where only fluorescent markers for which the coherence condition is satisfied produce self-interference patterns. Alternatively, TD-FPM can employ phase-shifting methods to extract the amplitude of self-interference patterns at various z positions of the fluorescent source and reconstruct the source autocorrelation function. The profile formed by these amplitudes determines the mesoscopic axial PSF—and hence the optical depth sectioning capabilities—of the FPM system. In addition, TD-FPM can make use of phase-shifting techniques to retrieve the phase maps of $I(x,y,z)$ in Eq. (18.1)

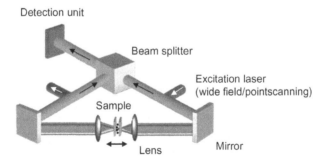

Figure 18.2

Experimental setup of TD-FPM. Fluorescence waves emitted from a fluorescent sample positioned in a self-referencing interferometer comprising two matched opposing lenses are directed by mirrors to a beam splitter where they recombine. The sample can be excited by either a wide-field or a point-scanning illumination and is scanned to generate the interference pattern on the detection unit (e.g., a camera). Source: *This figure is reproduced from figure 3(a) of Ref. [40] with permission of John Wiley & Sons Ltd.*

with high precision, thereby enabling the precise localization of the source along the z axis. In general, the axial localization precision is determined by the detected signal-to-noise ratio (SNR) and roughly increases inversely with its square root [4]. Finally, it is worth mentioning that FPM based on phase-shifting methods has recently showed localization of single fluorescent markers at <10 nm precision in 3D, thereby demonstrating the power of FPM for 3D optical nanoscopy [2,4].

The TD-FPM system consists of three main components: (i) excitation optics, (ii) a self-referencing interferometer, and (iii) a detection module. The excitation optics comprises a wide-field or point-scanning illumination depending on the application. Next, the self-referencing interferometer has two opposing matched lenses, which collect fluorescent light from both sides of the specimen, and two mirrors that direct the light to a beam splitter where the fluorescent fields are recombined. Low-numerical-aperture lenses are employed when a large depth of field is required and a moderate transversal resolution is sufficient. Moderately high-numerical-aperture lenses may be used for providing improved optical sectioning capabilities without scarifying transversal resolution but at the expense of the confocal length. Finally, the detection module comprises imaging and focusing optics as well as a photodetector, for example, a two-dimensional (2D) charge-coupled device (CCD) camera or a point detector, depending on the application.

It is important to point out that 4Pi and I5M microscopy systems combine the fluorescence self-interference process with high-numerical-aperture microscope objectives to achieve high axial resolution of approximately 100 nm [34–37], whereas TD-FPM uses the resolving power of optical path-length difference measurements to image fluorophores with mesoscopic resolution for applications in depth-resolved optical imaging of thick specimens or to precisely localize fluorescent emitters for applications in optical nanoscopy. We also note that in contrast to 4Pi and I5M microscopy systems, TD-FPM requires the translation of the sample along the optical axis and the demodulation of the recorded data to obtain the depth information from the sample.

18.3.2 Spectral-Domain Fluorescence Phase Microscopy

An alternative setup of FPM can be realized in the spectral (Fourier) domain by using spectrally resolved interferometry. An SD-FPM system is shown in Figure 18.3.
The detector output in the case of a single fluorescent point source is expressed using the Wiener–Khintchine theorem [43] as:

$$I(x,y,k) = I_0(x,y)S(k)\{1 + \cos(2kz)\} \quad (18.2)$$

Here, k denotes the wavenumber of the source, $I_0(x,y)$ is the transversal image of the source, $S(k)$ is the power spectral density (PSD) of the source which is closely related to the Fourier transform of $\gamma(\cdot)$, and z is the position of the source (relative to the zero

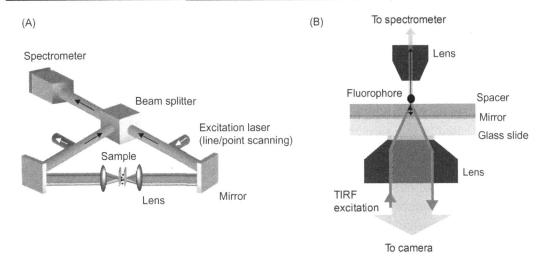

Figure 18.3
Experimental realizations of SD-FPM. (A) Line-focus or point-scanning illumination and a spectrometer are employed to spectrally resolve fluorescence self-interferences from the sample. The axial (depth) profile of the fluorophore distribution at each transversal point is obtained by performing the inverse discrete Fourier transform. (B) A TIRF-based SD-FPM employing a common-path-folded interferometer. The lateral localization is obtained by analyzing the diffraction-limited image of individual emitters, whereas the axial localization is retrieved through Fourier analysis. Source: *(A) This figure is reproduced from figure 1 of Ref. [41] with permission of the Optical Society of America.*

differential optical path-length point z_0 of the self-referencing interferometer). As seen in Eq. (18.2), the location of a fluorophore along a particular axial line is encoded by the interferometric frequency modulation of the emission spectrum, where the frequency is proportional to the fluorophore's distance from z_0. The axial (depth) position of the source (z) can be retrieved with mesoscopic accuracy by analyzing the periodicity of the self-interference-induced modulation of the source PSD ($S(k)\cos(2kz)$) in the spectral or spatial domains. The spatial analysis can be executed by identifying the z position that corresponds to the peak value in the modulus of the Fourier transform of Eq. (18.2). Similarly to TD-FPM, the z position of the source (z) can be extracted with nanometer precision. This ability is achieved either by examining the differential phase shifts of the modulated spectrum $S(k)\cos(2kz)$ or by analyzing the minute axial shifts of the source autocorrelation function $\gamma(\cdot)$ following the Fourier transform of Eq. (18.2). As in TD-FPM, the axial localization precision is determined by the detected SNR and approximately increases inversely with its square root.

The main components of the SD-FPM system are (i) excitation optics, (ii) a self-referencing interferometer, and (iii) a 1D/2D spectrometer. Typically, the excitation optics comprises a scanning point/line-focus or a wide-field illumination depending on the application. While a line-focus illumination allows one to excite a single plane within thick sample and as a result parallelizes the image-collection process, a point focus excites only a single axial line in the specimen. On the other hand, for moderate excitation power levels, a point-scanning illumination enables confocal detection that further improves the rejection of scattered fluorescent light from undesired points within the sample. We note that scanning of the illumination can be avoided by using 3D imaging spectrometers such as a computed-tomography-based imaging spectrometer. The self-referencing interferometer may consist of a two-opposing-lenses interferometer or a common-path-folded interferometer employing a reflecting surface depending on the application. Low-aperture lenses/mirrors are preferred when a large depth of field is required and moderate transversal resolutions are sufficient, whereas higher lens/mirror apertures can be used for obtaining a higher transversal resolution with a smaller confocal length. Importantly, phase-sensitive fluorescence measurements by planar reflectors [33] and two-opposing-lenses interferometers [2,4,32] have been performed with nanometer-level sensitivity. Advantages of the planar reflectors include simple alignment and high sensitivity due to the common-path characteristic of the interference process, though with limited transversal resolution. To circumvent the transversal resolution problem, the two-opposing-lenses interferometers have been employed. However, experiments with these interferometers require the implementation of careful and complex alignment procedures.

18.3.3 Time Domain/Spectral Domain-FPM—Capabilities and Limitations

TD-FPM and SD-FPM can be considered as a hypothetical optical low-coherence interferometric system [18–23,26,28] in which the sample under observation also acts as a spatially incoherent source with low-temporal coherence. Since the light source in FPM is spatially incoherent, coherent crosstalk that degrades measurement quality in low-coherence interferometry (LCI) is suppressed in FPM. However, unlike LCI, FPM suffers from the relatively limited light-collection efficiency (dictated by the aperture of the objective lenses) because it employs self-referencing interferometry where the reference signal results from fluorophores in the sample itself and not from a separate strong reference signal as in LCI. Therefore, the heterodyne gain commonly employed by LCI to place the detection system in the shot-noise limited regime cannot be utilized by FPM, which consequently requires the use of low-noise, high-sensitivity CCD cameras and relatively bright fluorescent tags. Moreover, the absence of heterodyne gain results in an SNR curve that monotonically increases with fluorescence power, in fundamental contrast to SNR curves of LCI that achieves a global maximum at a particular reference power level.

Unlike TD-FPM, SD-FPM can acquire the entire fluorescence profile along a specific depth without any mirror or sample scanning. Moreover, when operating in the shot-noise- or intensity-noise-limited detection regime, SD-FPM should theoretically provide an increased

SNR compared to that obtained using the TD scheme due to noise decorrelation in the Fourier domain. This improved detection sensitivity may be beneficial for the localization of dim fluorophores such as fluorescent proteins. An additional important aspect of TD-FPM and SD-FPM is related to artifacts due to the sample motion during data acquisition. While in TD-FPM the sample motion at a given time will affect only the particular sampling volume that is being acquired, the effect of motion in SD-FPM is likely to be more severe and complex because the signal is integrated over time and is obtained by the Fourier transform integral. In general, the magnitude of motion artifacts in SD-FPM will be governed by the total axial or transverse displacement during a single axial line signal acquisition time. Finally, for a given sample motion, as the temporal resolution improves, the axial and transverse displacements are decreased and so are the motion artifacts in both TD-FPM and SD-FPM.

The detection sensitivity of FPM is determined by the distribution of the excited fluorophores. In general, the detection sensitivity is degraded for a continuous distribution of fluorophores that extends along the axial dimension over a large range ($\sim\lambda/2$) because fringes produced by the fluorophores are linearly combined. Therefore, the use of FPM for imaging and ranging applications is mostly appropriate when using discrete distributions of fluorophores. Interestingly, this characteristic of FPM is similar to that of LCI; the latter is sensitive to well-defined specularly reflecting interfaces (produced by discontinuities in the scattering potential). Importantly, when TD-FPM and SD-FPM systems operate in the shot-noise- or intensity-noise- detection limited regime, a bright fluorophore at any point inside the sample will increase the noise floor, thereby making it difficult to detect weaker fluorophores located along the same axial line. To circumvent these limitations, FPM could be potentially combined with fluorescence lifetime imaging techniques [44]. Similarly to spectrally resolved interferometry, the phase ambiguity that occurs for fluorophores located at positive and negative distances from the zero differential optical path-length point of the interferometer (z_0) can be readily resolved by recording the complex spectral density, thereby doubling the maximal imaging depth [45].

18.4 Applications of FPM

As mentioned in the previous sections, the application of FPM is particularly attractive for (i) optical sectioning imaging with mesoscopic resolution and (ii) metrology with nanometer-level axial localization precision. To investigate the optical sectioning ability of FPM, we implemented FPM in both SD and TDs. The SD-FPM setup used either a wide-field excitation and a folded dielectric-mirror-based interferometer or a scanning line-focus illumination and a two-opposing-lenses interferometer, depending on the application. The TD-FPM system employed a wide-field illumination and a two-opposing-lenses interferometer.

18.4.1 Experimental Characterization of the Fluorescence Phase Microscopes

The Fluorescent Axial (Depth) PSF

The full-width at half-maximum (FWHM) of the axial PSF of the FPM systems is an adequate measure for the axial accuracy with which individual fluorophores or a dense aggregate of fluorophores can be determined. The axial PSF FWHM can be obtained by estimating the extent of the amplitude profile of the self-interference signal. To characterize the axial PSF of the FPM systems, we used a single thin fluorescent layer made by drying fluorescent nanoparticles on a coverslip and measured the spectrum or autocorrelation function of the systems. Sharp axial PSFs with FWHM of approximately 3 μm were reconstructed by the SD-FPM setup shown in Figure 18.3A at shot-noise limited SNR levels of 15–30 dB as shown in Figure 18.4 [39,41]. Similar FWHM values were also obtained for the TD configuration [39]. A first-order approximation to the extent of a Gaussian axial PSF is given by $2 \ln 2 / \pi \lambda_0^2 / \Delta\lambda$ where λ_0 and $\Delta\lambda$ are the peak wavelength and bandwidth of the fluorescence emission spectrum, respectively. Under the Gaussian PSF profile assumption, common fluorescent markers, such as cyan fluorescent protein, 4′,6-diamidino-2-phenylindole, and dimethylsulfoxide, would provide theoretical axial resolution levels of 1.8 and 0.9 μm, respectively. We note, however, that non-Gaussian profiles will result in the presence of sidelobes in the axial PSF that generate spurious structures in the acquired images and mask weak fluorophores located near a strong fluorophore. In addition, we point out that in contrast to the TD-FPM configuration, the SD-FPM system resulted in a lower SNR level when the fluorescent layer was positioned farther away from the zero differential optical path-length point of the interferometer (z_0). The reason for this effect is the finite resolution of the spectrometer that averages the

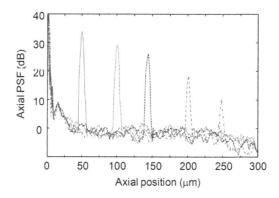

Figure 18.4
Axial PSFs of a single layer of 100 nm fluorescent beads versus axial position of the layer. Sharp axial PSFs with FWHM on the order of a few micrometers are clearly observed. Source: *This figure is reproduced from figure 4(a) of Ref. [41] with permission of the Optical Society of America.*

spectral fringes. The severeness of this effect increases for larger oscillation periods of the spectral fringe, i.e., for increased distances between the fluorescent layer and the z_0.

Axial Localization Precision of Individual Fluorophores

As discussed in the previous sections, FPM can localize individual fluorophores with nanometer-level accuracy in all 3D. The lateral position can be retrieved by adequately analyzing the acquired diffraction-limited PSF image. In general, lateral localization can be performed by fitting a 2D Gaussian to the image of individual fluorophores. However, to incorporate the effect of the dielectric mirror in the total-internal-reflection-fluorescence (TIRF)-based FPM system on the recorded PSF images, we developed a novel PSF model for fluorescent dipoles, which expresses the emission pattern as a superposition of three orthogonal dipoles with different radiation weights (see Ref. [39] for more details). Using this approach, we achieved a lateral localization precision below 10 nm [42]. The axial localization can be obtained by detecting the phase of the self-interference signal. To measure the 3D localization precision of FPM, we dried a low concentration of fluorescent QDs on a layered custom-made slide comprising a glass slip, dielectric mirror, and SiO_2 spacer [39,42]. The image and the self-interference spectrum of a single QD were acquired by a homemade objective-type TIRF microscope combined with a spectrometer as shown in Figure 18.3B. Figure 18.5 (top-left panel) presents a conventional measured image of a single QD. The resulting diffraction-limited PSF of the single QD was next analyzed to evaluate its centroid and consequently its lateral position with high precision. A typical cluster of multiple lateral position determinations (referred as localizations) resulting from repetitive localization of the QD is shown in the top-right panel of Figure 18.5. The standard deviation of this cluster was computed to be ~8 nm suggesting a lateral resolution below 10 nm in FWHM. The bottom-left panel in Figure 18.5 shows the measured (solid line) and fitted (dashed line) self-interference spectrum of a single QD. The spectral measurements lasted for a few seconds to a minute depending on the resolution and the lens aperture of the spectrometer. The resulting spectral interference pattern together with the a priori knowledge about the spacer thickness were next Fourier analyzed to yield an axial localization precision below 10 nm as shown in the bottom-right panel of Figure 18.5. These 3D localization accuracy levels are in accordance with other methods employing fluorescence self-interference for the development of novel optical imaging systems with a high resolution in all 3D [2,4].

18.4.2 Optical Sectioning Imaging with Mesoscopic Resolution by FPM

One of the unique abilities of FPM is to acquire images deep inside the sample and across a wide field of view with a conventional collinear excitation and detection geometry at mesoscopic resolution. To demonstrate this capability of FPM, we first used a calibrated sample comprising a dual-layered fluorescent sample and recorded the tomogram of the

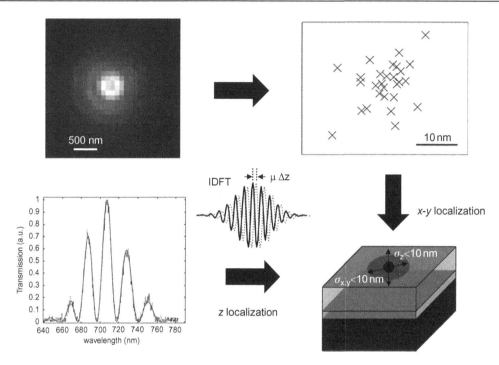

Figure 18.5
3D localization of a single fluorescent QD. A typical fluorescence image of a single QD is shown in the top-left panel. Repetitive lateral localizations of the single QD are retrieved by analyzing the recorded PSF as depicted in the top-right panel. Typical measured (dashed line) and fitted (solid line) spectra of an individual QD are presented in the bottom-right panel. The axial localization of the QD is obtained by Fourier analysis of the fitted spectra. As a result, an individual QD can be localized in all three dimensions with a precision below 10 nm as illustrated in the bottom-right panel. Source: *This figure is partially reproduced from figures 2(a) and 3(a) of Ref. [42] with permission of the Optical Society of America.*

sample (Figure 18.6A) by the SD-FPM of Figure 18.3A [41]. To reduce the noise floor fluctuations, these measurements were performed by averaging several consecutive images. The two layers can be clearly observed over a wide transverse field (>1 mm). The mean distance between the layers was measured to be 120.1 μm and closely matched the calibrated 120 μm separation of the layers. Importantly, the acquisition of the tomogram by the SD-FPM system was performed without any scanning of the sample or focus—a key advantage of SD-FPM which does not require any moving components to resolve axial (depth) information. We carried out similar measurements by using the wide-field TD-FPM setup of Figure 18.2 and reconstructed a 3D image of the sample as presented in Figure 18.6B. We note that in contrast to SD-FPM the wide-field TD-FPM system requires axial scanning of the sample to detect axial (depth) information; yet, no transversal scanning is needed to reconstruct a 3D image of the specimen.

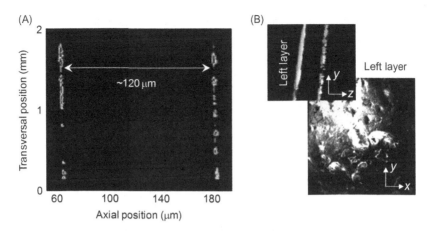

Figure 18.6
(A) A scanningless SD-FPM measurement of a tomogram of a dual-layer fluorescent sample.
(B) Two views of the dual-layer fluorescent sample as captured by the TD-FPM setup. Both lateral and depth distributions of the fluorophores can be imaged effectively. Source: *(A) This figure is reproduced from figure 6(c) of Ref. [41] with permission of the Optical Society of America.*

Next, SD-FPM and TD-FPM were used to image fixed biological samples consisting of the *Drosophila* embryonic nervous system labeled with immunofluorescence [39]. As shown in Figure 18.7A, the SD-FPM modality outlined the $y-z$ profile of the nervous system (top panel) and assessed the fluorescence gradients (bottom panel) inside the *Drosophila* embryo in a single shot of a few seconds without scanning. This capability might be useful in large-scale analyses, such as those involved in the formation of morphogenetic protein gradients, where imaging with mesoscopic resolution is sufficient and the superabundance of image data resulting from imaging at submicroscopic resolution is to be avoided [46]. Furthermore, 3D images of the nervous system of the *Drosophila* embryo were reconstructed by the wide-field TD-FPM system (Figure 18.2) as shown in Figure 18.7B. Finally, to further investigate the capabilities of TD-FPM for wide-field fluorescence imaging, we imaged fluorescently labeled cells (FluoCells, Invitrogen) with a conventional epifluorescence microscope and with the wide-field TD-FPM system as shown in Figure 18.7C. The two modalities employed identical low-aperture objectives. It is clearly observed that the coherence gate offered by the TD-FPM system assisted in the rejection of out-of-focus fluorescence and yielded sharper image details even with low-aperture objectives.

Metrology with Nanometer-Level Axial Localization Precision

To demonstrate the ability of FPM to perform measurements at the nanometer scale, we characterized the polymer swelling process in nanometer-thick polymer bilayers prior to and following exposure to water [42]. This characterization was carried out by the

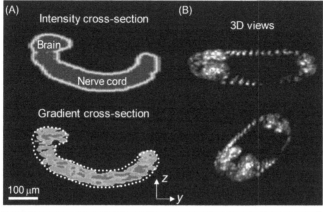
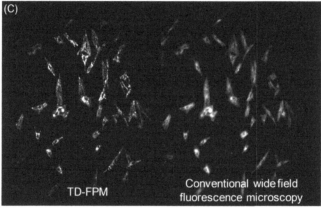

Figure 18.7
(A) A single shot depth profile of the fluorescently labeled embryonic nervous system of *Drosophila* measured by SD-FPM. Intensity and gradient images are shown. (B) 3D views of the fluorescently labeled nervous system of the *Drosophila* embryo imaged by the TD-FPM system. (C) Wide-field images of fluorescently labeled endothelial cells captured by the TD-FPM system (left) and by a conventional wide-field fluorescence microscope (right) employing identical low-aperture objectives. Sharper details can be detected by TD-FPM due to the coherence gate mechanism which assists in rejecting out-of-focus fluorescence.

TIRF-based SD-FPM system shown in Figure 18.3B which localizes single QDs attached to the bilayers with nanometer-level accuracy in all 3D. Figure 18.8A shows that QDs attached to a six-bilayer substrate underwent an axial displacement of 20 ± 5 nm, whereas the lateral displacement was determined with a lateral localization precision below 10 nm. The axial displacement was comparable to that measured by surface plasmon resonance (SPR); yet, the TIRF-based SD-FPM system is advantageous over SPR systems as it can also identify lateral nano-displacements across a large field of view.

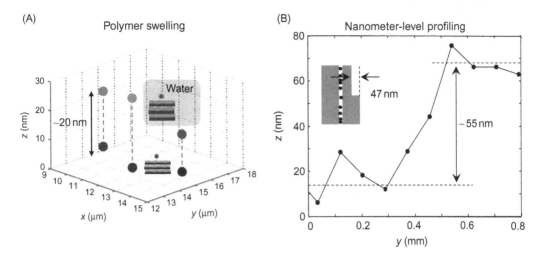

Figure 18.8
(A) Single QD localization measurements before (dark gray) and after swelling (light gray) of a six-bilayer polymeric substrate. Adding water resulted in an increase of the substrate thickness by 20 ± 5 nm. No lateral shifts of QDs were observed across a large field of view. (B) Nanometer-level profiling of a transparent nano-etched surface by SD-FPM. Source: *(A) This figure is reproduced from figure 5(a) of Ref. [42] with permission of the Optical Society of America. (B) This figure is reproduced from figure 6(b) of Ref. [40] with permission of John Wiley & Sons Ltd.*

In a different experiment, we measured the thickness of a calibrated sample by using the SD-FPM system shown in Figure 18.3A [39,40]. The sample comprised a step pattern etched on one surface of a coverslip where fluorescent beads were dried on the second surface. The step profile was obtained by computing the differential spatial phase of excited fluorophores along a single axial cross-section of the sample as expressed by $z|(x,y) \propto <I_{peak}(x,y) - <I_{peak}(x=x_0, y=y_0)$, where $z|(x,y)$ represents the axial displacement at a given transversal coordinate (x,y) and (x_0,y_0) is a reference coordinate. The surface profile was obtained by averaging several consecutive profiles (each recorded in a single shot) to increase the measurement accuracy. As shown in Figure 18.8B, this measurement predicts a step profile with a thickness (depth) of ~55 nm which is comparable to the 47 nm thickness measured independently by SD phase microscopy [26].

18.5 Conclusions and Outlook

The phase of fluorescent light waves is largely unexplored in the development of new tools of fluorescence microscopy. However, it comprises a novel source of information about the wave and can be used to yield new forms of optical imaging technologies with nanoscopic and mesoscopic resolution scales. In this chapter, we presented two new methods that employ concepts of white-light interferometry to retrieve this phase information and demonstrated their capabilities for optical sectioning imaging with mesoscopic resolution and metrology with nanometer-level axial localization precision. In particular, we showed that FPM can be realized in the Fourier domain and is useful for outlining in a single shot (i.e., without scanning) fluorescence gradients in semitransparent samples. Importantly, our recent report on image formation in fluorescence coherence-gated imaging suggested that the coherence gating mechanism may provide a unique approach for the rejection of out-of-focus fluorescence in thick scattering samples [47]. In addition, we discussed the usefulness of FPM for the precise localization of individual fluorescent emitters in all 3D. In closing, the potential next steps in the development of FPM would include explorations of its ability to track dynamical processes as well as its extension to other types of luminescence.

References

[1] A.G. York, A. Ghitani, A. Vaziri, M.W. Davidson, H. Shroff, Confined activation and subdiffractive localization enables whole-cell PALM with genetically expressed probes, Nat. Methods 8 (2011) 327–333.

[2] D. Aquino, A. Schönle, C. Geisler, C.V. Middendorff, C.A. Wurm, Y. Okamura, et al., Two-color nanoscopy of three-dimensional volumes by 4Pi detection of stochastically switched fluorophores, Nat. Methods 8 (2011) 353–359.

[3] J. Tang, J. Akerboom, A. Vaziri, L.L. Looger, C.V. Shank, Near-isotropic 3D optical nanoscopy with photon-limited chromophores, Proc. Natl. Acad. Sci. USA. 107 (2010) 10068–10073.

[4] G. Shtengel, J.A. Galbraith, C.G. Galbraith, J. Lippincott-Schwartz, J.M. Gillette, S. Manley, et al., Interferometric fluorescent super-resolution microscopy resolves 3D cellular ultrastructure, Proc. Natl. Acad. Sci. USA. 106 (2009) 3125–3130.

[5] S.R.P. Pavani, M.A. Thompson, J.S. Biteen, S.J. Lord, N. Liu, R.J. Twieg, et al., Three-dimensional, single-molecule fluorescence imaging beyond the diffraction limit by using a double-helix point spread function, Proc. Natl. Acad. Sci. USA. 106 (2009) 2995–2999.

[6] D. Wildanger, R. Medda, L. Kastrup, S.W. Hell, A compact STED microscope providing 3D nanoscale resolution, J. Microsc. 236 (2009) 35–43.

[7] J. Fölling, V. Belov, D. Riedel, A. Schdiaeresisonle, A. Egner, C. Eggeling, et al., Fluorescence nanoscopy with optical sectioning by two-photon induced molecular switching using continuous-wave lasers, Chem. Phys. Chem. 9 (2008) 321–326.

[8] A. Vaziri, J.Y. Tang, H. Shroff, C.V. Shank, Multilayer three-dimensional super resolution imaging of thick biological samples, Proc. Natl. Acad. Sci. USA. 105 (2008) 20221–20226.

[9] M.F. Juette, T.J. Gould, M.D. Lessard, M.J. Mlodzianoski, B.S. Nagpure, B.T. Bennett, et al., Three-dimensional sub-100 nm resolution fluorescence microscopy of thick samples, Nat. Methods 5 (2008) 527–529.

[10] B. Huang, W. Wang, M. Bates, X. Zhuang, Three-dimensional super-resolution imaging by stochastic optical reconstruction microscopy, Science 319 (2008) 810–813.

[11] L. Schermelleh, P.M. Carlton, S. Haase, L. Shao, L. Winoto, P. Kner, et al., Subdiffraction multicolor imaging of the nuclear periphery with 3D structured illumination microscopy, Science 320 (2008) 1332–1336.

[12] V. Ntziachristos, Going deeper than microscopy: the optical imaging frontier in biology, Nat. Methods 7 (2010) 603–614.

[13] P.J. Keller, A.D. Schmidt, J. Wittbrodt, E.H. Stelzer, Reconstruction of zebrafish early embryonic development by scanned light sheet microscopy, Science 322 (2008) 1065–1069.

[14] C. Vinegoni, C. Pitsouli, D. Razansky, N. Perrimon, V. Ntziachristos, In vivo imaging of *Drosophila melanogaster* pupae with mesoscopic fluorescence tomography, Nat. Methods 5 (2008) 45–47.

[15] H.U. Dodt, U. Leischner, A. Schierloh, N. Jährling, C.P. Mauch, K. Deininger, et al., Ultramicroscopy: three-dimensional visualization of neuronal networks in the whole mouse brain, Nat. Methods 4 (2007) 331–336.

[16] J. Huisken, J. Swoger, F. Del Bene, J. Wittbrodt, E.H.K. Stelzer, Optical sectioning deep inside live embryos by selective plane illumination microscopy, Science 305 (2004) 1007–1009.

[17] J. Sharpe, U. Ahlgren, P. Perry, B. Hill, A. Ross, J. Hecksher-Sørensen, et al., Optical projection tomography as a tool for 3D microscopy and gene expression studies, Science 296 (2002) 541–545.

[18] B.J. Vakoc, R.M. Lanning, J.A. Tyrrell, T.P. Padera, L.A. Bartlett, T. Stylianopoulos, et al., Three-dimensional microscopy of the tumor microenvironment in vivo using optical frequency domain imaging, Nat. Med. 15 (2009) 1219–1223.

[19] R. Huber, D.C. Adler, V.J. Srinivasan, J.G. Fujimoto, Fourier domain mode locking at 1050 nm for ultra-high-speed optical coherence tomography of the human retina at 236,000 axial scans per second, Opt. Lett. 32 (2007) 2049–2051.

[20] S.H. Yun, G.J. Tearney, B.J. Vakoc, M. Shishkov, W.Y. Oh, A.E. Desjardins, et al., Comprehensive volumetric optical microscopy in vivo, Nat. Med. 12 (2006) 1429–1433.

[21] C. Yang, Molecular contrast optical coherence tomography: a review, Photochem. Photobiol. 81 (2005) 215.

[22] E. Beaurepaire, A.C. Boccara, M. Lebec, L. Blanchot, H. Saint-Jalmes, Full-field optical coherence microscopy, Opt. Lett. 23 (1998) 244–246.

[23] D. Huang, E.A. Swanson, C.P. Lin, J.S. Schuman, W.G. Stinson, W. Chang, et al., Optical coherence tomography, Science 254 (1991) 1178–1181.

[24] C. Fang-Yen, W. Choi, Y. Sung, C.J. Holbrow, R.R. Dasari, M.S. Feld, Video-rate tomographic phase microscopy, J. Biomed. Opt. 16 (2011) 011005:1–5.

[25] Z. Wang, L. Millet, M. Mir, H. Ding, S. Unarunotai, J. Rogers, et al., Spatial light interference microscopy (SLIM), Opt. Express 17 (2011) 1016–1026.

[26] C. Joo, C.L. Evans, T. Stepinac, T. Hasan, J.F. de Boer, Diffusive and directional intracellular dynamics measured by field-based dynamic light scattering, Opt. Express 18 (2010) 2858–2871.

[27] T. Ikeda, G. Popescu, R.R. Dasari, M.S. Feld, Hilbert phase microscopy for investigating fast dynamics in transparent systems, Opt. Lett. 30 (2005) 1165–1167.

[28] C. Yang, A. Wax, M.S. Hahn, K. Badizadegan, R.R. Dasari, M.S. Feld, Phase-referenced interferometer with subwavelength and subhertz sensitivity applied to the study of cell membrane dynamics, Opt. Lett. 26 (2001) 1271–1273.

[29] K.E. Drabe, G. Cnossen, D.A. Wiersma, Localization of spontaneous emission in front of a mirror, Opt. Commun. 73 (1989) 91–95.

[30] K.H. Drexhage, Interaction of light with monomolecular dye layers, Prog. Opt. 12 (1974) 163–192.

[31] A. Lambacher, P. Fromherz, Luminescence of dye molecules on oxidized silicon and fluorescence interference contrast microscopy of biomembranes, J. Opt. Soc. Am. B 19 (2002) 1435–1453.

[32] M. Dogan, A. Yalcin, S. Jain, M.B. Goldberg, A.K. Swan, M.S. Ünlü, et al., Spectral self-interference fluorescence microscopy for subcellular imaging, IEEE J. Sel. Top. Quantum Electron. 14 (2008) 217–225.

[33] L. Moiseev, M.S. Unlü, A.K. Swan, B.B. Goldberg, C.R. Cantor, DNA conformation on surfaces measured by fluorescence self-interference, Proc. Natl. Acad. Sci. USA. 103 (2006) 2623–2628.

[34] M. Lang, T. Müller, J. Engelhardt, S.W. Hell, 4Pi microscopy of type A with 1-photon excitation in biological fluorescence imaging, Opt. Express 15 (2007) 2459–2467.

[35] M.G.L. Gustafsson, D.A. Agard, J.W. Sedat, I5M: 3D widefield light microscopy with better than 100 nm axial resolution, J. Microsc. 195 (1999) 10–16.

[36] M.G.L. Gustafsson, D.A. Agard, J.W. Sedat, Sevenfold improvement of axial resolution in 3D widefield microscopy using two objective lenses, Proc. SPIE 2412 (1995) 147–156.

[37] S.W. Hell, E.H.K. Stelzer, Properties of a 4Pi-confocal fluorescence microscope, J. Opt. Soc. Am. A 9 (1992) 2159–2166.

[38] X. Brokmann, M. Bawendi, L. Coolen, J.-P. Hermier, Photon-correlation Fourier spectroscopy, Opt. Express 14 (2006) 6333–6341.
[39] A. Bilenca, I. Maerki, B.E. Bouma, G.J. Tearney, T. Lasser, Low-level light interferometry: principles and applications in the life sciences, SPIE BIOS Conference (2009).
[40] A. Bilenca, J. Cao, M. Colice, A. Ozcan, B.E. Bouma, L. Raftery, G.J. Tearney, Fluorescence interferometry: principles and applications in biology, Ann. N. Y. Acad. Sci. 1130 (2008) 68–77.
[41] A. Bilenca, A. Ozcan, B.E. Bouma, G.J. Tearney, Fluorescence coherence tomography, Opt. Express 14 (2006) 7134–7143.
[42] I. Märki, N.L. Bocchio, S. Geissbuehler, F. Aguet, A. Bilenca, T. Lasser, Three-dimensional nano-localization of single fluorescent emitters, Opt. Express 18 (2010) 20263–20272.
[43] G.A. Korn, T.M. Korn, Mathematical Handbook for Scientists and Engineers, Dover Publications, Inc., 1968.
[44] E.B. van Munster, T.W. Gadella, Fluorescence lifetime imaging microscopy (FLIM), Adv. Biochem. Eng. Biotechnol. 95 (2005) 143–175.
[45] M. Wojtowski, A. Kowalczyk, R. Leitgeb, A.F. Fercher, Full range complex spectral optical coherence tomography technique in eye imaging, Opt. Lett. 27 (2002) 1415–1417.
[46] N. Kasthuri, J.W. Lichtman, The rise of the "projectome", Nat. Methods 4 (2007) 307–308.
[47] A. Bilenca, T. Lasser, A. Ozcan, R.A. Leitgeb, B.E. Bouma, G.J. Tearney, Image formation in fluorescence coherence-gated imaging through scattering media, Opt. Express 15 (2007) 2810–2821.

Index

Note: Page numbers followed by "*f*" and "*t*" refer to figures and tables, respectively.

A

Abbe's theory, 342–343
Acetic acid, 250
Actin, 225
Adaptive optics microscopy, 53, 61–62
Airy disk, 26
Allium cepa bulb scale, inner epidermal nucleus from, 56*f*
Analyzer, 331
Angular spectrum method, 132–133
Anisotropy, uniaxial, 331
Apoptosis and necrosis, imaging of, 114–115
Arabidopsis root sections, adaptive optics in, 63*f*
Artificial insemination, 154–155
Autocorrelations and spectral densities for DLS, 221–222, 222*t*
Autofocusing, 106–108
Azimuth, 331

B

Babinet-Soleil compensator, 29
Becke line in optical microscopy, 57–58
Berek compensator, 29
Bias optimization, 27–31
Birefringence, 312*f*, 317–318, 331–332
Birefringent lens, 20–21, 313, 317, 328
Born approximation, 233, 236
vs. Rytov approximations, 245, 245*f*

Bovine spermatozoa, 196–197
Bräce-Köhler compensator, 27, 317–319
Brightfield transmitted light microscopy, 4
Bulk image shifts (BISs), 271

C

Caenorhabditis elegans, 240
Calcium imaging, 85–86, 87*f*
Cardboard box, microscope enclosure made from, 46*f*
Carl Zeiss, 326
Cell cycle, 79–80
Cell death, 43
  neuronal, 85–86
  Stx1-induced, 112–113, 113*f*
Cell imaging and quantitative phase signal interpretation, 76–78
Cell membrane fluctuations (CMF), 82–84
Cell tracking in 3D environments, 121–124
Charge-coupled device (CCD) cameras, 74, 99–100, 174, 284–285, 321–322, 358, 366
Colchicine, 225
Color filter, 13
Commission Internationale d'Eclairage (CIE), 302–303
Compensator, 317, 332
  Babinet-Soleil, 29
  Berek, 29
  Bräce-Köhler, 27, 317–319
  Ehringhaus, 29

Complementary metal oxide semiconductor (CMOS) camera, 174, 236–237
Computer-assisted semen analysis (CASA) systems, 155
Confocal microscopes, 17
Contrast, 3
Correlative live cell–fixed cell observations, 51–52
Crane fly spermatocyte orientation-independent DIC (OI-DIC) image of, 35–37, 36*f*
CRi Inc., 322
Cucurbita, cross-section of, 66*f*
Curvature correction, 133–134
Cytochalasin D, 225

D

Dark field, 54
Deformable membrane mirror (DMM), 53
Dichroic material, 334
Dichroism, 332
Differential detection DIC (D-DIC) with polarizing beamsplitter, 22
Differential interference contrast (DIC) microscopy, 17, 19, 43, 53, 58–60, 59*f*, 60*f*, 61*f*, 71–72, 231–232, 287, 322
  bias optimization, 27–31
  measuring shear angle of DIC prism, 22–27
  OI-DIC microscope with fast modulation of bias and shear direction, 31–37

Differential interference contrast (DIC) microscopy (*Continued*)
  and fluorescence imaging, 39–40
  and orientation-independent polarization imaging, 37–39
  with two switchable beam-shearing DIC assemblies, 36f
  principles of, 19–22
  setup, 20f
Diffraction gratings, 355
Diffraction tomography
  and filtered back-projection algorithm, 244f
  optical, *see* Optical diffraction tomography (ODT)
Digital holographic microscopy (DHM), 69, 97–98, 231–232
  biological applications, 78–87
    cell membrane fluctuations, 82–84
    dry mass and cell cycle, 79–80
    extended depth of focus and 3D tracking, 79
    future perspectives, 86–87
    hematology, biophysical parameters in, 80–82
    neuronal cell death, 85–86
    neuroscience, 84–85
  cell tracking in 3D environments, 121–124
  with microfluidics, combining drive and analyse functions, 194–201
    experimental results, 200–201
    experimental setup, 196–200
    trapping theory, 195–196
  morphology and thickness changes, analysis of, 111–115
    apoptosis and necrosis, imaging of, 114–115
    cell reaction on chemical substances, 112–114
    vacuole formation, imaging of, 114
  quantitative cell division monitoring, 116–121
  recording and numerical evaluation of digital holograms, 101–108
    intensity distribution in hologram plane, 101–102
    quantitative phase imaging, 103–106
    spatial phase shifting-based reconstruction of, 102–103
    subsequent refocusing and autofocusing, 106–108
  refractive index determination of cells in suspension, 109–111
  setups, for live cell imaging, 98–101
    fiber optic modular DHM, 99–100
    self-interference DHM, 100–101
  technical introduction, 74–78
    classical holography, 74–78
Digital holography (DH), 174, 194, 215, 282
Digital in-line holographic microscopy (DIHM), 174–175
Digital off-axis holograms evaluation by spatial phase shifting, 104f
Drosophila embryo imaged by the TD-FPM system, 400f
Dry mass, 79–80
Dry mass surface density (DMSD), 80
Dual-interference channel quantitative phase microscopy (DQPM), 282–286
Dual-wavelength digital holography apparatus, 130–131
Dual-wavelength phase imaging
  linear regression method, 145–150
  synthetic wavelength, 142–144
Dynamic light scattering (DLS) spectroscopy, 212, 219–223

E

Ehringhaus compensator, 29
Evanescent waves, 355–356
Ewald sphere, 234, 349–350
Extinction coefficient, 332
Extinction position, 315–316
Extinction, 332

F

Fast axis, 318, 333
Fast Fourier transform (FFT), 300–301
FertilMARQ, 155
Fiber optic modular DHM, 99–100
Filtered back-projection algorithm, 237–238, 244f
  and diffraction tomography, 244f
Fisher Permount mounting medium, 30
Flickering, *see* Cell membrane fluctuations (CMF)
Fluorescein isothiocyanate (FITC), 327
Fluorescence imaging, 320
  combining polarized light with DIC and, 325–327
  phase contrast images with, 17
Fluorescence microscopy
  with orientation-independent DIC (OI-DIC) microscope, 39–40
Fluorescence phase microscopy (FPM)
  applications of, 395–401
  experimental characterization of, 396–397
  experimental setups of, 390–395
    spectral-domain FPM, 392–394
    TD/SD-FPM, capabilities and limitations of, 394–395
    time-domain FPM, 391–392
  fluorescence self-interference process, 389–390
  fluorescent axial PSF, 396–397
  individual fluorophores, axial localization precision of, 397

optical sectioning imaging with mesoscopic resolution by, 397–401
nanometer-level axial localization precision, metrology with, 399–401
Fluorescence self-interference process, 389–390
Fluorescent and nonfluorescent polystyrene beads, 301–304
Fluorescent axial (depth) PSF, 396–397
Focus stability, 44
Fourier diffraction theorem, 235
Fourier transform (FT), recording, 358
Fourier transform holographic interferometry (FTHI), 353
Fourier transform holography (FTH), 353–355
Fourier-domain OCM (FD-OCM), 263–264
Fresnel transformation, 102–103
Fresnel–Huygens principle, 102
Fresnel–Kirchhoff integral, 132
Fringe pitch measurement process, 364–365
Full width at half maximum (FWHM), 238–239, 396–397

G

Gabor, Dennis, 173
Gabor's regime
SALDHM inside, 177–182
SALDHM outside, 182–188
Gamma-ray microscope, 342
Global phase fluctuations (GPFs), 271
Glutamate, 84–85, 86f
Gold nanoparticles, detection of, 275–277
Gradient force, 195–196
Gratings
illumination by evanescent waves, 355–356
pitch determination, 364–365
Green fluorescent protein (GFP), 43, 329–330
Green's theorem, 235

H

HeLa cell, 239
Hematology, biophysical parameters in, 80–82
Hemoglobin, 81
High-extinction optics, 320
High-throughput on-chip semen analysis results, 165–168
Hoffman modulation contrast (HMC) microscope, 53, 60–61, 62f
Hologram plane, intensity distribution in, 101–102
Holograms formation, at nanoscale, 364–368
Holographic interferometric metrology, 98
at the nanoscale, 353–355
Holographic motility contrast imaging (MCI) of live tissues
dynamic light scattering (DLS) spectroscopy, 219–223
motility contrast imaging, 216–219
optical coherence imaging (OCI), 213–215
tissue dynamics spectroscopy (TDS), 223–226
Holography techniques, 72, 173
digital, see Digital holographic microscopy (DHM)
principle, 74–78
HT29 cells
3D tomogram of, 246f
Human brain microvascular endothelia cells (HBMECs), 107–108

I

Illumination imaging, 345–349
Image reconstruction, 343–345
Image shift, 368
In vivo observations, nano-holographic interferometry for, 353
Individual fluorophores, axial localization precision of, 397
Interferometric microscopy, 62–65

Inverse Radon Transform, 234
Ion–water relationship, illustrating, 84–85
Iterative constraint algorithm, 243f

K

Kohler illumination, 13, 44–45, 48–50

L

LC-PolScope, 313, 320–322
Lensless holographic on-chip microscope, 153–154, 154f
Lensless on-chip holographic microscopy (LOHM), 176
Light scattering spectroscopy, 258
Linear regression phase unwrapping, 145–150
Liquid crystal phase modulator (LCPM), 62–64
Live cardiomyocyte measurements, 291f
Live cell–fixed cell observations, 51–52
Live human cells, long-term recordings of, 43
correlative live cell–fixed cell observations, 51–52
microscope, 44–51
illumination, 44–45
observation chambers, 48–50
preparing chambers, 50–51
temperature control, 45–47
Low-coherence interferometry (LCI), 394

M

Mach–Zehnder interferometer, 99, 106
quantitative phase imaging arrangements based on, 100
Magnetic resonance imaging of plants showing the flow component, 65–67
Magnification, 3
Male fertility tests, 154–155
Maxwell equations, Fourier solution of, 354
Maxwell–Neumann equations, 363

MDA-MB-468 human breast cancer cell, 285f
Mean corpuscular hemoglobin (MCH), 82
Mean corpuscular hemoglobin concentration (MCHC), 80–81
Mean corpuscular volume (MCV), 80–81
Membrane fluctuation, 82–84
Metrological information extraction, from recorded images, 368–370
　change of optical path through observed object, 369–370
　image shift, 368
Michel-Lévy chart, 30–31, 314–315
Michelson interferometer-based DHM arrangement setup, 100–101
Microfluidics, digital holographic microscopy with, 193
Microscope, 44–51
　illumination, 44–45
　observation chambers, 48–50
　preparing chambers, 50–51
　temperature control, 45–47
Microscope enclosure made from cardboard box, 46f
Microscope objective (MO), 75–76, 197
Microscopic images reconstruction and phase recovery process, 161–162
Microvasculature within retina, visualizing, 271
Modulation transfer function (MTF), 345–347, 347f
Moiré effect, 345, 346f
Molecular order, 317, 331
Motility contrast imaging, 216–219
Mouse cells, 196–197
Multi-k vector fields, 359–364
Multimodality microscopy, 85

N

Nano Fourier transform holography, theoretical basis of, 353–355
Nanocrystals thickness, determination of, 374–377
Nano-holographic interferometry, for in vivo observations, 353
Nanometer-level axial localization precision, metrology with, 399–401
Nano-objects, observation of, 356–359
Nano-sized objects, observation of, 372–382
　polystyrene nanospheres, 377–382
　prismatic NaCl nanocrystals, 372–374
　thickness, determination of, 374–377
Near-field scanning optical microscopy (NSOM), 342–343
Nematode growth medium (NGM) buffer, 240
Nephrotoma suturalis, 38
Nerve displacement during action potential, 272–273
Neuronal cell death, 85–86
Neurosciences, 84–85
Newton's interference colors, 30–31
N-Methyl-D-aspartate (NMDA) receptor, 84–85
Nocodazole, 217–219
Nomarski prism, 21, 58–59, 287, 326
Nonfluorescent polystyrene beads, 301–304
Nonlinear phase dispersion spectroscopy (NLDS), 299–306
　fluorescent and nonfluorescent polystyrene beads, 301–304
　red blood cells, 304–306
Normalized standard deviation (NSD), 217
Numerical Aperture (NA) optics, 335
Numerical super-resolution methods, 383
Nycodenz, 77–78

O

Off-axis digital holography, 75–76
Off-axis geometry, 74
On-chip imaging of sperms using partially coherent lensfree in-line holography, 162–165
Onion cells, unwrapping, 141f
Optic axis, 333
Optical anisotropy, 311
Optical coherence imaging (OCI), 213–215
Optical coherence microscopy (OCM)
　phase-sensitive, *see* Phase-sensitive optical coherence microscopy
　principles, 262–266
Optical coherence tomography (OCT), 388
Optical diffraction tomography (ODT), 231, 234–236
Optical microscopy, Becke line in, 57–58
Optical path change, evaluation of, 374–377
Optical path delay (OPD), 261
Optical path difference, 23–24
Optical path length (OPL), 10, 97–98, 287
Optical resonator, 366
Ordinary refractive index, 331
Orientation-independent DIC (OI-DIC) image, 22
　of live crane fly spermatocyte, 35–37, 36f
Orientation-independent DIC (OI-DIC) microscope
　combining fluorescence microscopy with, 39–40
　with fast modulation of bias and shear direction, 31–37
　and orientation-independent polarization imaging, 37–39
　with two switchable beam-shearing DIC assemblies, 36f
Orientation-independent differential polarization (OI-Pol) system, 37–38

Index

Out of focus imaging, 54
Outer retina, imaging, 268–269
Out-of-focus haze, in structured illumination microscopy (SIM), 22

P

"Pacing cells," 292
Partially coherent in-line holography on chip, 156–161
 on-chip imaging of sperms using, 162–165
Particle tracking in 3D, 201–206
 experimental results, 204–206
 modeling for 3D tracking, 202–204
Passive focus stability, 44
Pf-RBCs, 3D refractive index maps of, 253*f*
Phase contrast (PhC), 71–72
 basic overview, 4–10
 experimental uses, 15–17
  cell culture models, 15
  fluorescence overlay, 17
  image analysis, 15–17
 halo artifacts, 12
 image, 10–11
 limitations of phase optics, 17–18
 positive and negative phase, 11–12
 principles of, 1
 working of, 4–12
Phase contrast image, formation of, 55*f*
Phase contrast images, acquiring, 12–15
 adjustability, 14–15
 components in phase contrast system, 12–13
 general alignment, 13
 phase ring alignment, 13–14
Phase contrast microscopy, 43, 231–232
Phase contrast techniques, 54–57
Phase imaging, 343
Phase optics, 17
 limitations of, 17–18
Phase ring alignment, 13–14

Phase shift, 5–6, 71–72, 75, 86*f*
Phase unwrapping problems
 dual-wavelength phase imaging
  linear regression method, 145–150
  synthetic wavelength, 142–144
 experimental techniques
  additional phase background removal, 134–137
  angular spectrum method, 132–133
  curvature correction, 133–134
  dual-wavelength digital holography apparatus, 130–131
  in phase microscopy of biological cells, 129
  varying reconstruction distance method, 137–142
Phase-sensitive optical coherence microscopy, 261, 266–267
 Gold Nanoparticles, detection of, 275–277
 microvasculature within retina, visualizing, 271
 nerve displacement during action potential, 272–273
 outer retina, imaging, 268–269
 Red Blood Cells, measurements of, 273–275
 RI measurements, 272
 rodent cerebral cortex, motion correction of, 270–271
 skin, determining elastic properties of, 269–270
Phase-shifting DIC (PS-DIC), 22
Phase-shifting interferometric microscopy, 231–232
Phosphate buffered saline (PBS), 47–48
Photo multiplier tube (PMT), 17
Plant cells and tissues, phase imaging in, 53
 adaptive optics microscopy, 61–62
 Becke line in optical microscopy, 57–58
 dark field, 54
 DIC microscopy, 58–60, 59*f*, 60*f*, 61*f*

Hoffmann modulation contrast (HMC), 60–61, 62*f*
interferometric microscopy, 62–65
magnetic resonance imaging of plants showing the flow component, 65–67
out of focus imaging, 54
phase contrast techniques, 54–57
second harmonic imaging microscopy (SHIM), 65
Plasmodium falciparum, 253
Plexiglas enclosures, 45–46
Pluta Interference microscope, 64
Point-spread function (PSF), 387–388
Polarization microscopy, 311
 in three dimensions, 327–329
 LC-PolScope, 320–322
 practical considerations, 322–327
  choice of optics, 322–324
  combining polarized light with DIC and fluorescence imaging, 325–327
  specimen preparation, 324–325
 traditional polarized light microscopy, 314–320, 316*f*
  basic setup, 315–317
  birefringence, 317–318
  retardance, 317–320
  slow axis, 318
Polarization modulation DIC (PM-DIC), 22
Polarization optical train, 324
Polarization rectifiers, 320
Polarized light, 333
 circularly, 333
 elliptically, 333–334
 linearly, 333
Polarizers, 315–316, 334
Polskie Zakłady Optyczne (PZO), 64
Polystyrene nanospheres, 377–382
Poplar stem sections, 66*f*

Q

Quantitative cell division monitoring, 116–121

Quantitative phase measurement, 74
Quantitative phase microscopy (QPM), 72
   dual-interference channel, 282–286
   nonlinear phase dispersion spectroscopy, 299–306
   fluorescent and nonfluorescent polystyrene beads, 301–304
   red blood cells, 304–306
   spectral-domain differential interference contrast microscopy, 286–292
   live cell imaging with rat cardiomyocytes, 290–292
   principles of, 287–289
   USAF resolution target, characterization of, 289–290
   spectral multiplexing by RGB color channels, 292–299
   of quantitative phase and fluorescence biomarkers, 296–299
   RGB camera multiplexing for phase unwrapping, 293–296
Quantitative phase-imaging techniques, 103–106, 281
Quasi-elastic light scattering (QELS), 223
Quasi-transparent specimens, 73

R

Rat ventricular cardiomyocytes, 298f
RAW cells, 139f, 140f
Reconstruction distance, phase unwrapping by varying, 137–142
Recorded image, system of fringes in, 359–364
Red blood cells (RBCs), 79, 231
   biophysical parameters of, 81–82
   cell membrane fluctuations, 82–84
   measurements of, 273–275

Red plate, 30
Refocusing, 106–108
Refractive index, 231–232
   determination of cells in suspension, 109–111
   of a medium, 5
   tomogram
      of HeLa cell, 239f
      of HT29 cell, 252f
Re-pixelation, 383
Retardance, 24, 317–318, 334–335
   quantitative analysis of, 318–320
Retardation modulation DIC (RM-DIC), 22
Retardation, 317–318
Retarder, 335
RGB color channels, spectral multiplexing by, 292–299
   for phase unwrapping, 293–296
   of quantitative phase and fluorescence biomarkers, 296–299
RI measurements, 272
Rodent cerebral cortex, motion correction of, 270–271
Ronchi test, 379
Rotating-beam geometry, 233
Rytov approximation, 233, 236
   vs. Born approximations, 245, 245f

S

Scale parameter for the FT plane, 370–372
Scattering force, 195–196
Schizosaccharomyces pombe, 80
SD-optical coherence tomography (SD-OCT), 287
Second harmonic imaging microscopy (SHIM), 54, 65
Self-interference DHM, 100–101
Self-interference off-axis holograms, evaluation of, 105f
Semen analysis results, high-throughput on-chip, 165–168
Senarmont method, 30

Senarmont technique, 30
Sensitive tint plate, 30
Shack–Hartmann detectors, 72
Shade-off, 12
Shear angle measurement of DIC prism, 22–27
Signal to noise ratio (SNR), 263, 391–392
Silicone grease, 51
Simulated cells, background removal for, 134–135, 136f
Skin, determining elastic properties of, 269–270
SKOV-3 ovarian cancer cells, 144f
Slow axis, 318, 336
"Smith double-focus system," 20
"Smith double-refracting interference shearing microscope system," 20
Space-bandwidth product, 341–342
Sparsity, 341–342
Spatial decoding, 75
Spatial light interference microscopy (SLIM), 53, 62–64, 63f, 64f
Spatial light modulators, 53
Spatial modulation tomography, 248f
Spatial phase shifting-based reconstruction of digital holograms, 102–103
Spatial resolution in phase microscopy, 339
   Abbe's theory, 342–343
   parallel full-field linear imaging, 343
   reconstruction from diffraction patterns, 343–345
   structured illumination microscopy, 345–349
   three-dimensional phase objects, imaging, 349–350
Speckle interferometry and holography, 212
Speckle, 211–212
Spectral domain phase microscopy (SDPM) system, 273–274, 299
Spectral multiplexing by RGB color channels, 292–299

for phase unwrapping, 293–296
of quantitative phase and fluorescence biomarkers, 296–299
Spectral-domain differential interference contrast microscopy (SD-DIC), 286–292
  live cell imaging with rat cardiomyocytes, 290–292
  principles of, 287–289
  USAF resolution target, characterization of, 289–290
Spectral-domain fluorescence phase microscopy (SD-FPM), 392–394
  capabilities and limitations, 394–395
  components of, 393–394
  experimental realizations of, 393f
Spectral-domain low-coherence interferometry (SD-LCI), 287
Spectral-domain OCM (SD-OCM), 264
Spectroscopic OCT (SOCT), 299
SpermCheck, 155
Sperms imaging
  using partially coherent lensfree in-line holography, 162–165
  using SALDHM, 165f
Structured illumination microscopy (SIM), 345–349
Super-oscillation, 341
"Super-resolution," 341
Surface plasmon resonance (SPR), 399–400
Swept-source OCM (SS-OCM), 264
Synthetic aperture lensless digital holographic microscopy (SALDHM), 176–177
  inside the Gabor's regime, 177–182
  outside the Gabor's regime, 182–188
  for superresolved biological imaging, 173

Synthetic wavelength, phase unwrapping, 142–144
Sysmex KX-21, 82

T

Temperature control of specimen, 45–47
Temporal decoding, 75
Temporal speckle contrast, see Normalized standard deviation (NSD)
"Theoretical" phase pattern, 374–375
Thickness measurements, 374–377
Three-dimensional imaging techniques, 231
3D index tomogram, 239
Three-dimensional phase objects, imaging, 349–350
3D refractive index maps, of Pf-RBCs, 253f
3D tracking, modeling for, 202–204
Threshold-based image analysis, 15–17
Time-domain fluorescence phase microscopy (TD-FPM), 391–392
  capabilities and limitations, 394–395
  experimental setup of, 391f
Time-domain OCM (TDOCM), 263–264
Time-lapse imaging, 44
Tissue dynamics spectroscopy (TDS), 223–226
Tomographic phase microscopy (TPM), 229
  biological applications, 253–258
    assessing light scattering of intracellular organelles in single intact living cells, 255–258
    study malaria-infected RBCs, 253–255
  data processing by the ODT, 240–247
  experimental setup, 236–237
  inverse radon transform, 234
    data processing by, 237–240

optical diffraction tomography, 234–236
video-rate TPM, 247–252
  experimental scheme, 247–250
  results, 250–252
Total-internal-reflection-fluorescence (TIRF) based FPM system, 397
Transmitted light microscopy, 15–17
Trapping theory, 195–196
Two-dimensional (2D) imaging techniques, 231

U

U-DICT, 26, 29–30
U-DICTH, 27
U-DICTHR, 26, 29–30
Uniaxial anisotropy, 331
Unit retardation plate, 30
USAF (United States Air Force) resolution target
  characterization of, in SD-DIC, 289–290
  phase images of, 143f
  two wavelength hologram of, 131f, 132
  using SALDHM, 181f, 185f
  with/without curvature correction, 135f

V

Vacuolated HBMECs, DHM analysis of, 115f
Vacuole formation, imaging of, 114
Video-enhanced-DIC (VE-DIC) microscopy, 27

W

Waveplate, see Retarder
Whispering-gallery modes (WGM), 377–378
Wide-field digital interferometry (WFDI) system, 275
Wiener–Khinchin theorem, 221
Wiener–Khintchine theorem, 392
Wollaston prism, 19, 20–21, 326

Printed in the United States
By Bookmasters